International Development

International Development is a comprehensive inquiry into the field of socio-economic development founded on an understanding that economic advancement involves the transformation of society. It explores successful developmental strategies but also tries to identify factors behind failed endeavors and the human costs associated with them. The book evaluates the role played by influential agents of development, such as the state and its institutions, authoritarian leaders, international organizations, donor agencies, non-governmental organizations, civil society activists, and private business actors.

Key features:

- A multi-disciplinary approach taking into account politics, economics, sociology, cultural aspects, and history of development;
- Examines a breadth of different theoretical approaches and their practical applications;
- Presents both mainstream and critical viewpoints;
- Addresses such complex issues as governability processes, rights of the poor, colonial legacy, armed conflict, environmental sustainability, gender relations, foreign aid, urbanization, rural development, and international trade;
- Suggested further reading list at the end of each chapter.

This well-balanced book will be a key text for students and practitioners working in the area of socio-economic development and more broadly in development studies, the politics of development and international political economy.

Anna Lanoszka is Associate Professor of International Relations at the University of Windsor in Canada. She specializes in international political economy and teaches courses on international trade and development.

"This excellent book provides an innovative conceptual analysis of international development using a human agency based approach. Professor Lanoszka's expert analysis moves beyond the free markets or state directed development models to a richer understanding that highlights the crucial role of individual agency. This is a powerful, often passionate study of international development. It convinces us of the need to place people at the centre of international development strategies."

Donna Lee, *Manchester Metropolitan University, UK*

"Anna Lanoszka has written an impressive introduction to the complex subject of development. It is grounded in key historical and theoretical debates, while retaining a lively and accessible style throughout. It makes excellent use of case studies to bring its subject to life. Lanoszka's passion for the subject is clearly projected, and will make this book a lively and insightful guide to the persistent and pivotal challenges of global development."

David Black, *Dalhousie University, Halifax, Canada*

"The study of development is a multi-disciplinary exercise and this lucid and informative book should be of interest to a wide variety of disciplines. Anna Lanoszka has written an accessible introduction to the study of international development that should appeal to both teachers and students. *International Development* is an historically informed, theoretically engaged text that places the individual at the centre of the process of development without at the same time neglecting the importance of structural and institutional factors."

Marc Williams, *University of New South Wales, Sydney, Australia*

International Development

Socio-Economic Theories, Legacies, and Strategies

ANNA LANOSZKA

Routledge
Taylor & Francis Group

LONDON AND NEW YORK

First published 2018
by Routledge
2 Park Square, Milton Park, Abingdon, Oxon OX14 4RN

and by Routledge
711 Third Avenue, New York, NY 10017

Routledge is an imprint of the Taylor & Francis Group, an informa business

British Library Cataloguing in Publication Data
A catalogue record for this book is available from the British Library

Library of Congress Cataloging in Publication Data
Names: Lanoszka, Anna, author.
Title: International development : socio-economic theories, legacies and strategies / Anna Lanoszka.
Description: New York : Routledge, 2018. | Includes bibliographical references and index.
Identifiers: LCCN 2017039541 | ISBN 9781138670341 (hardback : alk. paper) | ISBN 9781138670358 (pbk. : alk. paper) | ISBN 9781315617671 (ebook)
Subjects: LCSH: Economic development -Developing countries. | Sustainable development -Developing countries. | Urbanization -Developing countries. | Developing countries -Foreign economic relations.
Classification: LCC HD75 .L36 2018 | DDC 338.9009172/4 -dc23
LC record available at https://lccn.loc.gov/2017039541

ISBN: 978-1-138-67034-1 (hbk)
ISBN: 978-1-138-67035-8 (pbk)
ISBN: 978-1-315-61767-1 (ebk)

Typeset in Avenir and Dante
by Wearset Ltd, Boldon, Tyne and Wear

To all my students

Contents

Figures

Tables

Abbreviations

AAY	Antyodaya Anna Yojana
Btu	British thermal units
CAP	Common Agricultural Policy
CCP	Chinese Communist Party
CDP	Committee for Development Policy
CIAT	International Center for Tropical Agriculture
Comecon	Council for Mutual Economic Assistance
DSU	Dispute settlement understanding – WTO
EIA	Energy Information Administration
EVI	Economic Vulnerability Index
FAO	Food and Agriculture Organization – UN
FPR	Front Patriotique Rwandais (Rwandan Patriotic Front)
GAD	Gender and development approach
GATS	General Agreement on Trade in Services – WTO
GATT	General Agreement on Tariffs and Trade
GDI	Gender-related Development Index
GDP	Gross domestic product
GEM	Gender Empowerment Measure
GII	Gender Inequality Index
GMO	Genetically modified organism(s)
GNI	Gross national income
GNP	Gross national product
GVC	Global value chain(s)
GWh	Gigawatt hours
HAI	Human Assets Index
HDI	Human Development Index
HLPF	High-level Political Forum on Sustainable Development – UN
IBRD	International Bank for Reconstruction of Development – World Bank
ICA	International Coffee Agreement

ICSID	International Center for Settlement of Investment Disputes – World Bank
ICTR	International Criminal Tribunal for Rwanda
IDA	International Development Association – World Bank
IEA	International Energy Agency
IFC	International Finance Corporation – World Bank
IGO	Intergovernmental organization
ILO	International Labor Organization
IMF	International Monetary Fund
IOM	International Organization for Migration
IPC	International Patent Classification
IPR	Intellectual property right(s)
ISI	Import Substitution Industrialization
ITO	International Trade Organization
LDC	Least developed country
mb/d	Million barrels a day
MFN	Most Favored Nation
MGE	Mainstreaming Gender Equality
MIGA	Multilateral Investment Guarantee Agency – World Bank
MPLA	Popular Movement for the Liberation of Angola
MRND	National Republican Movement for Democracy and Development – Rwanda
NAFTA	North American Free Trade Agreement
nGDI	New Gender Development Index
NGO	Non-governmental organization
ODA	Official development assistance
OECD	Organization for Economic Cooperation and Development
OPEC	Organization of Petroleum Exporting Countries
PDS	Public distribution system
PPP	Purchasing power parity
SOE	State-owned enterprise
TKDL	Traditional Knowledge Digital Library
TKRC	Traditional Knowledge Resource Classification
TRIPS	Agreement on Trade-Related Aspects of Intellectual Property Rights – WTO
UN	United Nations
UNAVEM	United Nations Angola Verification Mission
UNCTAD	United Nations Conference on Trade and Development
UNDP	United Nations Development Programme
UNFPA	United Nations Population Fund
UNHCR	United Nations High Commissioner for Refugees
UNICEF	United Nations International Children's Emergency Fund
UNITA	National Union for the Total Independence of Angola
UNODC	United Nations Office on Drugs and Crime
UPOV	International Union for the Protection of New Varieties of Plants
USAID	United States Agency for International Development

USPTO	United States Patent and Trademark Office
WAD	Women and development
WHO	World Health Organization
WID	Women in development
WIPO	World Intellectual Property Organization
WTO	World Trade Organization

Introduction

Development and transforming societies

I Conceptualizing development

Economic development escapes easy definition, but is habitually associated with the idea of material progress. Developmental agencies share the conviction that we can improve the welfare of distant societies by intentional efforts. The goal is to eradicate poverty and minimize the inequality among the rich and the poor. But poverty transcends a condition of unfulfilled material needs. The meagerness and economic deprivation bring about social exclusion and feelings of hopelessness. The poor are routinely politically and socially marginalized and they hope for solutions that go beyond ensuring basic endurance in the form of food and shelter. This is why developmental initiatives that fail to enhance individual agency miss the point. People in need not only want to survive, they seek opportunities to live secured and dignified lives.

The task, of course, is incredibly complex. Individual agency, which manifests itself in our capacity to act, is shaped by a number of factors embedded in the political, social, cultural, and economic norms, and institutions that signify the environment where a person's home is. In many dysfunctional or oppressive countries, these norms and institutions only serve a narrow elite connected to the government. In other countries, the imposed norms perpetuate the existence of institutions that maintain the system characterized by inequality and the marginalization of minorities. Then again there are countries where institutions and norms have completely broken down because of the ongoing violent struggles for power, land, or resources.

The book begins with an understanding that the well-being of communities is closely related to the freedoms, rights, and opportunities that individuals have access to. A possibility for the advancement of people stems from the democratic accountability of the decision makers who formulate and implement the developmental policies that impact those communities. Offering a constructive helping hand to the disadvantaged societies involves conceptualizing development by thinking simultaneously about its two dimensions: the individual dimension and the macro-societal dimension. Too many developmental plans did not produce the expected results because they neglected to properly evaluate the requirements of people in need and overlooked the range of constraints surrounding them. The stakes are high. The failure of a misguided developmental project, at minimum, has a disabling effect. Incomplete or unsuccessful reforms lead to economic instability, deepening inequality, and, as a result, can destroy the social fabric of societies.

Perhaps the most notable claim about the necessity to place people at the center of the discourse about development comes from the literary world. In one heartbreaking novel that questions the impossibility of friendship between individuals who are socially unequal, we get a glimpse of a nation that endures a chronic economic instability. *The Kite Runner*[1] is a story about a series of botched experiments on how to govern a country and how these experiments impact the lives of ordinary people. The novel takes us on a journey throughout the recent history of Afghanistan. The main protagonist of the novel, Amir, is born to a privileged Pashtun family at the time when the country is still a monarchy. He enjoys the company of Hassan who lives in the same household but he may as well be on another planet. Hassan is a servant who has no rights and is not even allowed to go to school. As a member of an ethnic and religious minority, the Hazaras, Hassan is powerless, subject to abuse, and ultimately is killed during a period of political upheaval. While the novel allows us to become a witness to the divergent lives lived by Hassan and Amir, we also observe how Afghanistan goes from a monarchy, to authoritarian rule, only to be overtaken by a foreign army and then "rescued" by the destructive Taliban. The book ends on a positive note after yet another western intervention, which resulted in the December 2001 Bonn conference, tasked with reconstruction of Afghanistan under the leadership of Hamid Karzai. Sadly, over a decade later, the country remains deeply socially divided and economically unbalanced.

The turbulent history of Afghanistan provides many powerful insights about a chronically unstable country. A country marked by the hardship and suffering of its people, who are trapped inside the environment over which they have no control. I would like to share with you two passages from the novel that I believe to be remarkably salient in contemplating development. Here is the first one through the words of Amir:[2]

> I told you how we all celebrated in 1996 when the Taliban rolled in and put the end to the daily fighting. I remember coming home that night and finding Hassan in the kitchen, listening to the radio. He had a sober look in his eyes. I asked him what was wrong, and he just shook his head. "God help the Hazaras now, Rahim Khan sahib," he said. "The war is over, Hassan," I said. "There's going to be peace, *Inshallah*, and happiness and calm. No more rockets, no more killing, no more funerals!" But he turned off the radio and asked if he could get me anything before he went to bed. A few weeks later, the Taliban banned kite fighting. And two years later, in 1998, they massacred the Hazaras in Mazar-i-Sharif.

This quote demonstrates how critically vulnerable are poor people without rights. They live in fear and are particularly defenseless during political and economic transitions.

The second quote:[3]

> That same night, I wrote my first short story. It took me 30 minutes. It was a dark little tale about a man who found a magic cup and learned that if he wept into the cup, his tears turned into pearls. But even though he had always been poor, he was a happy man and rarely shed a tear. So he found ways to make himself sad so that his tears could make him rich. As the pearls piled up, so did his greed grow. The story ended with the man sitting on a mountain of pearls, knife in hand, weeping helplessly into the cup with his beloved wife's slain body in his arms.

The second quote reminds us about the corruptive appeal of wealth. Money itself is just a means that can be used to different ends. From the economic point of view, money can serve as a medium of exchange, a unit of accounting, or a storage of value. Money also offers stability, but it cannot guarantee happiness. If money falls short on providing people with protection, long-term prospects to live better lives, and the sense of accomplishment, it becomes meaningless. The purpose of eradicating poverty must go beyond supplying material possessions.

Given the complexity of the task, it is impossible to define development. The clues how to approach the field come from multiple sources and ongoing debates. Scholars and practitioners from different backgrounds contribute to the richness of development studies. However, one incredible woman, Wangari Maathai, captured the essence of thinking about development in a meaningful way. Let her insights guide our inquiry:

> "Development" doesn't only entail the acquisition of material things, although everyone should have enough to live with dignity and without fear of starvation or becoming homeless. Instead, it is a means of achieving a quality of life that is sustainable, and of allowing the expression of the full range of creativity and humanity.
>
> In trying to explain my work and my philosophy in the wake of being awarded the Nobel Peace Prize, I was reminded of the traditional African stool, which is comprised of a seat and three legs. The first leg represents democratic space, where rights, – whether human, women's, children's, or environmental – are respected. The second leg symbolizes the sustainable and accountable management of natural resources both for those living today and for those in the future, in a manner that is just and fair, including for people on the margins of society. The third leg stands for what I term "cultures of peace." These take the form of fairness, respect, compassion, forgiveness, recompense, and justice.
>
> Just as the African stool is made out of a single block of wood, each leg, or pillar is reinforced by the others and formed from the same grain, so the issues must be addressed together and simultaneously....
>
> The three legs of the stool support the seat, which in this conception represents the milieu in which development can take place.[4]

II Problems with measuring progress and underdevelopment

Development is a contested term, especially when used to imply that those who consider themselves to be developed have a right to instruct those who are seen as underdeveloped. Further confusion arises because of the way developmental advancement is thought to be possible through the source of deliberate and subjective action.[5] The belief that humanity is developing steadily towards better times became popular around the time of the Enlightenment. The scientific discoveries of the sixteenth century inspired a generation of European thinkers to celebrate progress, understood in terms of individual liberty, secular reasoning, and material improvements. To scholars such as David Hume, John Stuart Mill, Immanuel Kant, Francis Bacon, and Thomas Jefferson, the future looked decisively attractive, especially if contrasted with the oppressive past ruled by the Church and the King. But in their enthusiastic embrace of science and the enlightened self-interest, those thinkers paved the way for inventions that did not always serve us well. Some scholars worry about so-called progress traps, which happen when technological advancements threaten to do permanent damage to our planet.[6]

Consider, for example, the dangerous side of the military technology and nuclear weapons, with their potential to obliterate our planet in an instant. There is also a case to be made with respect to hazardous environmental degradation caused by rapid industrialization. Presently, in the fast-growing Chinese capital of Beijing, the quality of life is endangered by the hazardous level of pollution. When in 2015 the city issued its first "red alert" over the smog levels, all schools were closed, outside activities were discouraged, public transportation curtailed, and private cars were restricted. Such red alerts became more frequent throughout 2016 and took place in several Chinese cities known for experiencing unprecedented air pollution. These are the same cities that also represent marvels of new architectural designs and have become a symbol of China's economic progress.

China is an interesting example of a transforming country. Over last 30 years, China has become a powerful trading nation and lifted millions of people out of poverty. Its economic progress as measured in terms of GDP (gross domestic product) has been unquestionably impressive since the late 1980s. GDP is a useful statistic but has its limitations. GDP is measured in terms of aggregate domestic spending, which includes consumer, business and government spending, investment spending, and the country's net exports: GDP $= C$ (consumer spending) $+ G$ (government spending) $+ I$ (investment spending) $+ NX$ (net exports). As a result, this important economic indicator can provide only a limited, if not distorted, view about the state of a country's economy. Think about the situation when the government adopts a policy of reckless deficit spending, or the case when easy access to credit stimulates insatiable private consumption. Both scenarios can have long-term negative consequences for the health of the economy, but in the short-term they push the GDP higher, possibly inflating its value. The GDP measure also underestimates the true size of those economies that are more knowledge based and rely on globalized production networks. To remedy these problems, the new "inclusive wealth" statistic has been developed by the UN. It calculates a country's assets by adding: 1. manufactured capital (roads, buildings, machines, and equipment); 2. human capital (skills, education, health); 3. natural capital (sub-soil resources, ecosystems, the atmosphere). The evidence suggests that when measured this way, China's inclusive wealth is not that remarkable, and 4.5 times smaller than the wealth of the United States.[7] But again, the growing inequality compounded by the high level of public debt in the United States also questions the strength of the American economy.[8]

In recognizing how difficult it is to determine progress and measure underdevelopment, several indicators have been developed to help categorize different countries for the purpose of aid eligibility. The criteria have been standardized by the UN Economic and Social Council, whose Committee for Development Policy (CDP) is responsible for monitoring and reviewing the status of least developed countries (LDCs). There are three ways of assessing whether a country qualifies for LDC status. The most common, and that used by the World Bank, is the income criterion measured by the gross national income (GNI) per capita. GNI is equal to the GDP less primary incomes payable to non-residents plus primary incomes receivable from non-residents, and is expressed in US$. A country, with a GNI over a three-year-period averaging US$1,035 per capita[9] qualifies as an LDC. The second marker is the Human Assets Index (HAI), which measures the level of human capital by examining the following four indicators: percentage of population undernourished, mortality rate for children five years or younger, gross secondary school enrolment ratio, and adult literacy rate. In 2014, the CDP permanently fixed the LDC inclusion threshold at 60. This value corresponded to the third quartile in the distribution of HAI values of a reference group, which consisted of all LDCs and other low-income countries. The graduation threshold was set at 66 (10 percent higher). The last criterion is the Economic Vulnerability Index (EVI) designed to measure the structural vulnerability of countries to external shocks. The EVI consists of eight indicators that assess a country's level of exposure to economic and environmental shocks given its geography, agricultural and manufacturing makeup, and population. In 2014, the CDP permanently fixed the LDC inclusion threshold at 36. This value corresponded to the first quartile in the distribution of EVI values of a reference group, which consisted of all LDCs and other low-income countries. The graduation threshold was set at 32 (10 percent lower). Although there is no common understanding as to what happens when a country graduates to the next level and what it means to be a developing country.

Based on the above three criteria, there were 48 countries designated as LDCs in 2015: 1 in the Americas, 34 in Africa, and 13 in Asia and the Pacific region.[10] Fulfilling the requirements of being the world's poorest countries triggers their eligibility for different assistance programs. The status of being an LDC indicates problems, but it tells us little about the politics and institutional arrangements inside these countries. We know, however, that many people suffer the consequences of poverty. The numbers are overwhelming. According to the UN, in 2014 the population of LDCs was 898 million people or one-eighth of the world's total.[11] The developmental agencies have been trying to aid these countries for some time. The doctrine of foreign aid is attributed to American president Harry S. Truman, who announced in his 1949 inaugural address the plan of a program aimed at relieving the suffering of people in underdeveloped areas.

This doctrine of development has been strongly criticized as Eurocentric and principally invented to justify the promotion of practices of the advanced industrialized world. First used to explain attempts to institute social order during the industrial revolution in

Europe, when rapid growth of capitalism dislocated many people, development later came to signify superficial divisions between the developed and the underdeveloped societies.[12] Such an instrumental view of development served to induce progress in those countries perceived as immature, and hence deemed to be in need of external help. By extension, this view also informed programs of the newly created post-World War II developmental organizations and aid agencies. Any critical assessment of the foreign aid industry, however, should not stop us from responding to those communities in need. When the help is asked for, some form of outside assistance can be beneficial, especially if it empowers the individuals on the receiving end.

III Social dimension of development economics

Development economics is evolving, both in theory and practice. Some of the better-known economists in the field recognize this adaptation, characterized as branching out beyond the classical and neo-classical approaches to include political economy and new institutional analyses. The authors conclude that development economics "in addition to being concerned with the efficient allocation of existing scarce (or idle) productive resources and with their sustained growth over time, ... must also deal with the economic, social, political and institutional mechanisms, both public and private."[13] This welcome turn towards the social dimension of development economics has led to new insights and policy recommendations.

Perhaps the most ground-breaking impact on the developmental discourse came in the writings of the Nobel Prize winning economist Amartya Sen. In his book *Development as Freedom*, Sen explores the mutually reinforcing linkages between political and economic freedoms and development. His approach sees expanding individual freedom as "the primarily end and as the principal means of development."[14] Rather than classical economics' preoccupation with income and wealth creation, Sen concentrates on capability deprivation and the removal of obstacles to instrumental freedoms including: political freedoms, economic facilities, social opportunities, transparency guarantees, and protective security. Sen's view:[15]

> permits simultaneous appreciation of the vital roles, in the process of development, of many different institutions, including markets and market-related organizations, governments and local authorities, political parties and other civic institutions, educational arrangements and opportunities of open dialogue and debate (including the role of the media and other means of communication).

Another economist, Dani Rodrik, looks at development in the context of globalizing economic networks and growing interconnectedness among individuals and states. Such globalizing trends are having a transformative effect on social norms and expectations. Rodrik is interested in to what extent global pressures constrain national processes of economic policy making. As an economist grounded in the classical tradition, he sees economic growth as the most powerful instrument for reducing poverty. But Rodrik also pays attention to social aspects of economic advancement and believes that governments can play a positive role in stimulating domestic economies. Worried about the deepening scope of international economic organizations, Rodrik supports global regulatory coordination among states rather than outright harmonization, and speaks against hyperglobalization, defined as economic globalization that becomes an end in itself.[16] Instead, he argues for prioritizing the genuine developmental needs of people that for too long have been marginalized by the focus on economic globalization.

The social dimension of development economics is now decisively established. It slowly leads to the integration of approaches to development previously separated for underdeveloped and developed countries. In the past, the idea that poor countries are incapable to conceive their own developmental strategies gave excuses to aid agencies to advance a technocratic model of development. This antiquated model, imposed from the outside, attempted to address developmental issues, mostly in terms of technical solutions and generalized blueprints. The model ignored the historical knowledge about the countries while it favored grand untested designs. Created by external experts, such designs were left to be implemented by the authoritarian leaders of the aid receiving countries. In most cases, the model reinforced existing social divisions, maintained autocrats in power, and did very little to eradicate poverty. To break away from this ineffective approach to development, a persuasive new argument by William Easterly highlights the importance of learning from history while it focuses on the rights of individuals and supports the move towards creating the enabling environment for spontaneous solutions.[17]

In examining attempts to reconcile the micro and the macro approaches to development, the book draws upon a range of insights from several disciplines. The field of economics provides important knowledge, but other fields in social sciences also contribute in a significant way. For example, it is constructive to think about the importance of good democratic governance that involves trust (sound banks, insurance companies, transparent regulations), law and order (rule of law, enforceable contracts), security and protection (all individuals and property), social and labor mobility, competitive elections, just to name a few. Such concepts alert us to the critical importance of state institutions that can either facilitate or hinder processes of socio-economic advancement. Good institutions enable the environment where individuals can grow and prosper in a safe society. A society where people can express their needs and voice their dissatisfaction if their rights are violated. A society where property rights and contracts are respected according to the legitimate rules and where the developmental policy is made in a transparent and accountable way to benefit all citizens. Indeed "we must not let caring about material suffering of the poor change the subject from caring about the rights of the poor."[18]

IV Classical scholars and continuing debates

Over the years different approaches to development have emerged to shape the economic policies of the time. Although many scholars have contributed to the study of socio-economic issues, a look back at the foundational years shows that certain individuals stand exceptionally tall. The ideas espoused by them not only survived the passage of time, but also inspired scholars and practitioners for generations to come. Adam Smith, Karl Marx, John M. Keynes and Friedrich Hayek became known for shaking the establishment and suggesting alternatives to the prevailing attitudes of their contemporaries. Most likely these intellectuals would not be called developmental economists by today's standards. In spite of that, it is difficult to find a book or an article on matters related to development that is not somehow connected to their ideas.

Adam Smith (1723–1790)

It is often said that much of what is covered in economic textbooks can be traced back to Adam Smith. This eighteenth-century Scottish social philosopher, who never used a mathematical equation in his writings, is widely considered a founding father of economic liberalism. Adam Smith was born in the mercantilist's world dominated by powerful landowners and aristocracy. It was a world of conflictual economic relations. Just as the rich landowners wanted to perpetually oppress the peasants, the powerful nations wanted to dominate the weaker nations. The notion of self-interest meant disregard for others. The wealth was associated with territorial gain and accumulation of precious metals like gold and silver. International exchange of any kind was viewed with suspicion, and foreign imports were considered a threat to the stability of monopolistic enterprises at home and favored by the elites. Protectionism was the policy of choice, and international trade was only encouraged on the exports side in order not to deplete the accumulated coins. It was a hierarchical order with very little option for social mobility. Mercantilism resisted innovation, protected inefficient monopolies, and hindered human potential. And Adam Smith despised it.

After securing a professorship at Glasgow University when he was still in his twenties, Smith wrote *The Theory of Moral Sentiments* (1759), which explored the nature of human interactions. In this philosophical treatise, he outlined a vision for mutual cooperation between people, which reconciled his positive view of human nature with the awareness of competition among individuals. Smith observed that we all are naturally guided by self-interest but we are also prone to explore ways for mutual cooperation. His core revolutionary idea was the assertion that humans had developed natural empathy towards each other that allowed the development of a more peaceful and open society.

The concept most closely associated with Smith is the *invisible hand* or *laissez-faire capitalism*, which describes an economic system where the government's involvement in the economy is kept at minimum. Smith advances an organic view of a society based on free-exchange of ideas where cooperation improves people's material well-being. Cooperation thus becomes a matter of self-interest. In Smith's view, any society can function well if there is not too much interference into its natural processes of human interactions. As a result, he warns against a ruthless political leader (or a government) who:

> seems to imagine that he can arrange the different members of a great society with as much ease as the hand arranges the different pieces upon a chess-board. He does

not consider that in the great chess-board of human society, every single piece has a principle of motion of its own, altogether different from that which the legislature might choose to impress upon it.[19]

Smith's most famous work, *An Inquiry into the Nature and Causes of the Wealth of Nations*, was published in 1776. Here he translated his ideas into an economic analysis. Smith maintains that any economic system should be left alone in order to function properly; let the "invisible hand" of free markets work with minimum intervention from the government. In his view, governments should just focus on ensuring peace and order, with taxes collected mainly for defense purposes. Smith concluded that the welfare of the whole society is advanced when people are free to pursue their own interest. It is a powerful thought that expresses Smith's unswerving skepticism about the role of political leadership in managing the economy.

Smith outlined the advantages resulting from the division of labor, such as greater specialization, efficiency, incentives for new inventions, less burdensome work, and skill enhancement. His liberal discourse relies on spontaneous private actors to organize most economic activity through free markets. According to Smith, free markets self-regulate themselves because people are naturally creative, and if they are allowed to freely compete among themselves such entrepreneurs tend to improve their products and services. In an open environment, enterprises have the incentives to provide the right quantity of what is needed by the population.

Smith's writings continue to stimulate contemporary theorists of limited government and practitioners who want to deregulate the economy in favor of free markets. The critics, however, point out the weaknesses of Smith's views, especially when it comes to his reliance on self-regulation of all economic activities. They observe that at least some economic sectors – for example, the resource extraction industry, weapons trade, and health care – should be under close scrutiny of the government's regulators. Because if they become purely profit oriented, these activities can result in illegal activities or disastrous externalities. There is also a danger of collusive business behavior. It is quite possible that Smith himself would be sympathetic to putting in place limited regulatory laws to prevent the establishment of monopolies. A fact sometimes forgotten by those who associate Smith with unrestrained privatization is that he worried about dominant groups – governmental or private elites – who can unfairly dominate societies and distort their natural functions. For him, the test of a well-operating liberal order was not only a small government but, most importantly, free competition. Smith could be disturbed by a number of practices taking place in present-day liberal democracies, such as the political power of big business, corporate subsidies, and the influence of interest groups on policy making. Adam Smith warned against the danger of monopoly.

A monopoly granted either to an individual or to a trading company has the same effect as a secret in trade or manufactures. The monopolists, by keeping the market constantly under-stocked, by never fully supplying the effectual demand, sell their commodities much above the natural price, and raise their emoluments, whether they consist in wages or profit, greatly above their natural rate. The price of

> monopoly is upon every occasion the highest which can be got. The natural price, or the price of free competition, on the contrary, is the lowest which can be taken.[20]

There is a liberating energy in Smith's writing. Smith's liberalism sees people as capable individuals who waste their potential in a hierarchically organized and tightly controlled political system. And he wants to set them free. He believes that only liberated people can create a prosperous and harmonious economic order.[21]

Karl Marx (1818–1883)

In his lecture about Marx and his theory, Dennis Dalton paints a somber picture of a determined activist faced with life-altering poverty. Marx was a German economic philosopher, radical politician, and the author of two influential books, *Das Kapital* and *The Communist Manifesto* (with Friedrich Engels). He hoped to eliminate economic inequality through revolutionary transformations. By the time Marx and his family settled in London in 1849 – after being expelled from several European cities – he had gained a reputation of a dangerous communist agitator. Marx continued to live in absolute poverty to the end of his life and despised it. Not only did he lament the horrible existence of many industrial workers, he was also marked by personal tragedy when three of his children died of preventable diseases.[22]

Marx is perhaps best known for his dialectical understanding of history, stipulating that humans are historical creatures, "who are simultaneously the producers and the products of historical processes."[23] Marx believed that the historical development of capitalism and the corresponding rise of the bourgeoisie "has left remaining no other nexus between man and man than naked self-interest, than callous 'cash payment.'" Marx viewed all social relations as defined by capitalism.

His critique of capitalism is uncompromising.

> In Western Europe, ... the capitalist regime has either directly conquered the whole domain of national production, or, where economic conditions are less developed, it, at least, indirectly controls those strata of society which, though belonging to the antiquated mode of production, continue to exist side by side with it in gradual decay.... The capitalist mode of production and accumulation, and therefore capitalist private property, have for their fundamental condition the annihilation of self-earned private property; in other words, the expropriation of the laborer.[24]

According to Marx, capitalism has become a stage of history when the commodification of social life has reduced every societal relation "to a mere money relation."[25] This is why he viewed capitalist society as inherently unstable and prone to crises, positing that the widening inequality between the property-holders (the bourgeoisie) and workers with no property (the proletariat) was going to undermine the existing order; the internal contradictions of capitalism would lead to intensifying class struggles, which instability would eventually explode in a massive revolution and the bourgeoisie class would be overthrown. Marx hypothesized that, after the revolution, capitalism would be replaced by socialism (dictatorship of the proletariat). With time socialism was expected to evolve into communism, where there is no private property as everything is owned by all.

Marx also talked about alienation to explain how individuals in capitalist society have lost their understanding and control of circumstances around them. Alienation was caused by the exploitative nature of industrialized societies that distorted the natural social development of communities. It turned human beings into *the most wretched of commodities*, deprived of their true identities and subject to class conflicts. Marx's writings became the intellectual foundation of the 1917 revolution in Russia, which triggered a transformation of the country into a centrally run authoritarian regime. One of its main leaders, Vladimir Ilich Lenin, used Marx's ideas to explain the era of conquest as fueled by relentless capitalist accumulation. According to his law of imperialism, the dominant capitalist countries in their search for higher profit will have to seize colonies and create dependencies to serve as markets, outlets of investment, and sources of food and raw materials.[26] More recent neo-Marxist writers continue to argue that capitalism shapes the structures of the world economy in such a way that poor countries are permanently disadvantaged.

Marxism stands on the opposite side of liberalism, which celebrates free markets and self-interest. Liberalism socializes us into entrepreneurial individuals who want to cooperate, create, and advance. Marxism, however, sees liberalism as an illusion that hides the real conflictual societal relations defined by capitalism. The disabling impact of capitalist society is turning human beings into alienated creatures who operate on the logic of survival and greed. Only under communism, people can return to their natural nature because with no private property there will be no conflicting rights claims. Demand for unnecessary products and services will cease to exist and the world will be guided by the principle "from each according to his ability to each according to his needs."[27] But this is an unrealistic assumption. Even if we define the human needs in a very basic survival way, individual needs will always exceed available resources. In any society, adjudicating competing claims is inescapable and always involves opportunity costs. A world free of conflicts that normally arise from scarcity is not possible. In fact, managing scarcity is endemic to collective human life.[28] This important insight signals the major limitation of Marx's theory and the extent to which it can be applied when thinking about economic development.

John Maynard Keynes (1883–1946)

Keynes was a British economist and diplomat. He contested many underlying principles of classical economics. He also dismissed Marx.[29] A pragmatist by admission, Keynes attempted to revolutionize economics into a policy oriented field to make it more relevant in a world changed by the industrial revolution. As an envoy working for the British government, he was not afraid of controversy when he resigned from the public office in protest over the text of the 1919 Treaty of Versailles. Keynes objected to the punitive nature of the treaty and predicted it would have dire consequences for Europe.

The explanations behind his opposition were outlined in his first bestselling book *The Economic Consequences of the Peace*. Keynes argued that the reparation demands and other measures were unrealistically harsh for Germany and would push the population towards drastic solutions. Keynes was proven right, and most historians now agree that the treaty ultimately led to the rise of Nazism and predatory economic nationalism in Germany, paving a way to World War II.

Keynes was a practitioner economist, but he is principally remembered for his classic work, *The General Theory of Employment, Interest and Money* written in 1936. This book became the foundational work of macroeconomics. In contrast to microeconomics, which concerns itself with individual firms and consumers, macroeconomics examines the economy as a whole by focusing on such statistics as unemployment, growth rate, savings, inflation, price level, and national income. Keynes in fact created some of the most recognized macro indicators like GDP. In doing so, he demonstrated his preference for an activist government by means of using fiscal and monetary policy to help eliminate recessions and control economic booms. Presently governments use estimates and predictions derived from sophisticated macroeconomic models to help them design and evaluate economic policies.

Since Keynes is an advocate of interventionist government policy, his ideas often conflict with those of his predecessor, Adam Smith. One particularly devastating period in the world's economic history challenged Keynes' faith in the classical theory. The idea that markets operate best when left alone was at odds with the grim realities of the Great Depression, triggered by the market crash of 1929. As unemployment rocketed to double-digits, factories closed, and hungry people were lining up in soup-kitchens, the government was expected to wait for the economy to eventually return to its equilibrium. The recession engulfed all the major economies, and by extension impacted most countries in the world. Despite the alarming international situation, there was no multilateral effort to coordinate a united response to the crisis. This non-action by governments was consistent with the principles of classical economics. Keynes disagreed. He thought the depression was a lesson that called for a different economic theory.

> Our criticism of the accepted classical theory of economics has consisted not so much in finding logical flaws in its analysis as in pointing out that its tacit assumptions are seldom or never satisfied, with the result that it cannot solve the economic problems of the actual world.[30]

For Keynes, the Great Depression was evidence that "the invisible hand" can sometimes err in catastrophic ways. He was convinced that free markets had their limits and were sometimes capable of dangerously exuberant behavior. If such behavior necessities a major correction, the growing level of unemployment can cause the economy to decline. Based on these observations, Keynes identified insufficient demand as a key problem during the time of economic crisis. He believed that since weak demand may plunge the economy into a recession, the government has responsibility to boost public spending to compensate for weak business and consumer spending.

> The enlargement of the functions of government, involved in the task of adjusting to one another the propensity to consume and the inducement to invest would seem … a terrific encroachment on individualism, I defend it, on the contrary, both as the only predictable means of avoiding the destruction of existing economic forms in the entirety and as the condition of the successful functioning of individual initiative.[31]

According to Keynes, interventionist governments are necessary to correct free markets and safeguard personal liberty from the excesses and defects of capitalistic greed. Keynes was controversial in suggesting that governments should stimulate the aggregate demand by an expansionary fiscal policy, even if it meant borrowing money. In short, Keynes would support budgetary deficits to restore full employment.[32] He famously said that waiting for the adjustment by market forces may take too long and "in the long-term we are all dead."

The theories formed by Keynes were so influential that with time all interventionist policies were referred as outcomes of Keynesian economics. After World War II, Keynes' was heavily involved, as the head of the British delegation, in the negotiations that established the Bretton Woods system of international institutions. Some of these institutions are still at the center of the developmental debate: International Monetary Fund (IMF), International Bank for Reconstruction and Development (World Bank). Keynes ideas informed the developmental policy making in many countries around the world. Although his influence diminished following the end of the Cold War and the subsequent global turn towards classical liberalism, his ideas remain eminent especially during the time of economic crises when government intervention is sought and justified.

Friedrich A. von Hayek (1899–1992)

Despite the post-war popularity of the Keynesian prescriptions, not everybody agreed with his approach. Friedrich von Hayek especially believed in the classical economics and he expanded the ideas of Adam Smith into the field of politics. For him a commanded economy translated into an oppressive authoritarian political system. Hayek did not think that it was possible for any government to effectively manage an economy. He argued that every economic system is volatile and unpredictable, and hence policy makers and central planners are not able to foresee the consequences of their actions. Hayek believed that governmental intervention in the economy hinders the natural operation of economic relations and results in crises. In his famous 1944 work *The Road to Serfdom*, Hayek claimed that socialism had a strong probability of leading towards totalitarianism, because central planning was in fact economic engineering by the government that could not be restricted to the economy and would necessarily impact social life as well.

Arguably the writings of Hayek, resurrected by the neoconservative western leaders in the 1980s, reinforced belief in the supremacy of free markets by providing philosophical justification against government involvement in the economy. The alternative between state-run economy versus free markets was presented as an epic battle of ideas for the global economy.[33] The message followed that the progressive forces of liberalism won the battle and the world finally emerged from the misguided era of central planning and Keynesian interventionism. But even Hayek, who stressed the importance of free competition, as contrasted with distortive totalitarian systems of planning, still recognized the need for the government. Says Hayek[34]:

> The successful use of competition as the principle of social organization precludes certain types of coercive interference with economic life but it admits of others which sometimes may very considerably assist its work, and even requires certain kinds of government action.... And it is essential that the entry into the different trades should be open to all on equal terms and that the law should not tolerate any attempts by individuals of groups to restrict this entry by open or concealed force.

Hayek idealized the free market competition, but he would enlist the help of the government's regulatory authority and rule of law to prevent any form of monopolistic and collusive behavior. In this crucial sentence, he validates the importance of limited state institutions. Unfortunately, Hayek never explained the logistics behind his approach, choosing to be better known as a passionate apostle of unrestricted free markets.

Like many scholars with strong ideas, Hayek's distrust of the state came from his personal experience. Having participated in a disastrous World War I and trying to live a fulfilling life in its aftermath, Hayek became dismayed over policy choices of the time. The war was pointless, but to add to the injury, governments continued to create havoc in the lives of millions of people in the post-war period. The experience of Austria's runaway inflation of the early 1920s became for Hayek a lesson in government's incompetence. The economy was collapsing and people were suffering. In contrast to Keynes, however, Hayek thought the remedy was to diminish the role of the state in the management of the economy. Soon the battle lines between these two economists were drawn, leading to the clash that famously defined modern economics:

> Keynes believed it was a government's duty to do what it could to make life easier, particularly for the unemployed. Hayek believed it was futile for governments to interfere with forces that were, in their own way, as immutable as natural forces.... Thus the two men came to represent two alternative views of life and government.[35]

Keynes would be known as an optimist who believed that the government can and should take an active role in the economy. Hayek would be known as a pessimist who believed that the economy, just like the laws of nature, should not be tampered with.

V Concluding remarks

At the center of the debates ignited by these four thinkers, is the question how to ensure the most effective use of scarce resources. Scarcity is the fundamental puzzle of economic development. Developmental practitioners wonder how the world can become a generous place for all its inhabitants. Can the world economy offer fair opportunities for advancement to every person on the planet? Should we leave production, innovation, and distribution to market forces or should we trust governments to manage economic systems for the betterment of societies? Is the government responsible for providing citizens with economic and social prospects that allow people to fulfill their dreams? Or rather, should it be up to individuals to fend for themselves and all governments should stay away from the economy? Just like the clash between Keynes and Hayek, the dichotomy of choices between unrestrained market forces and interventionist governments leads to polarizing options. It has been a sad feature of our civilization that at any given time of history a particular generation tends to exaggerate its preference towards one of these two choices. It is time to bridge the best elements of the two perspectives into an approach that recognizes the importance of state institutions, appreciates the role of free markets, and keeps an eye on individual rights when thinking about economic development.

Notes

1 Khaled Hosseini (2004) *The Kite Runner*, Anchor Canada: Toronto, Canada.
2 Ibid., pp. 224–225.
3 Ibid., p. 33.
4 Wangari Maathai (2009) *The Challenge for Africa*, New York, NY: Pantheon Books, pp. 56–57.
5 M.P. Cowen and R.W. Shenton (1996) *Doctrines of Development*, New York, NY: Routledge.
6 Ronald Wright (2006) *An Illustrated Short History of Progress*, Toronto, CA: House of Anansi Press.
7 Stephen G. Brooks and William C. Wohlforth (2016) "The Once and Future Superpower – Why China Won't Overtake the United States," *Foreign Affairs*, May–June, pp. 91–104.
8 Niall Ferguson (2013) *The Great Degeneration – How Institutions Decay and Economies Die*, New York, NY: Penguin Group.
9 As per the UN review conducted in 2015. Online, available at: www.un.org/development/desa/.
10 For a complete list of LDCs and detailed information about the three sets of criteria used to establish an LDC status, please go to the website of the UN Committee for Development Policy, online, available at: www.un.org/en/development/desa/policy/cdp/index.shtml.
11 UN Office of the High Representative for the Least Developed Countries, Landlocked Countries, and Small Island Developing Countries, online, available at: http://unohrlls.org/about-ldcs/facts-and-figures-2/ (accessed May 2016).
12 M.P. Cowen and R.W. Shenton (1996) *Doctrines of Development*, op. cit.
13 Michael P. Todaro and Stephen C. Smith (2003) *Economic Development* (8th edition), New York, NY: Addison Wesley, p. 8.
14 Amartya Sen (1999) *Development as Freedom*, New York, NY: Random House, p. xii.
15 Ibid., p. 9.
16 Dani Rodrik (2010) *The Globalization Paradox – Democracy and the Future of the World Economy*, New York, NY: W.W. Norton, p. xvii.
17 William Easterly (2013) *The Tyranny of Experts: Economists, Dictators, and the Forgotten Rights of the Poor*, New York, NY: Basic Books.
18 Ibid., p. 339.
19 Adam Smith (1759) *The Theory of Moral Sentiments*, Part VI, Section II, Chapter II, pp. 233–234, paragraph 17.

20 Adam Smith (1776) *An Inquiry into the Nature and Causes of the Wealth of Nations*, London, UK: Methuen & Company Ltd, p. 65.

21 For more information about Adam Smith and his works please visit the Adam Smith Institute, online, available at: www.adamsmith.org.

22 Dennis G. Dalton (1998) *Marx's Theory of Human Nature and Society*, Chantilly, VA: The Teaching Company.

23 Mark Rupert (2016) "Marxism," in *International Relations Theories – Discipline and Diversity*, Oxford, UK: Oxford University Press, p. 129.

24 Karl Marx (1967) *Capital Vol. 1*, chapter 33, online, available at: www.marxists.org./archive.

25 Karl Marx and Friedrich Engels (1848) *Manifesto of the Communist Party*, online, available at: www.marxists.org./archive.

26 Robert Gilpin (1987) *The Political Economy of International Relations*, Princeton, NJ: Princeton University Press, p. 39.

27 Karl Marx (1875) "Critique of the Gotha Programme," cited in Ian Shapiro (2016) *Politics Against Domination*, Cambridge, MA: Harvard University Press, p. 8.

28 Ian Shapiro (2016) *Politics Against Domination*, Cambridge, MA: Harvard University Press, pp. 6–9.

29 Claudio Sardoni (1997) "Keynes and Marx," in G.C. Harcourt and P. Riach (eds.), *The General Theory of Employment, Interest and Money* (2nd edition), London, UK and New York, NY: Routledge.

30 John Maynard Keynes (1936) *The General Theory of Employment, Interest and Money* (Illustrated and Extended with John M. Keynes Library), Electronic Edition, part iii of chapter 24 ("Concluding Notes on the Social Philosophy Towards Which the General Theory Might Lead"), online, available at: www.WealthOfNation.com, accessed November 7, 2012.

31 Ibid.

32 Ibid., part iii of chapter 8 ("The Propensity to Consume: The Objective Factors"). Electronic Edition.

33 Daniel Yergin and Joseph Stanislaw (2002) *The Commanding Heights – The Battle for the World Economy*, New York, NY: Simon and Schuster, pp. 123–127.

34 F.A. Hayek (1944) *Road to Serfdom: Text and Documents – The Definite Edition*, edited by Bruce Caldwell, Chicago, IL: University of Chicago Press (2007), p. 86.

35 Nicholas Wapshott (2011) *Keynes and Hayek – The Clash That Defined Modern Economics*, New York, NY: W.W. Norton & Company, pp. 43–44.

Suggested further reading

Bruce Caldwell (2005) *Hayek's Challenge: An Intellectual Biography of F.A. Hayek*, Chicago, IL: Chicago University Press.

David Harvey (2010) *A Companion to Marx's Capital*, London, UK and Brooklyn, NY: Verso.

Ehsan Masood (2016) *The Great Invention: The Story of GDP and the Making (and Unmaking) of the Modern World*, New York, NY: Pegasus Books.

Ingrid H. Rima (2009) *Development of Economic Analysis* (7th edition), London, UK: Routledge.

Agnar Sandmo (2011) *Economics Evolving: A History of Economic Thought*, Princeton, NJ: Princeton University Press.

Adam Simpson Ross (2010) *The Life of Adam Smith*, Oxford, UK: Oxford University Press.

Robert Skidelsky (2010) *Keynes: The Return of the Master*, New York, NY: Public Affairs.

Theories and approaches to development   1

I Conceptualizing theories

The intellectual trajectory of development studies has been shaped by historical events and enriched by insights from multiple disciplines. This is reflected in a sometimes uneasy relationship between diverse academic approaches and practical answers to questions about what needs to be done to improve the well-being of societies. To gain appreciation of those different approaches, we have to enter the realm of theory. Theory is a logically constructed framework, describing the behavior of a certain observable phenomenon. It has to be somehow testable, which distinguishes theory from a personal belief. Theory can also be generated by applying a method of deduction and logic. Scientific theory is formed and evaluated according to a scientific method, which involves experiments and hypothesis testing. When it comes to social theory, we look for patterns that allow the explanation of the conditions under which linkages between two or more events exist. Social theorists strive to be scientific when carefully evaluating the relevant facts, often with the help of statistical methods.

The first influential theories of development emerged in the 1950s. Since then, the growing field of development studies saw the retreat of some theories, revisions of others, and the ascendency of new ones. Different theories propose diverse ways of identifying the crucial aspects of a particular fact. Choosing which theory is the most appropriate is often determined on the assumptions we make about the nature of society and the role of the state in the economy. Theories put forward explanations of why events happened. Theories that stem from a philosophical system of positivism search for objective findings that are free from personal biases and opinions. The *Oxford Dictionary of the Social Sciences* defines positivism as:

> A philosophical and social scientific doctrine that upholds the primacy of sense experience and empirical evidence as the basis for knowledge and research. The term was coined by Auguste Comte to emphasize the doctrine's rejection of value judgments, its privileging of observable facts and relationships, and the application of knowledge gained by this approach to the improvement of human society.[1]

Post-modernist thinkers, however, question the very notion of objectivity. They also contest the neutrality of the relationship between knowledge and power because, as Foucault has argued, it is through discourse that the individual subject gets established. Post-modernism considers discourse as a social construct through which knowledge and power are connected. Discourse serves to control access and content of knowledge. Discourse is enabled by those who have power, understood as "a domain of strategic relations focusing on the behaviors of the other or others, and employing various procedures and techniques according to the case, the institutional frameworks, social groups, and historical periods in which they develop."[2] Such radically critical positions tender important philosophical insights, but in rejecting the possibility of objective knowledge, they are difficult to operationalize within positivist constraints of policy making.

Positivism believes in systematic empirical observation and objective rationalism. Accordingly, positivist thinkers tend to favor problem-solving theories when designing policies and developmental strategies. To clarify the term: problem-solving theory

> takes the world as it finds it, with the prevailing social and power relationships and the institutions into which they are organized, as the given framework for action. The general aim of problem-solving is to make these relationships and institutions work smoothly by dealing effectively with particular sources of trouble.

Our examination of theories of development would not be complete, however, without paying attention to a number of prominent critical theories. Critical theory

> is critical in the sense that it stands apart from the prevailing order of the world and asks how that order came about. Critical theory, unlike problem-solving theory, does not take institutions and social and power relations for granted but calls them into question by concerning itself with their origins and how and whether they might be in the process of changing.[3]

By learning different perspectives and understanding of both kinds of theories – problem-solving and critical – one is alerted to the strengths and weaknesses of the policies they inform or contest.

Theories can become powerful trendsetters. In any given historical period one theory can effectively stimulate an emerging paradigm of thought. A paradigm is a widely accepted dominant way of thinking about decisive world issues. Paradigms are customarily initiated, modified or phased out by major international events. For example, after the end of the Cold War, economic liberalism become a catalyst behind the globalization paradigm. Imposing paradigms can prompt international organizations to formulate a universal blueprint of policy prescriptions. At the height of the globalizing moment of the late 1990s, a growing number of developing countries started to question the set of liberalizing reforms they were expected to introduce. Their eventual objection to the universalized policy recommendations put forward by the major donor agencies led to a more thoughtful way of looking at the developing world. Related to the issue of paradigms is the question of how to learn from the past in order to bridge the lessons generated by diverse theoretical approaches. As a rule, theory comparisons always require an awareness of the historical context in which such theories are evaluated and applied.

II Early theories: from growth model to structural change models

The common themes running through the first post-war debates about development were economic growth and progressive industrialization. A dominant paradigm that emerged during this period was called developmentalism, which with time came to signify the promotion of western narratives of modernization. These narratives were grounded in the economic liberalism of Adam Smith, who described the workings of free market forces to advance the welfare of societies. Paradoxically, however, developmentalism failed to afford economic freedoms to "the helpless parts of the world"[4] by assuming that these poor and underdeveloped countries were incapable of developing on their own. Developmentalism needed a centralized state, not free markets, to implement the strategies articulated by the developmental agencies offering foreign aid.

Perhaps the most complete outline of developmentalism can be found in the 1970 report of the Commission on International Development, known as the Pearson Report.[5] It was a major document heralding the new approach to economic development. The report announced 30 major goals and recommendations categorized into 10 main categories: trade, foreign investment, economic growth, volume of aid, debt relief, aid administration, technical assistance, population control, aid to education and research, and multilateral aid. In a nutshell, the report stressed liberal priorities of expanding international trade and investment, while it explicitly tied any increases in foreign aid to targeted increases in developing countries' economic growth, measured in terms of GNP of at least 6 percent a year.[6] These goals and recommendations crystallized the main pillars of the developmentalist model, which for decades informed both theory and practice of development.

The report was named "Partners in Development," but the only partners mentioned on its pages were selective states: those states that were designing and providing foreign aid and the states receiving foreign aid. There was no attempt to examine the sociopolitical environments of targeted countries – they were referred to as a nameless group – and there was no attempt to anticipate the social consequences of the proposed recommendations. The International Development Association (IDA), a newly established sub-agency of the World Bank, was to provide worldwide institutional guidance for meeting the main goals of the report. The IDA was chosen because its mandate was to focus only on the economics. As the report noted:

> [The] IDA is in the best position to exert leadership in the effort to establish criteria for the allocation of aid which emphasize economic performance, rather than the political relationships and historical accidents which bear little or no relation to development needs or performance.[7]

In summary, the report legitimized the consensus among the western industrialized countries. It encouraged the selective involvement of foreign aid organizations in the economies of the South under the label of improving them.

In view of that, there is one major contradiction in the report. Despite promoting development based on ostensibly liberal economic principles, the report evidently relies on the state to turn policy prescriptions into practice. To cite one of the main assumptions of the report: "it is clear that the nation-state has particularly vital functions at this juncture in history."[8] Sure enough, the Keynesian economists who accepted the macroeconomic intervention in the developed world were ready to support centralized macroeconomic management in the developing countries. Yet by doing so, these economists overlooked the fact that many of the targeted states were authoritarian and oppressive. And Keynes himself recognized the importance of freedoms inside the state:

> the greater parts of goods may be carried from one end of the kingdom to the other, without requiring any permit or let-pass, without being subject to question, visit, or examination from the revenue officers. This freedom of interior commerce, the effect of the uniformity of the system of taxation, is perhaps one of the principal causes of the prosperity of Great Britain.[9]

Perhaps it was missed that Keynes, a liberal at heart, wrote his *General Theory* for an open economy, assuming that the government can be removed by its citizens if the macroeconomic policies fail.

Nevertheless, while we are not certain whether Keynes could accept the principle of the Pearson Report, we can be sure that Adam Smith would reject it. He would be puzzled by the embrace of a strong state as the engine of liberal economic policy making in the developing world. To sum up, Smith's liberalism was motivated by a belief in the creative human spirit, which if unhindered by oppressive governments and vested monopolies, can positively drive societies towards improvement and modernization. In the very same year that he published *The Wealth of Nations*, the United States was born. The American Declaration of Independence advanced novel claims that all men were born equal. The document was inspired by political changes in England. In 1623, English Parliament diminished the influence of the Crown by passing the Statute of Monopolies due to growing social discontent, claiming that monopolies (routinely given to the prominent families) were stifling the English economy. The significance of the statute was enormous as it prohibited the monarch from granting new monopolies. The decisive moment, however, came with the Glorious Revolution of 1688. It forced the King to negotiate a new constitution and established the supreme role of Parliament. This new political arrangement meant that the monarch could no longer arbitrarily change laws or call for new taxes. In 1707, Great Britain was created following the Act of the Union between England and Scotland, and Parliament became fully responsible for policies of the state.[10] John Locke's writings provided a philosophical justification of the revolution in England and encouraged the American Independence movement. It was just a matter of time before political rights and civil liberties would become more universally granted. Adam Smith took note of these developments with his uncompromising faith in the rationality of human actions; an idea inherited from the Enlightenment. As a testament to his work, the economic and political liberalism became a transformative force working towards freeing human potential and individualism.

The spread of the liberal paradigm motivated Joseph Schumpeter (1883–1950) to formulate the very first theory of development, which emphasized economic growth and technological innovation. Schumpeter's book *Theory of Economic Development*, written at the beginning of the twentieth century, used lessons from business and applied economics to assign new importance to the concept known as the free marketplace of ideas. At the center of his theory stands the entrepreneur, who enjoys the stimulus of free market competition and brings forward innovations. The resulting progress erodes old structures and gives rise to a powerful new force, which Schumpeter called *creative destruction*. Contemporary developmental economists see the processes described by Schumpeter as necessary prerequisites for establishing modern systems of democratic institutions that foster growth and more harmonious social relations. On the other hand, "fear of creative destruction is often at the root of the opposition to inclusive economic and political institutions."[11]

Schumpeter considered many economists to be too theoretical and detached from reality. He gained a reputation of being a fierce critic who disapproved of his colleagues, including John Maynard Keynes. What connected them, however, was the fact that they both favored problem-solving theory and hoped to offer concrete remedies for policy makers. Keynes, especially, became known for proposing a number of helpful statistical tools, such as GDP, to assist governments in monitoring the operation of their economies. For Keynes theory led to positive practical outcomes. Then again it is impossible to deny that Keynes understood the power of ideas. On the last page of his groundbreaking book *General Theory*, Keynes wrote:

> the ideas of economists and political philosophers, both when they are right and when they are wrong, are more powerful than is commonly understood. Indeed, the world is ruled by little else. Practical men, who believe themselves to be quite exempt from any intellectual influences, are usually the slaves of some defunct economist.[12]

As the head of the British delegation at the pivotal Bretton Woods Conference in July 1944, Keynes became instrumental in drafting the documents that created the IMF and the World Bank. World War II was coming to an end, and the allies in anticipation of their victory were ready to embrace liberalism as the underlying paradigm of the postwar world order. Keynes was prolific in developing a scheme for an international system of currency trading. On the American side, Harry Dexter White worked on a similar arrangement. They both agreed there was a need for multinational organizations to provide assistance for countries experiencing balance-of-payments problems. After three weeks of negotiations, 44 countries signed what become known as the Bretton Woods agreement. Keynes also became involved in managing the newly established organizations, but died suddenly less than two years later in 1946 of a heart attack.

While the IMF was responsible for managing the fixed exchange rate system, centered on the US dollar and pegged at the rate of US$35 per ounce of gold, the International

Bank for Reconstruction and Development (World Bank) was tasked with post-war European reconstruction. It was not until several years later, during the wave of the post-colonial independence movements, that the World Bank started to pay attention to countries in Africa, Asia, and Latin America. This early omission became the World Bank's institutional weakness. At the time of the Bank's establishment in 1944, there was no clear objective to help with the developmental projects in the developing world, which is primarily the reason for its existence today.

The de-colonization processes greatly influenced the field of economics. Sir W. Arthur Lewis (1915–1991), a pioneer researcher of economic development, personally experienced the legacy of colonialism. Lewis was born on the Caribbean island of St. Lucia. The island is small, but two major colonial powers, France and the United Kingdom, were fighting over it throughout much of the seventeenth and early eighteenth centuries. St. Lucia was eventually colonized by the British and would only become independent in 1979. Struggling with poverty and living with consequences of the colonial rule gave Lewis a unique perspective. Studious and determined, he was awarded a scholarship and moved to the United Kingdom to study at the London School of Economics. Upon completion of a doctorate degree, he devoted his life to study development intellectually fighting for racial equality and decolonization. He was the first black professor at a UK university (University of Manchester) and at Princeton University in the United States, where he worked for over 20 years until his retirement. In 1955, Lewis wrote a classic book, *The Theory of Economic Growth*. In it he formulated his famous two-sector model of a developing economy.[13]

Lewis theorized that traditional economies with surplus-labor, typically located in the developing world, were characterized by dualism. The dual economy consisted of a large agricultural sector and a newly industrializing urban center. Lewis lamented that the agricultural sector in many poor countries was neglected by governments in terms of technological innovation. This resulted in low productivity, demanding that many people continue to work for poor wages in agriculture. He believed that the government should stimulate innovation leading to the expansion of the manufacturing sector and the corresponding production for exports. In his own words:

> The fact that an expansion of manufacturing production does not require an expansion of agricultural production if it is backed by a growing export of manufactures is particularly important to those over-populated countries which cannot hope to increase their agricultural output for food as rapidly as their demand for food however much they may try. In such countries industrialization in no sense waits upon agricultural expansion, even though it remains true that they should give great attention to agricultural production. Such countries have therefore to give urgent attention to increasing the export market for their manufactures since, in the last analysis, it is the rate of growth of their exports which sets the limit to their internal expansion.[14]

Lewis was interested in the process of labor transition from agricultural to urbanized labor force and the growth of output and employment in the industrialized sector.

He observed that because they were surplus-labor economies with labor in infinite supply in the agricultural sector, the traditional labor had incentives to move to the urban center where wages were rising. Lewis maintained that this movement happened without compromising productivity in the agricultural sector. Competition among workers caused wages in the urban center to stabilize, leading to high profits in the industrialized sector, which financed its further expansion. This reallocation of labor would happen until the turning point was reached, defined as "the time when labor reallocation has outstripped population growth long enough for dualism to atrophy and the economy to become fully commercialized."[15] At this stage of economic development the economy becomes a one-sector industrialized economy.

Often overlooked is Lewis contribution to the concept of human capital within the development process. Lewis considered the processes of labor transition as necessary for the advancement of societies, but still mainly operating according to market forces. He warned that industrialization should never come at the cost of social development. The goal was to maintain the balance between agriculture and industry and between export and home consumption.[16] Although during his lifetime he was actively involved in policy making, Lewis believed that developmental policies could not be socially disruptive. In recognizing the importance of accountability, he rejected one-party rule, widely popular in Africa during the 1960s, and argued that competitive multi-party democracy was essential, given the social pluralism of African countries. For Lewis, the improvement in human capital, with workers having access to education and health care, was a necessary prerequisite for economic progress.[17]

Lewis was nevertheless criticized for simplistically equating development with growth and industrialization. Critics accused Lewis of perpetuating the view of newly decolonizing nations as the underdeveloped versions of western countries that would eventually resemble economies in the industrialized world. Despite the fact that his theory branched out of the classical liberal belief in free markets, the ideas Lewis espoused provided many governments in the developing world with justification to actively support industrialization in the cities at the expense of the rural areas. When theory was put into practice, rural workers attracted by the promise of better wages would abandon the agricultural sector to look for industrial jobs in the cities. Many such workers would be unable to find permanent manufacturing jobs, leading to the growth of shanty towns on the outskirts of major urban centers. Yet Lewis anticipated the problem and tried to warn policy-makers about this kind of outcome, known as the urban bias. The problem was caused by allocating a disproportionate amount of resources to the cities.[18] Industrial production could be short-lived if it relied on unpredictable government subsidies. In recognition for his work, Lewis was knighted in 1963, and in 1979 he became the first person born outside Europe or North America to be awarded the Nobel Prize in Economics.

Many economists who dominated the early post-war developmental debates subscribed to a Keynesian belief that governments can positively stimulate the economy. The intellectual paradigm of the day increasingly favored structural explanations, which assumed that all countries passed through the same stages of economic development. The assumption was that the underdeveloped Third World countries were merely at an earlier stage of the linear historical progress, while First World (the industrialized western economies) and Second World (Soviet Bloc) nations were at a later stage. The term Third World was coined by the French scholar Alfred Sauvy, in an article published in 1952. While commenting on poor countries in the South who were caught between the growing animosities of the Cold War, he wrote: "because at the end this ignored, exploited, scorned Third World like the Third Estate, wants to become something too."[19]

As a theoretical concept, the linear view of progress was developed by Walt W. Rostow (1916–2003) in his book *The Stages of Economic Growth*, to become known as a modernization theory.[20] According to Rostow's model, economic modernization takes place in five historical stages of different lengths: 1. traditional society (backward, unstable, underdeveloped); 2. establishment of preconditions for take-off (technological advances, more stable society); 3. take-off stage (rapid but uneven economic growth); 4. the drive to maturity (period of long sustained growth); 5. the age of mass consumption (income per capita at such a high level that consumers move beyond sustenance and purchase luxury items). Rostow set out a number of conditions that were expected to occur with respect to capital accumulation, consumption, use of technology, and social tendencies at each stage. He considered the take-off stage the most critical for economic advancement of societies. Still his summary of developments that occur at this stage sounds very optimistic, if not naive:

> We come now to the great watershed in the life of modern societies: the third stage in this sequence, the take-off.... During the take-off new industries expand rapidly, yielding profits a large proportion of which are reinvested in new plant; and these new industries, in turn, stimulate, through their rapidly expanding requirement for factory workers, the services to support them, and for other manufactured goods, a further expansion in urban areas and in other modern industrial plants. The whole process of expansion in the modern sector yields an increase of income in the hands of those who not only save at high rates but place their savings at the disposal of those engaged in modern sector activities. The new class of entrepreneurs expands; and it directs the enlarging flows of investment in the private sector. The economy exploits hitherto unused natural resources and methods of production.
>
> New techniques spread in agriculture as well as industry, as agriculture is commercialized, and increasing numbers of farmers are prepared to accept the new methods and the deep changes they bring to ways of life. The revolutionary changes in agricultural productivity are an essential condition for successful take-off; for modernization of a society increases radically its bill for agricultural products. In a decade or two both the basic structure of the economy and the social and political structure of the society are transformed in such a way that a steady rate of growth can be, thereafter, regularly sustained.[21]

Modernization theory inspired by Rostow was quite limited as it neglected to take into account the internal specificity of countries, such as their institutional fabric, social dynamics, and geography. Its main assumptions created theoretical obstacles to an in-depth analysis by pretending to be a-historical. For example, the theory alleged that the stages of growth could take place in different historical contexts, but then how one would be able to examine and compare distinct countries and societies across centuries? Modernization theory also ignored exogenous factors, which nevertheless could profoundly impact societies; factors such as international commerce and geopolitics.

Policy makers in the West nevertheless embraced Rostow's model, believing that economic progress in poor countries could be facilitated from the outside with the help of foreign aid and friendly governments. It was a conclusion widely accepted during the 1960s, given the popularity of Keynesian interventionists ideas. Rostow became a foreign policy advisor under President Kennedy, and then President Johnson, and modernization theory was put into practice. The Kennedy Administration was responsible for a number of critical initiatives concerning Third World policy, including the passing of the Foreign Assistance Act of 1961, a significant increase in the volume of foreign aid, the creation of the United States Agency for International Development (USAID) and the expansion of the Alliance for Progress to promote economic development in Latin America.

A proliferation of western organizations, foreign aid initiatives and advisory bodies, including the 1964 United Nations Conference on Trade and Development (UNCTAD), compelled political scientists to theorize international institutions. The Bretton Woods era produced institutions of all sorts, through which relationships between the developed and developing countries were being established and sustained. Initially, it was the theory of hegemonic stability that offered the most consistent explanation as to why certain institutions were created. The theory links the process of institutional formation with the ability of the internationally dominant state (hegemon) to exercise its authority by stabilizing the system of states. It stipulates that a hegemonic state tends to advocate its own model of development on the countries that fall under its sphere of influence "because it creates and enforces the rules of the game over each region it dominates."[22] The hegemonic stability theory was effective in responding to the theoretical questions of thinkers coming from two competing structural positions: neo-realists and neo-liberals. Both these theoretical schools liked it for pointing out that international order could be maintained by a single hegemonic state. Even after the collapse of the Soviet Union, the theory stayed relevant, as some scholars argued that the United States remained the only possible hegemon. Most recently, the spectacular growth of China prompted speculations over its hegemonic ambitions.

Different explanations for growing economic integration during the post-war period were offered by liberal institutionalism. This theory is derived from the writings of Immanuel Kant, who believed that lasting peace and global prosperity is possible. He managed to transfer the classical liberal ideas about individuals living in a society into the international context of inter-state relations. Just like humans, who are rational beings and inclined to cooperate which each other, Kant maintained that countries have natural propensities towards collaboration and institution building. Kant even proposed a world constitution based on the cosmopolitan law, with rules and norms that contained rights and obligations of the individual states.[23] Disciples of Kant believed that it was his take on liberal theory that inspired the creation of the UN, the IMF, and the World Bank as universal organizations for all nations. Liberal internationalists are known for a cosmopolitan view of the world; they see the world as a unified entity. This has implications for thinking about development because developed states are expected to assume responsibility for the poverty in the poor countries. As the Pearson Report stated: "there must be a great concern in all nations for the fate of all other nations, and that this must reflect itself in more effective co-operation, including co-operation for development."[24] Liberal internationalism and modernization theory tend to reinforce each other. Modernization theory recognizes the western economic model as the finest model that should be universalized. Liberal internationalism supports a proliferation of international organizations framed on universal norms, which critics consider to be predominantly based on western values.

Functionalism is another theoretical perspective that deserves equal consideration. Initially this approach was preoccupied with how to stimulate nations to work together when reconstructing war-torn Europe.[25] Later functionalism turned to the question on how to galvanize economic cooperation among all countries. David Mitrany, the founder of functionalism, believed that successful collaboration in one particular area would lead to further collaboration in related fields.[26] In other words, collaboration happens because there is need for it. Once people start to collaborate, collaborating units create functionally organic linkages among nations. And as the involved departments become more and more integrated, they create institutionalized networks beneficial to all. The anticipated result is a progressive process of global functional integration. Mitrany believed that he had already witnessed the positive impact of such processes in the areas of international transportation and communication. Theoretically speaking, functionalism coined what is known as *the cobweb model*. It theorized the creation of a transnational world society that would eventually include all developing countries. International institutions, which Mitrany called "functional schemes," are important global actors in this process because their "functional 'neutrality' was assumed and welcomed. In addition, functional arrangements have the virtue of technical self-determination, one of the main reasons that makes them more readily acceptable."[27] In contrast with other liberal perspectives, the functional alternative to global cooperation maintains that integration of poor countries into the world economy can happen in an organic way without any activist intervention of the industrialized countries or a hegemon.

Functionalism was distinct among the growing family of structural approaches to development, which were mainly interested in how post-colonial countries were transforming their economic structures from traditional to industrialized economies. The work by Seers and Mahbub ul Haq further enriched those perspectives. In 1969, Dudley Seers (1920–1983) published an article "The Meaning of Development," which argued for seeing development as a social phenomenon. He contested the primacy given to capital investment as a spur of growth and sought to broaden the measures of development from economic growth and GNP to measures for eliminating poverty, unemployment and inequality.[28] In 1976, Mahbub ul Haq (1934–1998) provided a critical analysis of modernization theory and policy in his 1976 book, *The Poverty Curtain: Choices for the Third World*. The work was enriched by his experiences as an economist in the Pakistani government and the World Bank. Mahbub ul Haq argued that a narrowly measured economic growth may actually be counterproductive to economic development since this indicator may mask reductions in social standards and increased income inequality.[29] To counter this, he suggested to rethink development by focusing on human needs. He called for new policy measures away from per-capita growth and towards indicators monitoring malnutrition, disease, illiteracy, unemployment, and inequality. During his tenure at the World Bank (1970–1982), he tried to transform it into an institution that focuses on people instead of on rigid economic indicators. In 1989, he moved to New York, where he served as special adviser to the UNDP (United Nations Development Programme) administrator until 1995. It was at this time that he created the Human Development Index (HDI), which ranks countries in terms of human development and quality of life. It has been used since 1993 by the UNDP in its annual report.

The HDI is a composite index (scale from 0 to 1) that measures the average achievements in a country in three basic dimensions of human development: 1. A long and healthy life – as measured by life expectancy at birth; 2. Knowledge – as measured by the adult literacy rate and the combined gross enrollment ratio for primary, secondary and tertiary schools; 3. A decent standard of living – as measured by GDP per capita in purchasing power parity (PPP) US$. The annual UN Development Report remains one of the most influential global reports and it will be discussed in depth later.

Both Seers and Mahbub ul Haq succeeded in broadening the understanding of economic development. Their legacy meant that structural issues such as population growth, inequality, urbanization, agricultural reforms, education, health, and unemployment, began to be reviewed on their own merits, and not merely as attachments to a prevailing economic growth and industrialization paradigm. Structuralism increasingly paid attention to the distinct structural problems of developing countries; it proposed that Third World nations were not merely backward versions of developed countries, but rather they had distinctive features of their own. As a result, this approach stressed the need for country-specific analysis of development.

During the Cold War, the Soviet Union as the hegemonic power in Central and Eastern Europe influenced the establishment of a Soviet style centrally controlled model of socio-economic development supported by unique institutional arrangements. Similarly, in the West, the idea of liberal institutions and free market economy that emanated from the United States influenced western developmental strategies in this part of the world. However, the growing perception that foreign agencies were interrupting rather than enabling developmental processes in the Third World resulted in strong intellectual critiques of western theory and policy. These critiques were influenced by Marxist thought, with dependency theory emerging as the most popular.

III The state as the agent of development: dependency movement

Karl Marx never tried to design a formal theory of development. Where he excelled was the strength of his ideas. His writings inspired theories that shared some transformational similarities with modernization theory, but radically differed from its main conclusions. The Marxist view of historical developmental process is dialectical, in contrast to modernization theory, where progress is linear. Dialectically understood development processes work to transform societies from traditional to modern and then to capitalism. The ultimate goal is a revolutionary destruction of the capitalist society so a new order can replace it. Far from mass consumption, this new world order is to be characterized by communist harmony and lack of private ownership. Skeptical about the liberal preference for free markets, Marx looked around and saw the exploitative asymmetry between the owners of the means of production and the alienated workers. Consequently, while western approaches to development took for granted the predominance of economic liberalism and market economy, the Marxist model of development fundamentally rejected the principles of free markets.

After its socialist revolution in 1917, the Soviet Union put into practice its own development model, rooted in Marxism. It was a self-reliant closed economy model, which supposed to take a society from a centrally planned socialism to ultimate communism. In the Soviet system markets did not exist, all production was nationalized, and private property was essentially prohibited. In summary: "supply and demand were irrelevant; they were exiled. Resources were allocated by bureaucratic decision rather than by the tens of millions of individual choices that add up to supply and demand."[30] In practice, the model translated to complete state control over economic activities, prohibitively high import tariffs, and the centrally planned production of goods and services. Soviet style Marxist economists maintained that the command economy could ensure the production of necessary goods and create an egalitarian society with all citizens enjoying the equal distribution of resources. It was believed that a centrally run system would avoid a crisis of overproduction and guarantee employment for all because it would not rely on the uncertainties of free markets. Until the early 1970s, the Soviet Union grew at an impressive rate as a massive concentration of resources, accelerated industrialization, and expansion of industrial-military complex drove the economy forward. The rigidity of the central allocation scheme gradually resulted in massive inefficiencies of all kinds. But these problems would not be revealed to the world until the Soviet economy entered an alarming crisis in the 1980s.

During the 1950s, the Soviet model looked very attractive. As the world become divided between the capitalist West and the socialist East, former colonies faced an uneasy choice upon independence. Both Cold War rivals pressed newly independent countries to adopt their version of a developmental model in the attempt to extend their sphere of influence. The statist model practiced by the Soviet Bloc looked particularly attractive to countries with a difficult colonial past. In addition, the western modernization model started to be contested by the new theories rooted in Marxism. The strongest criticism came from Latin America.

Raul Prebisch (1901–1986), a classically trained economist from Argentina, turned his attention to the centrally run economies and in the process encouraged a radically new theoretical perspective. It is said that having observed the gradual collapse of the Argentinean economy made him question the principles of international trade and free markets. From the 1860s to 1920s, Argentina was an economic giant, as it exported large amounts of beef and wheat to Europe. However, in the 1930s, Argentina was in the grip of the Great Depression and as the international trade collapsed the classical orthodox liberal imperative against state interference seemed out of place. Prebisch was unsettled to witness how the self-interested behavior of western countries led them to create subsidies for their farmers to further deteriorate the terms of trade for agricultural products.

Argentina went into a permanent decline as it was not able to compete with the growing economic dominance of the United States and American farmers who started to export subsidized beef and wheat.

In 1950, Prebisch released a document, "The Economic Development of Latin America and its Principal Problems,"[31] which laid the theoretical groundwork for dependency theory. The study took a structural view of the global economy in criticizing the outdated international division of labor as skewed in favor of the industrialized countries. Consistent with Marxist tradition, Prebisch considered the asymmetrical division of labor to serve mainly the interests of western economies under the leadership of the United States. The structural rift in the world economy relegated regions such as Latin America to the subordinate position of periphery and limited their engagement in the global system to be mere suppliers of agricultural products and raw materials for the industrialized center. According to the Singer–Prebisch thesis, the terms of trade between basic commodities – such as coal, wheat, coffee – and manufactured goods – such as cars, cranes – tends to deteriorate over time.[32] Consequently, poor countries who sell agricultural products and raw materials would be able to import increasingly less manufactured goods in exchange for their less valued exports.

In challenging the theory of comparative advantage, Prebisch maintained that international trade relations reflected the economic power of the industrialized nations. The core countries took advantage of their technological superiority and were exporting manufacturing goods to the traditionally agricultural South to make poor countries dependent on those exports. As a result, the developing world was kept in the position of periphery by the hierarchical international system, which kept reproducing itself over time. According to Prebisch, these structural tendencies contradicted classical liberal theory, which predicted that economic growth and increased productivity in leading industrialized countries would improve terms of trade for all trading nations. He observed that as incomes in the core industrial countries had risen more than productivity, the opposite took place in the periphery, resulting in a real decrease in price ratios between primary and manufactured goods.[33] Prebisch blamed the asymmetrical structure of the world economy for such developments.

The first conclusion of the Prebisch thesis suggested a need for an activist state with policies of self-sufficiency to address the structural asymmetry in the global economy. The second conclusion reinforced the view that that there was a fundamental problem with the capitalist market system and international trade.[34] Hence, Prebisch advocated that the poor countries in the periphery should become self-reliant and even should cut all the ties with international trade networks. Rather than exporting commodities and agricultural products while importing manufactured goods, developing countries should adopt a set of policies called *import-substitution industrialization* (ISI). Prebisch proposed a number of specific macroeconomic policies to address the structural inequalities he described. The priorities were given to the state-induced industrialization in order to steer the industrial production into self-sufficiency. Domestic industrial production was to become a substitute for imports from the core countries.

The ISI model became the predominant economic development paradigm in Latin America until the 1970s. Countries that pursued this statist model stayed outside the world trade system through different forms of protectionism. As the role of the state expanded, governments centrally controlled the production levels, manipulated domestic prices, introduced a range of domestic subsidies, and created state-owned enterprises (SOEs). Many commanding heights of the economy like coal, oil, electricity, and telecommunication sectors were nationalized, while a complicated system of regulations grew throughout the economy. Initially, the ISI model created a high level of economic growth based on domestic investment in manufacturing industries, especially heavy machinery, the auto sector, and chemicals. In Latin America, per capita income nearly doubled between 1950 and 1970. The systemic weaknesses of the ISI model were mostly hidden during this period. However, governments' inabilities to effectively plan the production of the required products led to growing shortages of food, basic goods, and mechanical components. The affected countries started to borrow foreign money to pay for supplies. The debt crisis that came in the 1980s revealed the disastrous state of Latin America's economies and placed a serious question mark on the ISI model.[35]

One obvious problem with dependency theory was its historical determinism. Poor countries were presented as having permanently stagnant socio-economic characteristics. The theory alleged that historically determined structures of the global economy worked to perpetually reproduce underdevelopment in the South. Dependency theory did not question the internal socio-political dynamics of the very countries it claimed to be preoccupied with.

The criticism led to modifications by dependency theorists such as Cardoso and Faletto, who did not suggest a complete break from the capitalist global system, and tried to identify domestic obstacles to development. They argued that dependency and underdevelopment were a result of historical forces and feudal societal structures, both within the nation and at the international level that created and reinforced existing power structures.[36] Even Raul Prebisch in his later work recognized that weak domestic institutions compromised developmental policy making. He admitted that the ISI approach could be abused and without "an intelligent state" would lead to stagnation.[37] New dependency theorists maintained that development was possible within the current capitalist system, but it would be unequal development both globally and nationally.[38] True development continued to be conceptualized as a sweeping societal transformation that altered internal and external power structures. But this true development, as contrasted with limited modernization, was only possible with "less dependency and self-sustained growth based on the local capital accumulation and on the dynamism of the industrial sector."[39]

Designed both as a critique of modernization and as a modification of dependency theory, the world-systems analysis by Immanuel Wallerstein aimed at examining the struggle of North versus South.[40] This theory asserts that capitalism is a necessary global mode of production that functions as an organic force, connecting all national and regional economies. Inspired by the French thinker Braudel and his long view of history, Wallerstein insists that "capitalism could only exist within the framework of a world economy, and that a world economy could only operate on capitalist principles."[41] The model stipulates that core industrial countries have obtained a historically generated advantage that allows them to control structures of the global economy. The core states exercise this control via patterns of commercial exchanges that benefit them to the detriment of poor countries in the periphery, and semi-developed countries in the semi-periphery. Consistent with Marxist tradition, the world-systems analysis perceives the relationship between the industrialized nations and the rest of the world primarily in exploitative terms. The model involves three unique spheres of the global economy – the core, the semi-periphery, and the periphery – reflecting the inequality of the new international division of labor. Asymmetrical labor becomes "a coordinating conjunction" among the three spheres, allowing the continuous reproduction of the world system as a whole possible.[42]

Drawing on the writings of Antonio Gramsci, the Italian Marxist of the interwar period, Robert Cox attempted to develop a more comprehensive explanation of the relationship between economic processes and political dynamics at the domestic and international levels. Cox has always maintained that any theory should be based "on changing practice and empirical-historical study, which are a proving ground for concepts and hypotheses"[43] because theory is always derived from some specific socio-political context. When theory detaches itself from reality and becomes too abstract it can lose its explanatory ability. Herein lies the significance of the Gramscian concept of hegemony.

Developed by Antonio Gramsci (1891–1937), the concept served to address some of the weaknesses of Marxism. The communist revolution took place in the predominantly agricultural Russia, but not in the western industrialized societies as predicted by Marx. This was a fundamental problem for many Marxist thinkers of the time. Gramsci noted that by focusing too much on economics and too little on the role of political and cultural factors employed by governments to control society, Marx did not fully explain the perseverance of capitalism. Gramsci found the answer by describing hegemony as a form of political control mitigated by the appearance of consent. Hegemony masked the underlying relationships of domination and allowed to maintain the structures of power. Gramscian hegemony became valuable for Marxists writing about international organizations for its ability to encompass both domination and consent as the two fundamental dimensions of system maintenance. Hegemonic control works to keep the order within the capitalist state and, ultimately, as Cox projects it, the global economic order. In applying the concept to the international level, Cox insisted that it is important: "to know both (a) that it functions mainly by consent in accordance with universalist principles, and (b) that it rests upon a certain structure of power and serves to maintain that structure."[44] Consequently, through the concept of Gramscian hegemony, Cox builds an understanding of the importance of ideas for international institution building. "The material aspect of power has its counterpart in ideas," he says.[45] Ideas lead to the creation of international norms and institutions, which, in turn, reinforce power relations within the global economy. This means that the only possible change to the system can come from some form of counter-hegemonic resistance based on different sets of norms and ideas.

Another Marxist economist, Samir Amin (1931–), also promoted the self-reliance of newly independent countries. His work was grounded in Lenin's theory of imperialism. Lenin was a Russian revolutionary and the first premier of the Soviet Union after the revolution of 1917. In his pamphlet "Imperialism, the Highest Stage of Capitalism" (1916), Lenin argued that mature capitalism was characterized by the expansion of capitalist monopolies into poor countries of the South in search for new markets and raw materials.

Samir Amin spent his life examining the practical consequences of these ideas. His theory of disconnection suggests that the developing world should disconnect itself from the industrialized capitalist countries and their model of development. For Amin, capitalism is inherently obsolete because it distorts the development of societies by fueling monopolistic production and ownership. In this destructive propensity monopolies accelerate the predatory dominance of powerful western countries over the global economy.[46] As predicted by Lenin, Amin believed that monopoly capitalism was aggressively looking for new markets, cheap commodities, and outlets for investment in the developing world. The Marxist theories of development lost their momentum with the collapse of the Soviet Union in 1991.

IV The state as an obstacle to development: neoliberal resurgence

From the end of World War II throughout most of the Cold War era, Hayek remained an obscure economist. As an uncompromising believer in unrestrained free markets, he was dismayed over the post-war influence of Keynesian interventionism over economic policy-making. Hayek detested every known state-directed model of development, especially the Soviet practiced socialism. He believed that socialism would lead to totalitarianism because central planning could not be restricted to the economy. Such view was not really popular during Hayek's lifetime. The tide was finally turning by the end of the 1970s. The budgetary deficits, inflation, and overproduction were increasingly difficult to handle by the overgrown western governments on both sides of the ocean.

When Margaret Thatcher – a studious daughter of a grocer and ambitious member of the British Conservative Party – decided to pursue a political career, she would famously carry in her briefcase Hayek's book.[47] She vowed to resurrect free markets by privatizing previously nationalized industries and by releasing the state from inefficient subsidies. Her determination would take her to become the first female prime minister of Britain. When Thatcher took power in 1979, she decided to change the country by implementing a set of policies aimed at privatization and the deregulation of the economy. Similar attitudes towards the economy would soon take hold in the United States with the presidential election of Ronald Reagan in 1980. These two politicians would end up closely working together during the most globally transformative years in recent history. The memorable decade started with the 1981 Solidarity movement in Poland, continued as the debt crisis engulfed the Latin American countries, to end up with the collapse of the Berlin Wall in 1989, and the Soviet Union in 1991. From the early 1980s, the signs of troubles in the state-run economies could no longer be ignored. The classical liberal theories of Smith and Hayek had slowly been making a comeback in the works of Deepak Lal and Jeffrey Sachs. These economists responded to the historical events by claiming that only a market-based liberal economic order was able to solve the problem of poverty and underdevelopment.

In his book *The Poverty of "Development Economics"* first published in 1983, Deepak Lal (1940–), an Indian born economist, took issue with what he believed was a doctrinaire approach to development.[48] Lal exposed the negative consequences of the so-called *dirigiste*[49] dogma; the state-centered model of development. In his view, the ills of unbalanced growth, dependency, etc. were all ascribed to excessive government *dirigiste*. In describing its various effects, he also showed how this dogma inevitably induced corruption by politicizing decision making and inviting rent-seeking. Lal argued that *dirigisme* captured the thinking of policy makers, journalists, and academics in both developing and developed countries, and made them blind and unable to comprehend the positive side of open markets. The major indictment was directed towards those development economists who, by supporting the *dirigisme*, contributed to: "'policy-induced' distortions which are more serious than any of the supposed distortions of the imperfect market economy it was designed to cure."[50] Lal concluded that the demise of this dogma would be beneficial for all the economies in the world.

A faithful disciple of the classical economic tradition, Lal claimed that the disastrous experience with central planning in the Soviet Union proved that Hayek was right. Observing the crisis situation as it was unfolding in the former Soviet Bloc countries where the state-run economies became unmanageable and financially bankrupt, Lal advocated a controversial big-bang solution. It called for a quick structural reform process involving the rapid liberalization of the troubled economic system. In his words:

> A big bang may therefore be desirable to smash the equilibrium of rent-seeking interest groups who have a stake in maintaining the past system of dirigisme. To stiffen the government's spine in this unenviable task, sweeteners which ease its fiscal problems, in the form of soft loans or grants from multilateral and bilateral foreign governments, may be desirable.[51]

When trying to create free markets in countries with a heavy interventionist past, Lal would quickly destroy the existing structures, hoping the new ones would materialize as fast. He would also support these efforts with foreign aid in the form of financing and advisory expertise. Lal's writings strengthened the emerging global neoliberal paradigm, focused on quick privatization, deregulation and free trade. Soon the belief in unrestrained free markets informed the prescriptive approach to development, consistent with a view that: "a market-based liberal economic order which promotes labor-intensive growth can cure the age-long problem of structural mass poverty."[52]

Another intellectual reinforcement behind the neoliberal paradigm came with the argument by Francis Fukuyama, who believed that the collapse of the Soviet Bloc proved the superiority of liberal democracy and rational free markets.[53] The belief that markets are always rational echoes the work of an early twentieth-century economist, Irving Fisher. His work on the predictable scientific behavior of stock markets was greatly influenced by Darwin's theory of evolution. The markets awarded the best economic endeavors and destroyed the superfluous ones by acting in an impassionate and logical way. It was then argued that markets should be completely deregulated because they operated in a way that individuals and governments could not because individuals and governments are influenced by political meddling and human emotions. In December 1928, Fisher wrote an essay "Will Stocks Stay Up in 1929?" in which he provided compelling (in his view) arguments that they absolutely would.[54] In October 1929, the markets collapsed, triggering the Great Depression. The ideas of Fisher and Hayek were strengthened by Fukuyama's writings and provided philosophical justification against government involvement in the economy. At the heart of these developments was the theoretical assumption "that the state has little role to play beyond that of creating a positive regulatory environment in a global economy dominated by firms."[55]

The neoliberal paradigm was put into practice by Jeffrey Sachs (1954–), an economic advisor to a number of Eastern European, Latin American and Asian governments on their economic transformations. A supporter of a big-bang way of introducing a set of structural reforms and capitalizing on his experience in the field, Sachs designed his own plan for eradicating poverty by 2025. In the book with a provoking title, *The End of Poverty*, Sachs expanded the principles of orthodox liberalism to argue that a key to economic development is a careful clinical diagnosis followed by consistent monitoring. He called this theory clinical economics.[56]

Sachs identified eight main reasons why countries failed to achieve economic growth: poverty trap, physical geography, fiscal trap, governance failure, cultural barriers, geopolitics, lack of innovation, and demographic trap. The poverty trap is particularly debilitating because poverty itself is a cause of economic stagnation. And just as clinical economics works with the help of professionals, poor countries need foreign aid to get started, because

> once they're on the first rung of the ladder of development, they'll start climbing just like the rest of the world ... I'm proposing that we help people help themselves. This can be done without legions of people rushing over to these countries to build houses and schools. This is what people in their own communities can do if we give them the resources to do it.[57]

In an engaging analysis, Paul Collier contests the very idea of a poverty trap. Rightly observing that all societies were once poor, he suggests that we would stay this way forever if poverty itself was such an impediment. Instead, he suggests that some countries remain stuck in perilous economic circumstances because of four traps: the conflict trap, the natural resources trap, the trap of being landlocked with bad neighbors, and the trap of bad governance in a small country.[58] Collier argues that poor countries are caught in one or another of these traps, but poverty itself is not a trap.

Sachs is also criticized for assuming the superiority of western economic liberalism and spending little time to explain what makes it work as a model for the developing world. Sachs continues to see the right path to development in poor countries mainly in economic terms and technical solutions. Namely, he wants to increase investment in people (health, education, nutrition, and family planning), the environment (water and sanitation, soils, forests, and biodiversity), and infrastructure (roads, power, and ports). Sachs observes that poor developing countries cannot afford these investments on their own, so rich countries must help. Together with foreign aid, he points out the necessity for transfers of technology and knowledge from the developed to the developing world. And while Sachs does mention the need for good governance, he does not elaborate how it can be achieved. In suggesting aggregate solutions, surprisingly little attention is paid to the institutional framework of the country. He equates the state with the government tasked with providing social and political stability in the country. The government is expected to create a business friendly environment and to maintain a judicial system responsible for protecting property rights and enforcing contracts.[59]

Perhaps the uncertainly over Sachs' prescriptions stems from his belief in Enlightened Globalization. He sees it as: "a globalization of democracies, multilaterals, science and technology, and a global economic system designed to meet human needs."[60] It is assumed that Enlightened Globalization emerges by following his ideas, but it is unclear what would be the operational principles of such a system. Only some clues are provided. The system would see rich countries addressing environmental degradation, shining a light on corporate responsibility, and advancing trade and investment via the World Trade Organization (WTO) Doha Round of negotiations. In short, Enlightened Globalization "would insist that the United States and other rich countries honor their commitments to help the poor escape from poverty."[61] In summary, Sachs wants to continue the promotion of western liberal values in the developing world, as supported by foreign aid and expertise of the industrialized nations.

It appears that the post-Cold War global push towards economic liberalization has lost its momentum after the financial crisis of 2008. The WTO has failed to complete its very first round of trade negotiations that started in 2001. Countries slowed down on making global liberalizing commitments from the fear of losing autonomy over their policy space. In his critique of globalization, Rodrik calls for the re-conceptualization of the international system:

> instead of viewing it as a system that requires a single set of institutions or one principal economic superpower, we should accept it as a collection of diverse nations whose interactions are regulated by a thin layer of simple, transparent, and commonsense traffic rules.[62]

It is an argument for a functional global integration that shifts the burden of responsibility on the states expected to do a balancing act between free markets and interventionism while focusing on their developmental priorities. It is a good idea notwithstanding that the state is not an unproblematic actor.

Liberal theory has always struggled with explaining the need for the state. Unable to offer a satisfactory account of centralized governance in the globalizing world, neoliberalism prefers to reject the state all together. The global economy may be fractured and asymmetrical but it still consists of a web of networks organized by the states. With its undivided focus on free markets, neoliberalism creates theoretical barriers to examining the state's potential. After all, it considers the state to be an obstacle to development.

V The state as an enabling environment: institutions and rights of the poor

There is something ostensibly reductionist in thinking about policy options for development as a choice between states and markets. Such a constraining view stems from a limited analysis of the state. Over the last 400 years, the concept of sovereignty provided a useful excuse for insufficient probing of the transformative potential of the state and its institutions. Finally, in moving beyond the schematic states versus markets dichotomy, a more nuanced set of economic ideas have been injected into the field of development.

The ideas of Amartya Sen have had a particularly transformative impact on development studies. The 2016 UNDP Human Development Report[63] takes as its philosophical foundation a human centered view of development refined by Sen. Amartya Sen was born in British India in a Bengali family but moved to the United Kingdom in the early 1950s to study economics at Cambridge. This 1998 winner of the Nobel Prize in Economics culminated his lifetime's work with *Development as Freedom*.[64] In this book he explores the mutually reinforcing relationship between political and economic freedoms and development. Sen considers development to be an integrated process of the expansion of substantive human freedoms that connect with one another.[65]

> Development can be seen, it is argued here, as a process of expanding the real freedoms that people enjoy. Focusing on human freedoms contrasts sharply with narrower views of development, such as identifying development with the growth of gross national product, or with the rise in personal incomes, or with industrialization, or with technological advance, or with social modernization.[66]

Sen's approach sees expanding individual freedom as "the primarily end and as the principal means of development."[67] His perspective on development emphasizes human agency. Sen sees the capacity of a person to *choose* to do one thing and not another as an essential ingredient of human welfare. As long as this choice is confined to the selection between options determined by other people, constrained by institutions, limited by oppression, destroyed by poverty, there is no real freedom and hence no possibility for meaningful development. In other words, so long as a person's capability set is determined by socio-economic and political arrangements in which s/he has no say, the path to advancement is closed off.

Rather than classical economics' preoccupation with income and wealth creation, Sen concentrates on capability deprivation and removal of obstacles to instrumental freedoms including:

1   Political freedoms (e.g., free elections, freedom to form opposition)
2   Economic facilities (e.g., freedom to open a business, ability to get a job)
3   Social opportunities (e.g., access to education and health care)
4   Transparency guarantees (e.g., predictability of regulations, no secret files)
5   Protective security (e.g., unemployment benefits, rule of law).

There are two reasons why these individual freedoms are so important for development. First, substantive individual freedoms are taken to be critical in the success or failure of a society. Second, each substantive freedom is a principal determinant of individual initiative and social effectiveness. Hence, Sen criticizes the limited application of traditional measures of development, like GDP, personal income, degree of industrialization, technological advance, etc. Instead, he offers a new measure of development, which is the removal of major sources of *unfreedom* such as poverty, political tyranny, poor economic opportunities, systematic social deprivation, neglect of public facilities, intolerance, denial of political liberty and basic civil rights. Sen argues that focusing on capability deprivation is a better measure of poverty than simple economic indicators because his approach can capture aspects of poverty hidden by statistical measures.

Another focus of Sen's work has been the role of women and minorities in development. Here he argues that, while improving their well-being is important, enhancing their agency is just as critical. Despite cultural and social differences across nations and communities, he sees all of us connected by the universal appeal of basic human rights and values, such as the need for respect, security, and independence. His critique of tradition questions the norms and customs that sometimes excuse conceivably oppressive practices in the name of cultural values. For Sen, the dominance of cultural values poses questions of legitimacy and authority. He asks who and on what grounds decides whether a traditional way of living can be given up to escape poverty. Just like with any other matters that impact our human agency, it is important for Sen that people are allowed to decide what traditions they wish to follow. Sen's freedom oriented perspective on development measures the progress in the context of human capacity to choose our own path towards advancement.[68]

The concept of development as freedom is lodged in the ideas of Adam Smith. Sen, however, acknowledges the inescapable role played by the state in economic policy making. He is cognizant about tensions within classical liberalism caused by the liberal embrace of free markets and liberal skepticism about mobilizing the state in advancing development. Sen reconciles these tensions by insisting on the importance of democracy. Democratic order becomes an essential prerequisite in creating an effective and legitimate governing framework, capable of removing obstacles restricting human agency. Only in democracy people become capable of shaping common norms and priorities: "Political rights, including freedom of expression and discussion, are not only pivotal in inducing social responses to economic needs, they are also central to the conceptualization of economic needs themselves."[69]

The people-centered view of development also informs the work of William Easterly. He puts it bluntly: "the real cause of poverty is the unchecked power of the state against poor people without rights."[70] Easterly rejects the conventional approach to development based on a technocratic illusion, which is easily adaptable to various strains of modernization theory. The technocratic approach blinds technical experts, economists, and policy advisors who unintentionally empower dictators by conferring undeserved legitimacy on the state expected to implement the prescribed technocratic solutions: "What used to be the divine rights of kings has in our time become the development right of dictators. The implicit vision of development today is that of well-intentioned autocrats advised by technical experts."[71] Easterly names it: *authoritarian development*. Authoritarian development ignores history by adopting the blank slate view of countries supposedly in need of grand solutions designed by the developmental agencies. In creating his own model, Easterly articulates three challenges in opposition to the authoritarian model, by insisting that history matters, individuals matter, and spontaneous solutions are preferable. In his words: "History and modern experience suggest that free individuals with political and economic rights – call it *free development* – make up a remarkably successful problem-solving system."[72]

The ideas of Sen and Easterly complement each other by locating multiple factors that shape the socio-economic development of people. Sen recognizes the universality of basic human needs that include the desire for justice as superior to the simplistic preoccupation with economic growth. Easterly's work pushes this further by examining why so many well-conceived developmental projects are never implemented or produced adverse results. He demonstrates that it does not matter whether the development community follows the steps of modernization theory or pursues neoliberal solutions, the expectations remain the same. By considering the state as a fairly unproblematic entity, aid agencies often put their faith in the hands of friendly dictators expected to implement the prescribed policies. It should come as no surprise that implementation rarely happened as planned. In conclusion, Easterly calls upon the development community to give up its authoritarian mind-set that leads to "the global double standard of rights for the rich and not for the poor."[73] Both approaches rely on effective democratic state institutions that serve to guarantee and protect individual rights and freedoms.

When researching democratization processes, Adam Przeworski tried to answer the question of what makes a democratic state effective and enduring. He observes:

> A democratic state "works," when it generates normatively desirable and politically desired effects, such as economic growth, material security, freedom from arbitrary violence, legal assurance that the contracts entered into will not be broken, and other conditions conducive to the full development of individuals. And democracy "lasts" when it can absorb and effectively regulate all major conflicts and when rules are changed only according to rules.[74]

Democracy, however, entails a complex system of governance supported by a sophisticated institutional framework. There are no two identical democracies around the world. Accountable government goes hand in hand with open society, where people with rights can democratically change their government. At its core, democracy rests upon several key principles: equality (guaranteed equal rights for all citizens), accountability (government answerable to its electorate and accepting responsibility for its actions), participation (ability of citizens to take part in the political processes), transparency (acting in an open and accessible way), and the rule of law (fair laws protect the rights of citizens and maintain order). Scholars of democratic transitions further underscore the need for separation between the executive, legislative and judicial power, ensured by relevant checks and balances. All these principles guard us against the *electoralist fallacy*, which assumes that a *necessary* condition of democracy, free elections, is also a *sufficient* condition of democracy.[75] Dysfunctional states come in different shapes and forms, but most of them display internal problems, perpetuated by extractive institutions and the absence of the rule of law. If left ignored, such problems consume the country's potential for positive change.

State institutions matter, and the quality of state institutions is the key difference between successful and failed economies. This is the argument advanced by Acemoglu and Robinson in their recent book. Domestic institutions are instrumental for the processes of policy making of the state, but in many countries state institutions work to the detriment of poor marginalized people. Institutions can be categorized as either *inclusive* and *extractive*. As defined by the authors "inclusive economic institutions foster economic activity, productivity growth, and economic prosperity."[76] Quite different are extractive institutions, which deserve to be called "extractive because such institutions are designed to extract incomes and wealth from one subset of society to benefit a different subset."[77] Extractive institutions perpetuate the existence of embedded inequality in a society by allowing dictators and their cronies to benefit from the system. On the other hand, inclusive institutions are capable of creating a framework for cooperation and to guarantee the political and economic rights of the people, allowing a majority of citizens to keep the government accountable. Once such inclusive institutions are established, their ability to improve and self-correct leads to the "virtuous circle" that paves the way to a well-operating political and economic system.[78] Another critical insight about state institutions comes from Hilton L. Root: "Economic models generally assume institutions already exist that provide regulations, oversight, and incentive driven by competition. But the emergence of such institutions is the very essence of development."[79]

In disagreeing that development should focus on individual rights and state institutions, Ha-Joon Chang favors a strong activist government. The state should play an active role in the economy, he believes, because there is no such thing as a truly free market. After all, governments exist and they all regulate different parts of the economy and this is why: "what is a necessary state intervention consistent with free-market capitalism is really a matter of opinion."[80] As a result, he advocates the existence of a large, and well-designed, welfare state based on universal programs that can shield people from the excesses of capitalism.

> People can accept the risk of unemployment and the need for occasional re-tooling of their skills more willingly when they know that these experiences are not going to destroy their lives. This is why a bigger government can make people more open to change and thus make the economy more dynamic.[81]

Ha-Joon Chang also maintains that asymmetries in the global market economy have put poor countries at a disadvantage. Consequently, if developing countries want "to leave poverty behind, they have *to defy the market*" – which means they should be allowed to "use extra policy tools for protection, subsidies, and regulation."[82] In his perhaps best known work, Ha-Joon Chang argues against economic globalization conducted in the name of liberal principles and open trade. He observes that pro-globalization policies advocated by the IMF and the World Bank forced many developing countries to lose their autonomy only to end up in a quagmire of economic mess. Instead, he supports state-directed development focused on infant industry protection, with the right tariffs and robust regulations in place. He even defends copyright violations by demonstrating that all such policies were at some point practiced by the industrialized countries when they were developing. Ha-Joon Chang also endorses certain SOEs. His argument in favor of interventionism is supported by examples from East Asia, where state policies positively transformed several previously backward economies.[83] However, Ha-Joon Chang's lack of attention to individual agency, his ambiguous view of political corruption,[84] and insufficient examination of state institutions, especially in ensuring common trust and accountability of rulers, is disappointing in this otherwise interesting work.

An important theory about the importance of government leadership in the economy was developed by Robert H. Wade. His examination of the economic successes in East Asia showed patterns of synergy between the governmental rules-setting, decision making and the private sector. The resulting governed market theory: "emphasizes developmental virtues of a hard or soft authoritarian state in corporatist relations with the private sector, able to confer enough autonomy on a centralized bureaucracy for it to influence resource allocation in line with a long-term national interest."[85] Such a complementary relationship between state and markets is essential for maintaining a well-functioning capitalist economy. One of the ideas Wade proposes for improving the effectiveness of the state in the developing world is the creation of effective institutions of political authority and corporatist institutions before the whole system is democratized. In a more recent article, Wade worries about the growing income inequality in the West. He also provides policy prescriptions for a government to play a positive role in remedying the shortcomings of free markets.[86]

Another scholar researching East Asia has shown how a cohesive–capitalist state can use its power to prioritize economic development as a political goal. A number of East Asian states achieved rapid progress via authoritarian control "by mobilizing resources, channeling them into priority areas, altering the socio-economic context within which firms operate, and even undertaking direct economic activities."[87] Critics of such cohesive–capitalist regimes maintain, however, that state-directed development can only go so far without listening to voices calling for democratic reforms, transparency, and accountability.

Ha-Joon Chang takes an issue with the argument that emphasizes the importance of institutions in development economics. It is possible that his criticism stems from a limited view of state institutions. Ha-Joon Chang contests the idea that institutions that "maximize market freedom and most strongly protect private property rights are the best for economic development."[88] Indeed, if state institutions are mainly considered in terms of performing such functions, the analysis is incomplete. Inclusive institutions go beyond protecting private property, they include democratic checks and balances over governmental decision making, as well as measures supporting a sustainable long-term growth. They also include institutionalized practices that secure peaceful cooperation and equal opportunities for all members of society.

Even Hayek, considered an apostle of free markets, believed that the state has a role to play as long as the government adheres to the rule of law. His view of state institutions, however, was very narrow. Hayek insisted on a minimum state, whose responsibility was to counter any form of monopolistic behavior and corruption.[89] He views the state as: "the rule of formal law, the absence of legal privileges of particular people designated by authority, which safeguards that equality before the law which is the opposite of arbitrary government."[90] Hayek still sees the state as the main unit of political authority in the international system. He worries, however, about the state run by an oligarchic government and privileged elites, who have a propensity to act in an arbitrary way, to the detriment of the society they are supposed to represent. Hayek's analysis, however, falls short in understanding that inclusive state institutions contribute to the democratic empowerment of society and this is why they are vital for supporting equality and sustaining the long-term socio-economic growth of a country.

Past theories of development assumed that the main problems of underdeveloped societies were their inabilities to generate fast growth rates or accumulate capital. In contrast, recent theories and developmental approaches have turned their attention towards individual agency, inequality, and the institutional context within which people's needs are articulated or repressed. And while domestic state institutions are often the primary focus of these approaches to development, states operate within the global milieu of interdependent economic networks. For that reason, it is important to understand the role of global institutional factors, because as one scholar observes:

> Such an understanding would lead us to take the unfulfillment of human rights abroad more seriously – provided we accept that persons involved in upholding coercive social institutions have a shared moral responsibility to ensure that these institutions satisfy at least the universal core criterion of basic justice by fulfilling, insofar as reasonably possible, the human rights of the persons whose conduct they regulate.[91]

In acknowledging the human centered approach to development, aid agencies increasingly concentrate on local engagement, while ignoring the state all together. This move misses the point. For example, instead of insisting on effective state institutions to provide a transparent and accountable aid management system, donors try to establish their own independent networks of managers and distributors, who are not necessarily more efficient or less corrupt. Such networks do nothing to improve the overall functioning of the country. The solution is not to bypass the state, but to deal with it, especially when it is ruled by autocrats. An institutionalized – within the state structures – aid management system constructed in partnership with the aid donor and the aid receiving state can be a significant step towards facilitating good democratic governance. The long-term strategy for individual communities cannot ignore the role of the state in attempting to create incentives for building inclusive institutions that work to guarantee peoples' rights and freedoms.

The optimum strategy for the state is to provide an enabling environment for good decision-making processes. By good, we mean effective and democratically accountable to the people, based on trust (sound banks, insurance companies, transparent regulations), law and order (rule of law, enforceable contracts), security and protection (all individuals and property), social and labor mobility, and competitive elections. The state is a complicated construct that needs to be further interrogated and better integrated with the discourses concerning economic development.

> The state is necessary for implementing a conception of justice ... because only an agency of that kind would be capable of discharging the many varied and demanding tasks involved: maintaining the developmental, institutional and material infrastructure that justice requires; establishing and adjusting the laws required to identify substantive and coherent basic liberties; ensuring that those liberties are resourced on the basis of any needed conventions and subsidies; and protecting people against the invasion of those liberties, whether in particular relationships or on a more general front. The tasks involved here are so complex, interconnected and dynamic that no abstract apparatus of rules could plausibly ensure their fulfillment. There can be no effective system of justice, so it appears, in the absence of a state.[92]

Notes

1 Craig Calhoun (ed.) (2002) *Dictionary of the Social Sciences*, Oxford, UK: Oxford University Press.
2 Michel Foucault (1997) "Subjectivity and Truth," in Paul Rabinow (ed.), *Ethics – Subjectivity and Truth, Essential Works by Foucault (1954–1984) Vol. I*, New York, NY: The New Press, p. 88.
3 Robert W. Cox (1981) "Social Forces, States and World Orders – Beyond International Relations Theory," *Millennium*, Vol. 10, No. 2, pp. 128–129.
4 Cited in William Easterly (2013) *The Tyranny of Experts: Economists, Dictators, and the Forgotten Rights of the Poor*, New York, NY: Basic Books, p. 49. Woodrow Wilson made this comment when justifying the transformation of former colonies into mandates under the League of Nations.
5 The commission consisted of seven individuals under the chairmanship of Lester B. Pearson. It was created at the request of Robert S. McNamara, the new president of the World Bank, in August 1968. Summaries of its main findings were published by the *UNESCO Courier* in February 1970 as "Partners in Development – The Pearson Report – A New Strategy for Development."
6 UNESCO (1970) "Partners in Development – The Pearson Report – A New Strategy for Development," *UNESCO Courier*, February Edition, pp. 15–17.
7 Ibid., p. 12.
8 Ibid., p. 8.
9 John Maynard Keynes (1936) *The General Theory of Employment, Interest and Money*. Electronic Edition: Location 24662 of 81062.
10 Daron Acemoglu and James A. Robinson (2012) *Why Nations Fail: The Origins of Power, Prosperity, and Poverty*, New York, NY: Crown Business, chapter 7.
11 Ibid., p. 84.
12 John Maynard Keynes (1936) *The General Theory of Employment, Interest and Money*, op. cit.
13 Robert L. Tignor (2005) *W. Arthur Lewis and the Birth of Development Economics*, Princeton, NJ: Princeton University Press.
14 W. Arthur Lewis (1955) *The Theory of Economic Growth*, Homewood, IL: Richard D. Irwin Inc., p. 278.
15 Gustav Ranis (2004) *Arthur Lewis' Contribution to Development Thinking and Policy*, Discussion Paper No. 891, New Haven, CT: Economic Growth Center, Yale University, p. 5.
16 Yoichi Mine (2006) "The Political Element in the Works of W. Arthur Lewis: The 1954 Model and African Development," *Developing Economies*, Vol. XLIV, No. 3, pp. 339–340.

17 Patsy Lewis (2005) "Grenada: A Testing Ground for Lewis's Balanced Development Perspectives," *Journal of Social and Economic Studies*, Vol. 54, No. 4, p. 206.

18 Yoichi Mine (2006) "The Political Element in the Works of W. Arthur Lewis: The 1954 Model and African Development," op. cit., pp. 344–345.

19 Alfred Sauvy (1952) *L'Observateur*, August 14. The Third Estate referred to the poor commoners during the French Revolution. The First and Second Estates consisted of the nobility and the clergy.

20 Walt W. Rostow (1960) *The Stages of Economic Growth: A Non-Communist Manifesto*, Cambridge, UK: Cambridge University Press, pp. 4–16.

21 Ibid., pp. 7–9.

22 Steven E. Lobell (2005) *The Challenge of Hegemony – Grand Strategy, Trade, and Domestic Politics*, Ann Arbor, MI: The University of Michigan Press, p. 8.

23 Bruce Russett (2016) "Liberalism," in Tim Dunne, Milja Kurki, and Steve Smith (eds.), *International Relations Theories – Discipline and Diversity* (4th edition), Oxford, UK: Oxford University Press.

24 UNESCO (1970 "Partners in Development – The Pearson Report – A New Strategy for Development," op. cit., p. 8.

25 David Mitrany (1948) "The Functional Approach to World Organization," *International Affairs*, Vol. 24, Issue 3, pp. 350–363.

26 David Mitrany (1975) "A Political Theory for the New Society," in A.J.R. Groom, and P. Taylor, *Functionalism*, London, UK: University of London Press, pp. 25–38.

27 David Mitrany (1948) "The Functional Approach to World Organization," op. cit., p. 358.

28 Dudley Seers (1971) "The Total Relationship," in Dudley Seers and Leonard Joy (eds.), *Development in a Divided World*, Baltimore, MD: Penguin Books, pp. 339–340.

29 Mahbub ul Haq (1976) *The Poverty Curtain: Choice for the Third World*. New York, NY: Columbia University Press, pp. 5–8.

30 Daniel Yergin and Joseph Stanislaw (2002) *The Commanding Heights – The Battle for the World Economy*, New York, NY: Simon and Schuster, pp. 280–282.

31 Raul Prebisch (1950) *The Economic Development of Latin America and its Principal Problems*, New York, NY: United Nations.

32 Hans Singer first developed this hypothesis, later expanded by Prebisch, in the late 1940s.

33 Raul Prebisch (1950) *The Economic Development of Latin America and its Principal Problems*, op. cit., pp. 8–10.

34 Edgar J. Dosman (2012) *Raul Prebisch and the XXI Century Development Challenges*, New York, NY: United Nations, ECLAC.

35 Daniel Yergin and Joseph Stanislaw (2002) *The Commanding Heights – The Battle for the World Economy*, op. cit., pp. 234–235.

36 Fernando Henrique Cardoso and Enzo Faletto (1979) *Dependency and Development in Latin America*, Berkeley, CA: University of California Press.

37 Edgar J. Dosman (2012) *Raul Prebisch and the XXI Century Development Challenges*, op. cit., p. 19.

38 Andrés Velasco (2002) "Dependency Theory," *Foreign Policy*, Vol. 45, No. 133, pp. 44–45.

39 Fernando Henrique Cardoso and Enzo Faletto (1979) *Dependency and Development in Latin America*, op. cit., p. 10.

40 Immanuel Wallerstein (2002) "The Itinerary of World Systems Analysis; or, How to Resist becoming a Theory?," in J. Berger and M. Zelditch Jr. (eds.), *New Directions in Contemporary Sociological Theory*, Lanham, MD: Rowman & Littlefield, pp. 358–376.

41 Ibid., pp. 361–362.

42 Immanuel Wallerstein (1979) *The Capitalist World Economy*, Cambridge, UK: Cambridge University Press, pp. 2–19.

43 Robert W. Cox (1996) "Special Forces, States, and World Orders: Beyond International Relations Theory," in Robert W. Cox with Timothy J. Sinclair, *Approaches to World Order*, Cambridge, UK: Cambridge University Press, p. 87.

44 Robert W. Cox (1996) "Realism, Positivism, Historicism," in Robert W. Cox with Timothy J. Sinclair, *Approaches to World Order*, op. cit., p. 56.

45 Robert W. Cox (with Harold K. Jacobson) (1996) "Decision Making," in Robert W. Cox with Timothy J. Sinclair, *Approaches to World Order*, op. cit., p. 362.

46 Samir Amin (2003) *Obsolescent Capitalism: Contemporary Politics and Global Disorder*, London, UK: Zed Books.

47 Daniel Yergin and Joseph Stanislaw (2002) *The Commanding Heights – The Battle for the World Economy*, op. cit., p. 89.

48 Deepak Lal (2002) *The Poverty of "Development Economics,"* London, UK: The Institute of Economic Affairs.

49 This term is rooted in a French word *diriger*, meaning "to direct."

50 Deepak Lal (2002) *The Poverty of "Development Economics,"* op. cit., p. 135.

51 Ibid., pp. 227–228.

52 Ibid., p. 256.

53 Francis Fukuyama (1992) *The End of History and the Last Man?*, New York, NY: The Free Press.

54 Justin Fox (2009) *The Myth of the Rational Market – A History of Risk, Reward, and Delusion on Wall Street*, New York, NY: Harper Collins, pp. 3–44.

55 Vivien A. Schmidt (2009) "Putting the Political Back into Political Economy by Bringing the State Back in yet Again," *World Politics*, Vol. 61, No. 3, p. 519.

56 Jeffrey D. Sachs (2005) *The End of Poverty – Economic Possibilities for our Time*, New York, NY: The Penguin Press.

57 Jeffrey D. Sachs interviewed by Onnesha Roychoudhuri for *Mother Jones* on May 6, 2005.

58 Paul Collier (2008) *The Bottom Billion – Why the Poorest Countries Are Failing and What Can Be Done About It*, Oxford, UK: Oxford University Press, p. 5.

59 Jeffrey D. Sachs (2005) *The End of Poverty – Economic Possibilities for our Time*, op. cit., pp. 59–60.

60 Ibid., p. 358.

61 Ibid., pp. 358–359.

62 Dani Rodrik (2010) *The Globalization Paradox – Democracy and the Future of the World Economy*, New York, NY: W.W. Norton, p. 280.

63 UNDP (2016) *Human Development Report 2016 – Human Development for Everyone*, New York, NY: UNDP.

64 Amartya Sen (1999) *Development as Freedom*, New York, NY: Random House.

65 Ibid.

66 Ibid., p. 3.

67 Ibid., p. xii.

68 Amartya Sen (1999) *Development as Freedom*, op. cit., pp. 31–32.

69 Ibid., p. 154.

70 William Easterly (2013) *The Tyranny of Experts: Economists, Dictators, and the Forgotten Rights of the Poor*, New York, NY: Basic Books, p. 6.

71 Ibid.

72 Ibid., p. 7.

73 Ibid., p. 350.

74 Adam Przeworski (1995) *Sustainable Democracy*, Cambridge, UK: Cambridge University Press, p. 11.

75 Juan J. Linz and Alfred Stepan (1996) *Problems of Democratic Transitions and Consolidation – Southern Europe, South America, and Post-Communist Europe*, Baltimore, MD: The Johns Hopkins University Press, p. 4.

76 Daron Acemoglu and James A. Robinson (2012) *Why Nations Fail: The Origins of Power, Prosperity, and Poverty*, New York, NY: Crown Business, p. 75.

77 Ibid., p. 76.

78 Ibid., p. 308.

79 Hilton L. Root (2006) *Capital and Collusion – The Political Logic of Global Economic Development*, Princeton, NJ: Princeton University Press, p. 16.

80 Ha-Joon Chang (2012) *23 Things They Don't Tell You About Capitalism*, New York, NY: Bloomsbury Press, p. 8.

81 Ibid., p. 230.
82 Ha-Joon Chang (2007) *Bad Samaritans – The Guilty Secrets of Rich Nations and The Threat to Global Prosperity*, London, UK: Random House Business Books, pp. 210, 218.
83 Ibid.
84 Ibid., p. 166.
85 Robert Wade (1990) *Governing the Market – Economic Theory and the Role of Government in East Asian Industrialization*, Princeton, NJ: Princeton University Press, p. 26.
86 Robert H. Wade (2013) "Capitalism and Democracy at Cross-Purposes," *Challenge*, Vol. 56, No. 6, November/December. Electronic Edition.
87 Atul Kohli (2005) *State-Directed Development – Political Power and Industrialization in the Global Periphery*, New York, NY: Cambridge University Press, p. 418.
88 Ha-Joon Chang (2011) "Institutions and Economic Development: Theory, Policy and History," *Journal of Institutional Economics*, Vol. 7, No. 4, pp. 473–498.
89 Daniel Yergin and Joseph Stanislaw (2002) *The Commanding Heights – The Battle for the World Economy*, New York, NY: Simon and Schuster, pp. 123–127.
90 F.A. Hayek (edited by Bruce Caldwell) (1944) *Road to Serfdom – Text and Documents – The Definitive Edition*, Chicago, IL: University of Chicago Press (2007), pp. 116–117.
91 Thomas Pogge (2008) *World Poverty and Human Rights* (2nd edition), Cambridge, UK: Polity Press, p. 55.
92 Philip Pettit (2012) *On the People's Terms – A Republican Theory and Model of Democracy*, Cambridge, UK: Cambridge University Press, p. 133.

Suggested further reading

Ha-Joon Chang and Ilene Grabel (2014) *Reclaiming Development: An Alternative Economic Policy Manual (Critique. Influence. Change)* (2nd edition), London, UK and New York, NY: Zed Books.
Edgar J. Dosman (2010) The *Life and Times of Raúl Prebisch, 1901–1986*, Montreal, Canada: McGill-Queen's University Press.
Khadija Haq (ed.) (2017) *Economic Growth with Social Justice: Collected Writings of Mahbub ul Haq*, Oxford, UK: Oxford University Press.
Nina Munk (2013) *The Idealist: Jeffrey Sachs and the Quest to End Poverty*, New York: NY: Doubleday.
Dani Rodrik (2015) *Economics Rules: The Rights and Wrongs of the Dismal Science*, New York, NY: W.W. Norton & Company Ltd.
Robert L. Tignor (2005) *W. Arthur Lewis and the Birth of Development Economics*, Princeton, NJ: Princeton University Press.
Michael P. Todaro and Stephen C. Smith (2014) *Economic Development* (12th edition), Toronto, Canada: Pearson.

Colonialism legacy of developing countries

2

I Conceptualizing historical legacy

Colonialism and poverty have a complex relationship. Multiple factors are responsible for the persistence of inequality today, but many socio-economic dispossessions in the developing world have deep historical roots. What started as curious explorations by the kingdoms of Spain and Portugal in the fifteenth and sixteenth centuries, became a well-organized conquest of the Americas, Africa, and Asia only a few decades later. The waves of imperial expansions interrupted the lives of people and imposed unfamiliar customs and attitudes on native communities in a way that permanently altered their indigenous existence.

This chapter looks at the challenges of economic development within a historical context of colonialism. The discussion begins with tracing the origins of European expansionism and examining the theoretical approaches that try to explain it. Next, the chapter reviews the establishment of colonies and the consolidation of colonial domination, only to ponder the complexity of the post-independence era. Our aim is to investigate to what extent key historical forces such as European imperialism, colonialism, and neo-colonialism, laid the foundation for at least some of the problems presently confronting a significant number of developing countries.

One of the most enduring consequences of the western conquest was the construction of colonies as states dominated by the European powers. The colonial states were frequently delineated by artificially demarcated borders, many of which remain contested today. Every one of them was, to a different degree, modeled after the European regime in charge. The colonies were subject to intense processes of adaptation with respect to laws, regulations, language programs, educational systems, religion, and administration. Under the premise of modernization, the colonists imposed new sets of rules in the areas of their influence. Such rules were grounded in European legal systems and traditions, but were completely foreign to the colonized people. In some cases, the development of indigenous administrative structures was allowed, in other cases any attempt towards building indigenous political capacity was actively discouraged.

There was one crucial aspect of historical legacy common to all colonial experiences: governing by creating a large bureaucratic state. It is particularly relevant to address this model of colonial rule in the context of contemporary debates about economic development. The socio-political dynamics stemming from the centralized model of authoritarian governing left permanent imprints on the post-independence generations. This is how one scholar describes it:

> a colonial regime is essentially one of bureaucratic authoritarianism, in which government is viewed as the initiator of all public policy, as well as the source of all amenities and of most good jobs. A colonial government rules through petition and administration, and not through competition and compromise. The public does not participate in the political process; it is "administered" by a bureaucratic elite, which by the system's definition knows what is best.[1]

The political leaders that later led their nations towards independence grew up in the environment arranged by bureaucratic authoritarianism. They were socialized to think about the state as a highly centralized structure. During the colonial period, this form of political entity became normalized as it was a top-down approach to policy making. The tenacity of such a model can somewhat explain why citizens of newly independent countries saw and accepted how their governments continued to rule from above without consultation with the public, showed limited attempts to build democratic institutions, and remained dependent on the international system for aid in order to stay in power – even if that system operated in a neocolonial way. The legacy of the colonial state can also partially explain why a pro-independence freedom fighter like Robert Mugabe of Zimbabwe became a tyrant who still ruled the country in 2017 as he celebrated his ninety-third birthday. Mugabe may suffer from "the conflicting loyalties of clashing cultural values." However, he most certainly "hails from a collective social system characterized by the expectation that people will not oppose those above them in the hierarchy, i.e., authoritarianism."[2]

Although the colonial period officially ended a long time ago, with former colonies becoming independent, it is important to ask, "what remains, at this turn of century, of the African quest for self-determination."[3] A recent article surveys a growing body of empirical evidence indicating the long-term effects of historic events on economic development. Without a doubt history matters, although how and why it matters remains the subject of ongoing studies. Not surprisingly, scholars point out the historical roots of domestic institutions and cultural norms of behavior.[4] The continent-wide post-traumatic stress resulting from the suffering under colonialism still lingers all over Africa, and its impact should not be discounted.[5] To comprehend at least some aspects of this continuing stress, we have to first understand the nature of the colonial relationship. As one scholar explains:

> The colonial relationship is based on the distinction between the animal and the domestic animal. Colonization as an enterprise of domestication includes at least three factors: the *appropriation* of the animal (the native) by the human (the colonist);

the *familiarization* of man (the colonist) and the animal (the native); and the *utilization* of the animal (the native) by the human (the colonist). One may think such a process was as arbitrary as it was one-dimensional, but that would be to forget that neither the colonist nor the colonized people emerge from this circle unharmed.[6]

II Indigenous civilizations and the European powers

European domination over the indigenous peoples of the Americas, Africa, and Asia was largely a negative experience, and its footprints still can be found on these continents. Although there is no precise date that marks the onset of the conquest, it is generally agreed that the turning point was 1492, when Christopher Columbus discovered the Americas, or to be precise, the year he discovered: San Salvador, the Bahamas, and Cuba. Columbus was sent by the Spanish Crown, worried about the growing influence of Portugal. The Portuguese were the pioneers of the overseas expansion. As early as 1415, Portugal conquered Ceuta, a prosperous Islamic city and a trade outpost, located in North Africa. Soon after, the Portuguese Crown claimed its first colonies in the Atlantic: Madeira and the Azores. By the beginning of the sixteenth century, imperial competition between Spain and Portugal was fast accelerating. In 1521, Cortez conquered the Aztec empire, paving the way for the eventual Spanish conquest of a territory known today as Mexico.[7] For the next 400 years Europe would strive to control most of the globe. According to a major study, in 1800 European powers controlled one-third of "the world's land surface," two-thirds by 1878, and four-fifths (or just over 84 percent) by 1914.[8]

The period of European domination lasted over several centuries. This is why it is helpful to clarify the terms used, such as imperialism, colonialism, and neocolonialism: "whereas colonialism means direct rule of people by a foreign state, imperialism refers to a general system of domination by a state (or states) of other states, regions or the whole world."[9] Neocolonialism is more elusive as a term. Following an observation of one scholar, it is a peculiar kind of relationship that translates to "inequitable ties" of former colonies with a world economic order. These ties, which long outlived independence, are the result of the colonial experience.[10] To put it more expressively:

> To African nationalists neocolonialism does not mean the imposition of new forms of alien control from outside Africa; rather, it is the persistence of residual European influence in Africa itself. Moreover, it is not direct political control or influence that is the irritant, but the omnipresent cultural influence of the West, which saturates all aspects of life.[11]

The above definitions serve to distinguish among the three stages of European expansionism. The first phase, characterized by imperialism, stretched from 1415 to 1776. The second phase involved the formal establishment of colonies. It lasted symbolically until 1945, when the end of World War II marked the beginning of the formal decolonization period. The last phase refers to the post-independence period, which is set apart by competing neocolonial tendencies originating in the major western countries.

During the first phase, the naval power of the Portuguese, Spanish, Dutch, British, and French was strategically used to conquer new lands for commercial gain and political prestige. It was during this time that the first multinational corporations were founded: the British East India Company in 1600, and the Dutch East India Company in 1602. These government-chartered joint stock companies were essentially trading monopolies obligated to confer certain rights to their respective governments in exchange for their monopoly privileges. With time, the companies would become powerful arms of the European powers with their own security forces (private armies) and banking systems.[12]

While the Dutch and the British sailed their ships to Asia and then to Africa, armed with superior weaponry, the Spanish and the Portuguese imperial powers conquered and ravaged the ancient empires they found flourishing in South America: the Aztecs and the Incas. The Incas developed a sophisticated civilization, located in South America on the territory of contemporary Peru. It reached its peak from 1438 until 1532, when it was destroyed by the Spanish conquistadors, led by Pizarro. The Inca empire was ruled by Atahualpa and stretched as far north as Colombia and as far south as Chile and Argentina. Pizarro defeated the Inca in the bloody Battle of Cajamarca on November 16, 1532. The fact that a small number of Spanish troops was able to defeat thousands of Inca warriors is attributable to many factors, among them the Spanish horses and lances, their steel blades, and cannons. The Incas were known for the beautiful textiles, art, and goods made of gold. None of these items survived the Spanish conquest. At Pizarro's request, all gold items were melted and bars of gold were sent to Spain. In addition, smallpox introduced by the Spanish had almost completely shattered the indigenous population.[13]

The Spanish and the Portuguese were later followed by a wave of British and French Settlers coming to North America. Upon their arrival, the imperial powers were prepared to engage into brutal conflicts with the native populations that almost exterminated many of the aboriginal tribes. While the Spanish, the British, and the French remained preoccupied with taking over massive territories for imperial control or settlement in the Americas, the Portuguese decided to unlock new frontiers of imperial expansion in the East. This imperial ambition was realized under Vasco da Gama – a Portuguese explorer who first sailed from Europe to India around the Cape of Good Hope in 1498. The new sea route gave the Portuguese enormous advantage, and soon parts of Africa, India, and East Asia fell under their sphere of influence.[14] The Dutch, the Swedes, the British, the French and the Danes also eventually arrived in Asia, rummaging through the continent in their bid to have access to the vast material resources of the region and to establish their own outposts.

Meanwhile, in the Americas, European imperialism was steadily evolving into the colonial phase with the formal settlements built on the conquered lands. Taking advantage of the warm southern climate, the European settlers were establishing sizeable plantations of cotton and sugar. As the Industrial Revolution was gaining momentum in Europe, the demand for cotton increased substantially. The production on the plantations could not keep up with the demand because of a severe shortage of workers. This scarcity of labor in the cotton and the sugar plantations of the Americas and the Caribbean and the need for cheap labor in the gold and silver mines of Latin America opened a new and more ominous dimension of European expansionism: the slave trade.

In order to satisfy the insatiable demand for labor in the colonies, the European governments turned a blind eye to the developments on the African continent, where a new troubling industry was taking shape. Africans were ruthlessly captured or bought and shipped to the Americas to become slave labor. Slavery existed in Africa to a certain extent before the European arrival, but it was a different kind of a system. Under the old tribal system, soldiers captured during wars or people who committed crimes were enslaved. In most places, such slaves were able to regain their freedoms upon paying a price or after working for several years. Still, the fact that the slavery was present on the continent turned out to be catastrophic for Africa. In the relatively short period of a few decades, the Portuguese exploited the system and became heavily involved in the increasingly profitable trade as the slave-buyers quickly multiplied. Slaves were shipped in millions to Brazil to work in mines and coffee plantations. Later came the demand from North America: "roughly one of every four slaves imported to work the cotton and tobacco plantations in the American South began his or her journey across the Atlantic from equatorial Africa."[15] Another significant place that slaves came from was the Windward Coast, which today just about comprises the Ivory Coast.

The European imperial powers proceeded to establish an elaborate intercontinental trading scheme, known as the *triangular* or the *transatlantic trade*. The system was perversely well organized. Traders on the African continent shipped millions of people to the Americas and the Caribbean. Once they were sold as commodities, the enslaved people were compelled to work in mines or to work on plantations to produce cotton, sugar and tobacco. The slave mode of production significantly lowered the cost of producing many commodities, which benefited the European owners, who reaped huge profits. Money was regularly sent to Europe. In turn, Europe produced clothes and guns, which were then shipped to Africa, where they were used to control more territory and enslave more people. For the African continent, the slave trade meant a huge socio-economic loss, since it involved the removal of the most productive sector of African society: young, able-bodied men and women. The whole commercial infrastructure was built to increase the ability to capture and sell people. The resulting disorder exacerbated the political and ethnic divisions on the already politically fragmented continent. The impact of slavery on Africa cannot be underestimated: "the export of about 11 million slaves from 1500 to 1800, including the astronomical increase between 1650 and 1800 in the Atlantic sector, could not have occurred without the transformation of the African political economy."[16]

The slave mode of production accelerated the economic performance of the colonies and accounted for much of the wealth during the European-settler years in the Americas and the Caribbean. By the nineteenth century, slave labor became a significant factor in maintaining the desired level of cotton production in the Southern states. However, as the number of slaves working on plantations multiplied, calls for abolishing slavery also intensified. Such calls were openly rejected by politicians. Citing economic considerations, South Carolina senator James H. Hammond famously named such proposals in 1858: *war upon cotton*.[17] He was known to represent powerful business interests, who maintained that cotton cultivation was the indispensable backbone of the southern economy and the source of vital profits.

The general pattern of economic imperialism in the first phase of European expansionism disproportionally benefited the dominant naval empires. By the time this phase was coming to an end in the eighteenth century, Africa and Asia had been incorporated into the asymmetrical world-trading network.

The second colonial phase roughly started in 1776. This phase was marked by the growing presence of the British empire in many regions of the world and the relative decline of other European powers. The literature, however, is mixed when assessing the strength of the British imperial dominance during this period.[18] Some scholars refer to two main events that serve as a manifestation of its slow decline. The first event was the American Revolution of 1776, which ended British rule in the most lucrative part of North America. The second event was, paradoxically, the rising influence of British economists, especially Adam Smith and David Ricardo, who argued for free trade among all nations and for the removal of government restrictions over domestic and international commerce. Partly in response to these liberal views, the British empire slowly began both to dismantle its system of protective tariffs and loosen its control over its imperial and colonial holdings, particularly in British North America (later, the Dominion of Canada) and in the Orange Free State and Transvaal in South Africa.[19] On the other hand, it is also argued that British imperial dominance became more pervasive during the second phase of expansionism. What is more, the industrial revolution gave the British unparalleled economic power. In the mid-nineteenth century, the British penetrated more regions in Africa, Australia, New Zealand and the western part of today's Canada (British Columbia). In India, rule by the East India Company was replaced by the establishment of direct Crown administration in 1858. The British exercised control over five main regions of Asia: the whole of the Indian peninsula, the northwest frontiers of India, Burma, the Malay peninsula, and the coasts of China.[20]

During the second colonial phase, the other imperial powers were consistently declining. In 1803, the United States purchased Louisiana from France after a series of failed attempts by the French to make it profitable.[21] With the final defeat of Napoleon in 1815, the French slowly lost their interest in further expansion, although they continued to reap benefits from African colonies and maintained their strong presence in Algeria. The Spanish and the Portuguese influence was effectively diminished by the revolt of their Latin American colonies in the 1820s. After Spain was forced into a peace deal with Morocco in 1860, it was noted that Spain became: "a lay figure in the diplomatic drama played out by the major powers,"[22] with the British leading the pack. Although we should not forget that from 1879 to 1884 Belgium under King Leopold II extended its political control over the Kingdom of Kongo in the heart of Africa. With the use of African mercenaries, Leopold created the most powerful foreign army on the continent, consisting of approximately 20,000 men. The costs of maintaining the army were substantial, it consumed more than half of the state's budget. Leopold, who treated the Kongo colony as a private venture, believed the army was necessary to quickly contain any possible rebellion over the territory exploited for ivory, rubber, and slaves.[23]

Despite losing the United States, the remnants of the British empire dominated the international economy with little threat to their position until 1914. Following World War I, victorious European nations continued to dominate overseas territories through colonial spheres of influence. Finally, after World War II was over, the calls for the self-determination of all people were acted upon. It would be a long and uneven process, punctuated by the lingering influence of the former colonial powers in Africa and Asia. Unlike the first and the second phases of European global expansionism, the third neocolonial phase was gradual and largely concealed. The central challenges when thinking about neocolonialism stem from its ambiguous perseverance. What started in the 1950s as awkward attempts by Europe to maintain relationships with the former colonies turned into confusing neocolonial knots. Writing in 1955, James S. Coleman provided pointed remarks about that decade in African history:[24]

> Colonialism in Africa has been the medium for the indiscriminate diffusion of Western ideas and institutions. Dominant among these is the modern state. Most current efforts to erect modern states in the heterogeneous cultural and racial milieu of African territories are producing situations that either defy peaceful solution or invite authoritarianism.... State formation and colonial liquidation in Tropical Africa are taking place in a climate of opinion and institutional environment quite different from that which prevailed in the other major areas formerly under European imperialism. Under the pressure of world opinion and an African nationalism saturated with mid-twentieth-century ideas of social democracy, the welfare state, and mass participation in politics, sobered and repentant colonial governments have pursued for nearly a decade various schemes of political, economic, and social development.... These special elements in the African situation are bound to produce a unique pattern of politics and Euro-African relationship in the post-colonial period.

III Theoretical explanations of imperialism and colonialism

In general, scholars agree that mercantilist ideas provided the intellectual foundation behind the first phase of European expansionism. Mercantilism can be defined as an economic philosophy that centers on maximizing the economic power of one's nation, often at the expense of other nations. In medieval Europe, economic prominence was traditionally associated with the accumulation of precious metals and having control over vast amounts of land. By the sixteenth century, mercantilism dominated economic thinking in Europe. Mercantilists supported the view that governments should promote economic nationalism by controlling the domestic economy. Mercantilist policies aimed to swell the state's coffers with gold and silver, obtained by ensuring large volume of exports while limiting imports to minimum.

In a more nuanced analysis of this approach, one scholar demonstrated how English mercantilism prioritized full employment and "a favorable trade balance."[25] To meet these objectives mercantilist leaders proposed the following commercial policy:

> duties on imports, with rebates on raw materials used in making exports; the prohibition of certain imported goods; the removal of export duties; subsidies and other

assistance to the export industries; monopoly grants to certain joint stock companies engage in foreign trade; a prohibition of the export of coin and bullion; and an aggressive foreign policy by which England would help its exporters capture markets from their competitors.[26]

This paragraph shows that even in its benevolent form, mercantilism still meant quite a protectionist policy mixture.

The era of exploration contributed to developing a less benevolent form of mercantilism. The inherent contradiction within mercantilist thought – rejection of free trade and yet acceptance of policies that would see the expanding of protectionist trade exports – was resolved by the colonial conquest. Mercantilism provided rationalization for the imperial expansion and the establishment of colonies viewed as indispensable because:

they afforded an opportunity to shut out commercial competition; they guaranteed exclusive access to untapped markets and sources of cheap materials (as well as, in some instances, direct sources of the precious materials themselves). Each state was determined to monopolize as many of these overseas mercantile opportunities as possible.[27]

Perhaps the most famous examples of British mercantilist trade policies were the Navigation Acts of the 1660s and the Staple Act of 1663. These laws gave undisputed preferences to the British commercial interests and promoted the establishment of colonial plantations in the southern part of North America. The Navigation Acts excluded all foreign merchants and their ships from carrying trade flows to and from British colonies. The Staple Act stipulated that the owners of plantations had to purchase goods from England. Even slaves had to be purchased from English slave traders. The colonies were only allowed access to English sources of capital and credit.[28] The Acts attempted to ensure that the British empire had definite monopoly over British controlled areas, including all trade activities.

From the beginning of the overseas exploration, mercantilist ideas were very attractive to European rulers, who had ambitions to conquer the world. The concept of peaceful international cooperation was foreign to the nations of Europe. After all, they were chronically fighting each other for centuries over money, prestige, and territory. The first escapades of Columbus, Pizarro, and Cortez were financed by the Spanish Crown hoping to become a powerful kingdom by finding large quantities of gold and silver. The conquistadors under Pizarro fulfilled these hopes, but in the process they destroyed the Incas. The last king of the Inca empire tried to buy his freedom from the invading Spaniards by offering Pizarro 13,420 pounds of 22 carat gold and 26,000 pounds of pure silver. This move did not make any difference. In 1533, Pizarro took the offering, executed the king, and allowed his soldiers to go after everything of value in the orgy of plundering that vanquished the Incas.[29]

History shapes the perceptions of policy makers, so there is no surprise that mercantilism was the widely accepted view of economic relations until Adam Smith's publication of his famous book in 1776, *The Wealth of Nations*. The book strongly supported the idea of free markets, but it also provided a convincing critique of mercantilism. Smith viewed mercantilism as a highly unstable approach to international economic relations. By the time these liberal ideas took hold, however, thousands of European explorers had sailed towards the New World. They were guided by mercantilist convictions that it was beneficial for a nation's wealth to engage in self-centered behavior, even if it meant predatory actions towards people living in distant lands.

Mercantilism continued to inform the second phase of European expansion. But as the empires pushed the frontiers of their territorial influence, the growing popularity of liberal arguments would change the character of their domination. Three distinct theoretical standpoints provide competing explanations as to why colonialism became such a powerful force in the global economy. The first is the theory inspired by Marxism and formulated in the writings of Lenin. The second is the view advanced by J.A. Hobson, who represented the attitudes of British liberals struggling to explain the continuation of colonialism. The third is the political argument of D.K. Fieldhouse who disagreed with both Lenin and Hobson, and suggested an alternative explanation.

Lenin, who was one of the principal architects of the Russian Bolshevik revolution of 1917, created a Marxist concept of colonial imperialism. In doing so, he achieved two goals with one theory. First: "Marxists required an economic explanation for the explosion, from the late nineteenth century on, of expansionist policies, on the part of all major advanced capitalist countries." Second, he used this theoretical reasoning to explain the absence of "the confidently expected socialist revolution" in the industrialized West as anticipated by Marx.[30] Lenin's theory declared:

> Imperialism is capitalism in that state of development in which the dominance of monopolies and finance capital has established itself; in which the export of capital has acquired pronounced importance; in which the division of the world among the international trusts has begun; in which the division of all territories of the globe among the great capitalist powers has been completed.[31]

Most likely, Lenin would have a problem with our way of periodizing European expansionism into three somehow distinct phases. Lenin's belief in historical materialism convinced him to see colonial imperialist expansion as part of a long process of global capitalist development.

Both theories of uneven development in the global economy – dependency and world-system theory – were built around this understanding of imperialism. Colonial expansion unleashed disruptive forces that permanently privileged the core industrialized countries within the world's capitalist system. Wallerstein, the author of the world-systems theory, specifically noted that:

> the three structural positions in a world economy – core, periphery, and semi-periphery – had become stabilized by about 1640.... The geographical expansion of the European world economy meant the elimination of other world-systems as well as the absorption of the remaining mini-systems.[32]

From the Marxist point of view, the colonial period was critical for the consolidation of the current inherently asymmetrical structures of the global economy.

On the other hand, Hobson viewed imperialism as caused by a combination of forces. He specially focused on the growing power of international finance playing a decisive role in the operations of the late nineteenth-century economies. In explaining the developments of this period, he rejected the proposition that European powers were driven by the logic of capitalism and simply engaged in the exploitative search for new markets. Rather, the colonies were sought because they provided logical opportunities for the investment of surplus savings. Hobson considered competition among European powers as the natural struggle for dominance. However, the colonial phase of this competition was unique to the extent that it was enabled by the transnational financier. He concluded:

> In view of the part which the non-economic factors of patriotism, adventure, military enterprise, political ambition, and philanthropy play in imperial expansion, it may appear that to impute to financiers so much power is to take a too narrowly economic view of history. And it is true that the motor-power of Imperialism is not chiefly financial: finance is rather the governor of the imperial engine.[33]

Furthermore, for Hobson there was nothing inevitable, or not-fixable, about colonial expansion "because its 'taproot' could be cut by social reform."[34] For suggesting such a straightforward solution to colonialism, Hobson was vilified by the Marxists, who "regarded him as a 'bourgeois revisionist.'"[35]

Fieldhouse did not think that either of these explanations for colonization were satisfactory. Going so far as calling some of the evidence provided by Hobson conspiratorial[36] and referring to Lenin as dogmatic,[37] Fieldhouse tried to champion a different view. His argument challenged the assumptions made by Lenin and Hobson in one principal way. To summarize, Fieldhouse stated that late nineteenth-century colonial expansion was a political phenomenon. He maintained that the costs of maintaining the overseas colonies were tremendous, and thus having colonies implied having special responsibilities. Accordingly, the decisions to intensify territorial control over overseas colonial lands came from "multiple problems arising on the frontiers of European activity in other continents." The full-scale colonization happened as a necessary response to these frontier problems during the period when the relations between the colonizers and the colonized "became fundamentally unstable."[38] For Hobson, it was an unfortunate situation that would eventually come to an end with the demise of colonial empires. He hoped for a natural satisfactory conclusion when: "Europe could withdraw the legions with reasonable confidence that the fundamental problems of international relations which had generated the imperialism of the late nineteenth century no longer existed."[39]

Fieldhouse's political theory of imperial colonialism was met with some grave disapproval, not only because he thought about it as a necessary "expedient" way of dealing with problems occurring in the "pre-capitalist periphery." He was also criticized for unquestionably praising some of the European actions during the colonial phase of the expansionism. Here is how one scholar summarizes the main critique of Fieldhouse's work:[40]

While conceding that colonial rule may have rankled Africans at times, Fieldhouse suggested that it was "historically the lesser of two evils facing most indigenous peoples in the later nineteenth century," with the other possible evil, he implied, having been to leave Africans to themselves. With its claims for prudent administration, firm control, and good intentions, Fieldhouse's description of colonial rule is one that most historians of Africa would argue against. Where Fieldhouse saw cool bureaucracy, systematic law codes, and coherent policies, historians have for some years been more likely to notice the randomness, incoherence, and unpredictable harshness of colonial "systems."

Although none of the theories discussed above provides a perfect explanation for the imperial colonial expansion, all are important attempts to comprehend this critical period of our world history. The Marxist perspective has proven to be particularly resilient in continuing to intellectually provoke many post-colonial conversations. Conversely, Fieldhouse's political argument is viewed with apprehension today. Especially given its assertion that colonialism simply ended, and its implicit rejection of neocolonialism.

Which brings this discussion to the third neocolonial phase of expansionism. There is a vast body of literature, and not all of it traced to Marxism, that deals with post-colonial theory. For all the complicated narratives that try to address a multiplicity of diverse forms of post-colonial relations between the West and the Rest, the following thoughts synthesize the role of postcolonial theory:[41]

postcolonial theory entails "the rejection of those modes of thinking which configure the third world in such irreducible essences as religiosity, underdevelopment, poverty, nationhood, [or] non-Westernness."[42] ... I take it to mean that the "postcolonial condition" is one in which the legacy of colonialism – practically, historically and theoretically – is ever-present, even in the attempt to think beyond it.

Post-colonial theory focuses on the question of how western practices of colonial domination impacted the processes of identity formation and political organization among indigenous communities. It asks to what extent the post-colonial identity reduces the capacity of individual and collective agency to be independent. It also examines the responses of postcolonial societies to offers of new engagements with the West in the form of cultural, social, or commercial relations. The analytical departure points of many postcolonial discourses come from diverse sources and focus on individual and communal experiences shaped by colonial institutions in the Americas, Asia, Africa, and the Pacific.

The crucial point made by the post-colonial thinkers concerns the global diffusion of economic liberalism. According to this view, liberalism's "universalizing ethos" contains justification for imperial ambitions.[43] Later in the book, we examine different actions by western governments, or agencies, undertaken in the name of liberal principles. Such actions often mean direct involvement in the politics of developing countries, and hence are considered by post-colonial thinkers to exemplify the continuation of neo-colonial ties. Post-colonial theory attempts to understand these ties and the subsequent trajectories of societies subjected to ongoing external pressures. In this regard, the label of a new imperial actor on the international stage is routinely attached to the United States. The previously examined hegemonic stability theory views American policies as created by a benign and well-intentioned hegemon. Neo-colonial critics would disagree with such an interpretation. When in 2003 the United States, and its coalition, decided to invade Iraq, the move was criticized for displaying signs of imperialistic behavior. One scholar offered a pointed explanation:[44]

> America's empire is not like empires of times past, built on colonies, conquest and the white man's burden. [It is] an empire lite, a global hegemony whose grace notes are free markets, human rights and democracy, enforced by the most awesome military power the world has ever known.

IV Colonialism and the creation of the Third World

By the end of the nineteenth century, the colonial powers had established spheres of influence in virtually every corner of the globe They were capable of exercising political domination through direct control made possible by the possession of superior military weaponry. As it turns out, Europeans had multiple advantages that allowed them to conquer native societies. Jared Diamond famously called these advantages: guns, germs, and steel. In terms of military technology, the difference was particularly staggering. While the Native Americans were using clubs and axes of stone and wood to fight the invaders, Europeans had steel swords, lances, daggers, and body armor, and were in the process of perfecting firearms. In addition, while Native American armies moved by foot, Europeans used horses, with a terrifying effect.[45]

Diamond attributes this superior military technology to a more favorable path of development enjoyed by Eurasian societies as conditioned by geography. The argument suggests that faster progress towards specialized settlements in the Eurasian region was initially triggered by a better climate and fertile terrain, which over the millennia created specific animals that could be domesticated and used as tools and sources of food and clothing. A more favorable geographical setting and better climate allowed the cultivation of cereal crops like wheat, barley, and rye. These factors permitted the establishment of permanent and specialized communities. With time the division of labor contributed to making Eurasian settlements more efficient, secured, and conducive for finding and employing new inventions.

By contrast, people living in a tropical inhospitable climate, full of predatory animals and with limited ability to grow crops, were compelled to spend most of their resources on trying to survive. A harsh climate and challenging terrain forced many indigenous communities to live as nomads, with very limited options to develop new technologies. According to Diamond, the historically traced geographical factors can explain differences in all aspects of technology between the colonizers and the native populations. And these differences are partially responsible for the success of the European conquest.[46] This geographical hypothesis has its fair share of critics because it can lead to a deterministic conclusion that the world's economic inequality originated with geographical differences.

The findings provided by Diamond offer an interesting interpretation of human history and complement the historical examination of the colonial period.

In critiquing the geography hypothesis, Acemoglu and Robinson claim that it is incapable of capturing the nuances of nations' historical trajectories. Some countries had been persistently unsuccessful, despite being situated in the area that provided abundant natural resources. While other countries had excelled where they should had failed as forecasted by their geography. Many nations around the world had gone through periods of prosperity only to experience acute crises later. The reverse scenario is also true. By disagreeing with the geography argument, Acemoglu and Robinson advance the institutional thesis.[47] It postulates that the global developmental inequality among nations emerged during the time of the Industrial Revolution in the late eighteenth century. The industrial progress propelled some countries forward by providing incredible incentives for technological innovations, political freedoms, and labor mobility. These developments paved the way for democracy. Rather than geography, the key for the successful development of societies is the quality of state institutions, the authors conclude. In their elaborate historical analysis, they show the benefits of inclusive democratic institutions on economic development and the disastrous consequences of the extractive institutions.

The institutional argument draws attention to critical junctures in the timelines of nations. The era of colonization is recognized as a crucial period for most colonized societies. During this time, the state became the most important construct transplanted to the overseas territories by the colonial powers. The state institutions varied according to the preferences of the colonial power and were imposed on societies without their consent. The authors note the intensification of extractive institutions during colonization.[48] Less convincing is their observation about the post-war de-colonization period, believed to create another set of critical junctures for many former colonies. Critical junctures can be defined as unique historical moments that offer opportunities for change. Unfortunately, as the authors lament: "the post-independence governments ... simply repeated and intensified the abuses of their predecessors, often severely narrowing the distribution of political power, dismantling constrains, and undermining the already meager incentives that economic institutions provided for investment and economic progress."[49] In many cases, the newly independent governments reinforced the extractive institutions instead of creating new ones. Sadly, the opportunity provided by gaining independence was not fully utilized into building inclusive institutions as new leaders struggled with multiple challenges of the post-colonial period. Such moves were often motivated by the desire to escape deprivation by remaining part of the governing elites. Processes of governability are never simple, especially when institutional structures that are supposed to constrain or stimulate such processes are weak or non-existent. A weak institutional environment, combined with limited state resources – further exacerbated by persistent poverty of a nation – elevates the level of urgency when it comes to the issue of political survival.

The independence movements aspired to alter the political configuration of post-colonial societies, but they were seldom fully successful. The states inherited from the colonial era were awkward creatures. Institutionally speaking, post-colonial states resembled centralized entities organized to ensure the implementation of colonial policies. Politically, they were oppressive organizations ready to employ violence if necessary to maintain control over the population. Many of these states were assigned superficial boundaries, drawn on the negoti-ating tables in the salons of colonial powers. The delineation of territorial boundaries fos-tered the effective administration of the conquered lands. In addition, these state-making colonial projects reflected Europe's conviction that the state should be replicated around the world as the most desirable form of political authority.

The concept of the state is traced back to the end of the Thirty Years' War in Western Europe and the signing of the Treaty of Westphalia in 1648. The treaty heralded the end of decades-long religious wars. It is an important historical benchmark, marking the origin of modern international relations by a way of establishing a sovereign state-system. After Westphalia, the state becomes the primary face of legitimate political authority exercised over a particular territory and not to be interfered with by other states.[50]

In the three centuries following the signing of the treaty, the increasingly powerful European states kept expanding the inter-state system around the world. The making of the colonies, however, was a paradoxical undertaking from the Westphalian point of view. The colonial states had borders, but the ultimate political authority rested with the colonial powers, contravening the most important principle of Westphalia: the principle of sovereignty. In fact, the paradox would not be resolved until the colonial independence movements contested the operation of colonies and the 1960 UN Declaration recognized all nations' rights to self-determination. Only then were all countries around the world acknowledged as sovereign states equal under international law.[51]

During the colonial era, which roughly coincided with the 300-year period following Westphalia, polities outside the European state system were considered according to one of three categories. The first category included European trade partners and their rivals, like China and the Ottoman empire. The second category were: "nonsovereign polities … that Europe treated as less than equal, but not as wholly subordinate."[52] Here were included some parts of Asia, the Middle East, and portions of Latin America. By the end of the nineteenth century, China and the Ottoman empire would fall into this category as well. The third category were the colonized nations, which were completely under the sovereign rule of European powers. All colonial states had governmental authorities installed by and answerable to European rulers. Colonies then constituted subordinated extensions of colonial powers under the tacit rule that no other European state could legitimately meddle in their affairs. Needless to say, colonial territories were the site of many armed conflicts between European powers.[53]

We can identify seven characteristics that all colonial states shared. First, there was an international political dimension, which meant the complex relationships between colonies and dominant countries, resulting from direct political control by European empires over colonized societies. Second, bureaucratic elitism and authoritarianism were the forms of governmental organizations in colonial states. Third, the use of "traditional" or customary authority figures in colonial society was employed by coercing and co-opting the local landlords or communal chiefs to serve as allies. Fourth, the use of force was regularly sanctioned to maintain order. Fifth, technological advantage gave Europe-ans not only the ability to control colonies, it also helped to engender the myth of their superiority. Sixth, statism, which translated to comprehensive command over the economy by the state. Seventh, hegemonic ideology reinforced by claims about the benevolent nature of the colonial rule by the invincible great powers.[54] These characteristics manifested themselves in different forms and intensities over the long period of colonialism. Upon the official completion of the decolonization processes, the UN Declaration prohibited countries from governing other countries as colonies. But becoming independent had its obvious limits. Only recognized colonial states were entitled to the UN recognized statehood. Nations or ethnic groups within those colonies were not.[55]

The European great powers were determined to spread their political domination over the world for almost 300 years. The most dramatic intensification of their efforts occurred between the 1880s and the start of World War I in Africa. Known for a rapid transition from measured engagement to direct military control, this period became known as the scramble for Africa. The accelerated attempts to extend the space of colonized territories threatened to spark a military conflict among the contending European powers. Anxious to resolve any destructive conflicts, Otto Von Bismarck – the ruler of the newly formed German nation – called a conference of the European colonial powers in Berlin. The Berlin conference took place from November 1884 to February 1885 and resulted in a peaceful partition of the African continent amongst its participants. It was all done with the "aim at instructing the natives and bringing home to them the blessings of civilization."[56] The borders were drawn with no consultation with African people, cutting through the lines of ethnic tensions and inserting Europeans inside competing local interests. For example, the conference divided the nation of Tutsis between the Belgian Congo and German East Africa, the region that would later become Rwanda and Burundi.[57] The division of the Tutsis was exploited by the colonizers, exacerbating ethnic tensions between them and another local ethnic group, the Hutu. Generations later the tensions exploded, leading to the genocide in Rwanda in 1994.

The maps shown in Figures 2.1 to 2.4 demonstrate the scope of colonial possessions and the formal process of independence. In their respective colonial territories, each

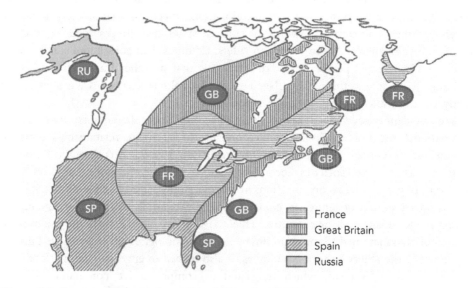

Figure 2.1 European colonization of North America, 1763

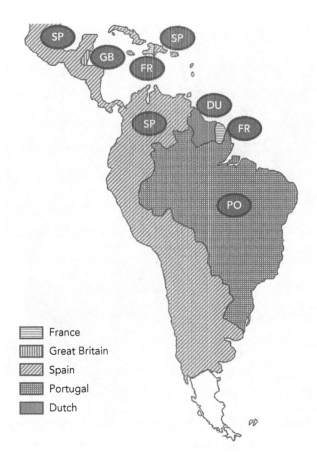

Figure 2.2 European colonization in eighteenth-century Latin America

European power went on to establish a political structure that was closely modeled on the institutions of the *Metropole*.

From an economic point of view, European powers exercised domination by organizing the colonial territories in a manner that best served their economic needs and interests. Through decree and coercion, the colonizers altered the existing modes of economic production in the conquered areas by establishing socio-economic structures that benefited them at the expense of the colonized people (see Box 2.1 below for the case study on a socio-economic and political profile of British rule in Kenya). What is more, most of the economic activities that emerged in the colonized areas were limited to the production of primary commodities, mainly in the agricultural and the mining sectors. The goal was to ensure supplies of cheap agricultural goods and raw materials for Europe's industrial consumption. While the European powers continued to develop their economies to become the core centers of industrial manufacturing, the colonies were kept at the subservient level of supplying basic commodities to Europe.

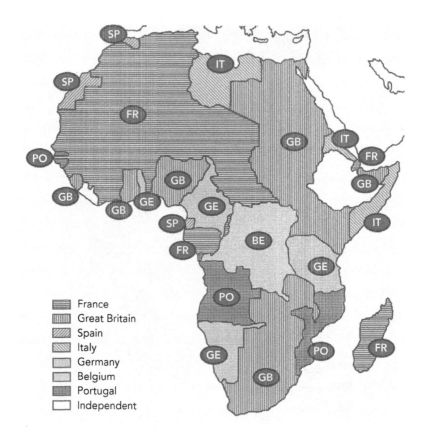

Figure 2.3 Colonization in Africa, 1914

France
Great Britain
Spain
Italy
Germany
Belgium
Portugal
Independent

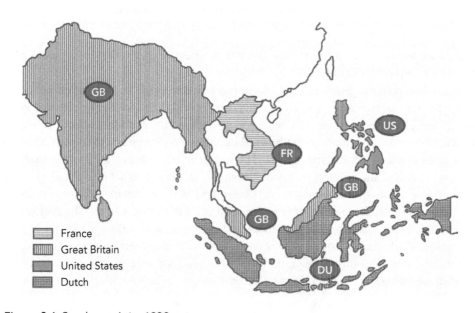

France
Great Britain
United States
Dutch

Figure 2.4 Southeast Asia, 1939

Box 2.1 Case study – Kenya under British rule: a socio-economic and political profile

I Building a white settler economy

In 1895 Kenya was declared an official part of the British East African Protectorate. From 1901 onwards, when there were only about 30 white settlers in the Protectorate, the new commissioner, Sir Charles Elliot, decided to actively encourage the establishment of European settlement in the highlands. He also recommended that other non-white settlements be confined to the lowlands. In pursuing this policy, Elliot was backed by the small numbers of white settlers who had already formed themselves into a political association in Kenya and successfully lobbied that the highlands should be reserved for Europeans only. The granting of land for settlement was governed by the 1902 Crown Lands Ordinance, which stated that "in all dealings with Crown land, regard shall be had for the rights and requirements of natives and in particular the commissioner shall not sell or lease any land in the actual occupation of the natives." The ordinance, while appearing to protect African land rights, actually entrenched the expropriation of large parts of the land belonging to the Kikuyu ethnic group. The ordinance empowered the commissioner to sell freehold estates on the land that was not under occupation by Africans without the consent of tribal chiefs. Such land was considered as "waste and unoccupied land." In the 1890s, the Kikuyu had suffered a series of disasters: between 1890 and 1899 there were severe droughts followed by an epidemic of smallpox, which had led to a 70 percent decrease in the Kikuyu population in the southern district of the Kikuyu land. As a result, much of the land in this area became unoccupied.

The Kikuyu did not regard the fact that land was unoccupied as meaning it no longer belonged to them. In addition, the Kikuyu practiced a system of shifting agriculture and their land was not always in use. The ordinance, however, allowed their rights to any unoccupied land to be ignored. Under the 1902 Crown Lands Ordinance, a number of pioneer white settlers were given titles to some enormous territories of land. In particular, the two most prominent settler leaders, Hugh Chalmondeley, later to become Lord Delamere, and Col. Ewert Grogan were key beneficiaries. Delamere was able to amass 115,000 acres, while Grogan bought and leased a total of 186,000 acres. By the end of 1905, over a million acres of Kikuyu land had been either leased or sold to the settlers by the British colonial authorities. No Asians or Blacks were allowed to lease or buy land in the fertile highlands.

II White settlers and political control

In 1905, the Colonial Offices took over responsibility for the Protectorate from the Foreign Office. In 1906, a new constitution was introduced, under which the commissioner became the Governor of the Protectorate. Later, the executive and legislative councils were established. Only Europeans were represented on the Legislative Council (LEGICO), which comprised of six government officials and two non-officials when it first met in 1907. The settlers slowly increased their power on the council and this gave them the means by which to advance policies favorable to them. By 1920, the settlers had won the right to elect members to the legislative council. In the same year, Kenya became a full British colony under Sir Edward Northey. While the settlers were over-represented in the council, the only outlet for African political aspirations was via the local native councils established in 1924. The lack of political rights was a fundamental problem for the Africans and later became a unifying force among the different ethnic groups in Kenya, thus providing an important basis for the development of African nationalism.

III African labor and African exclusion from the colonial economy

From the outset of the colonial conquest, Africans were compelled to participate in the unequal economy run by the Europeans. By the end of the second decade of white settlements, 25 percent of the Kikuyu population was living permanently on farms owned by white settlers as migrant laborers. Many Africans who had been landholders before, became squatters and laborers on these "white" farms when their lands were expropriated. Initially, the African laborers were able to do quite well on the land allotted to them by the white settler, in return for which they had to work for the settler farmer for three to four months a year. But after 1920, when European farming was established on a more formalized basis, the rights of African workers were undermined. White farmers, who were growing coffee, sisal, maize, and wheat for export, became more interested in a steady supply of resident laborers and as such, were less interested in allowing temporary African workers on their farms. What is more, the number of livestock Africans could keep and the amount of land granted to them for cultivation was reduced while the number of days they had to work for the white settler farmers increased. Progressively, the Africans were excluded from being key players in the economy and transformed from being independent producers into indentured agricultural laborers for the white settler farmers.

The process of creating favorable conditions for an increased supply of African labor was accelerated by an increase in the number of unemployed Africans. African taxation had also been introduced as a way of forcing Africans to work as farm laborers. Africans were prohibited from growing cash crops like coffee or tea. Gradually Africans found it impossible to accept these changes without resistance, and in 1929 a protest movement emerged. African anti-British protests continued throughout the 1930s and the 1940s. Further restrictions were placed on African rights following the intensification of agricultural production during World War II. As a result, the native communities, especially the Kikuyu whose land had been extensively expropriated by the white settlers, became increasingly impoverished.

IV World War I, the roots of resistance and the end of British rule

The Kenyan Africans who had served the British Empire during World War I bitterly resented the treatment they received when the war was over. As many as 160,000 Kenyan men served in the corps alongside the British soldiers, and they suffered extreme hardship during the war. Approximately 42,000 African porters died in the war, often as a result of the poor service conditions. Upon returning home, the British officers were rewarded, while the African servicemen received no deserving recognition. In fact, although these men had been promised compensation at the end of their service, many of them, including the relatives of those who died in service, never received it.

By the 1920s, the continued mistreatment of the Africans by the British had set in motion the conditions that eventually gave rise to the Mau Mau rebellion in the 1950s. The most burning grievance behind the rebellion was the expropriation of African land by the white settlers. Bitterness over the loss of land was particularly strong among the Kikuyu, who accounted for the majority of the freedom fighters during the Mau Mau struggle for African freedom. Apart from the land issue, Africans deeply resented their lack of political and economic rights. African anger increased throughout the first half of the century as their peaceful demands for equality and fairness were ignored. Ultimately, sustained internal resistance and a combination of external factors taking place in the international political system loosened the grip of British influence in Kenya, and on December 12, 1963, Kenya won her independence.

Excerpts from: James Bailey and Garth Bundeh (1993) *Kenya: The National Epic*, Nairobi, Kenya: Kenway Publishers.

Another important aspect of the colonial economy is that it accelerated the rise of urban centers, which started as outposts used by the colonial powers for both economic and administrative purposes. The colonial urban centers were mainly set up in locations that ensured political advantage, while at the same time these locations worked to facilitate good economic returns for the colonizing power. The physical infrastructure built during the colonial period also served to reinforce the extractive nature of the colonial economy. For example, most of the railways lines in colonial West Africa followed a direct route from the source of raw materials in the mainland to the coast; there were no roads and no other railway lines aimed at linking other areas of the colonial state, which were deemed to be of little or no significant economic importance for the colonial power.

At the cultural level, European domination manifested its potency through organized religion, the spread of continental languages, and new education systems. With the exception of a few areas like the Arab Middle East (where Islam and Arabic remained dominant), and China (where Mandarin or Cantonese were widely spoken), most of the other areas under colonial domination adopted the religion and the language of the colonizer. The colonial powers practiced the Christian religion, which became one of the most powerful tools of cultural disruption imposed on indigenous communities. The Christian religion, whether Protestant or Catholic, provided the Europeans with an aura

of superiority and a sense of mission. By refusing to learn the local languages and by prop-
agating their own cultures and educational curricula, the colonizers planted seeds of psychological and cultural inferiority in the minds of the native populations. Despite these
organized efforts to project the apparent invincibility of the colonial powers, European
colonialism began to crumble during the twentieth century. The formal beginning of self-
rule, however, failed to bring about a socio-economic and political stability and wealth for
a significant number of post-colonial states.

V Post-colonialism and developing countries

Most colonial states secured their independence by generating strong nationalistic feelings. Prior to independence, most of the areas under colonial domination had not been
governed in a manner that would instill a sense of national pride among often diverse
groups of peoples, who found themselves within the confines of the newly created colonial states. The native leaders who fought for independence still managed to galvanize
the support of the indigenous populations by buttressing nationalist ideology. The call for
native unity drew the oppressed together, and pitted them squarely against their common
enemy – the colonial powers. The rise of nationalism and the subsequent attainment of
self-rule in the post-colonial territories were made possible through the following three
main factors.[58] First, the weakening of European colonial states as a result of World War I
(1914–1918) and World War II (1939–1945) accelerated the rise of nationalism in the colonized lands. The two world wars, coupled with the Great Depression of the 1930s, significantly reduced the capacity of the European colonial powers to effectively administer
their overseas colonial possessions. Also, equally important and particularly more so at
the end of World War II, was the emergence of new global superpowers the United States
and the USSR. Neither the Soviets nor the Americans supported the continuation of the
colonial possessions by European countries.

The second factor, was the role played by ex-servicemen from the colonized areas that
had served together with their colonial masters in the two world wars. Before the wars,
the colonialists had displayed an aura of invincibility. However, the destruction and
carnage caused by the wars eroded the impregnable image of cultivated Europeans. As a
result, the colonized servicemen realized that their colonial masters had fears and weaknesses, and above all, could be defeated in a battle. The ex-servicemen became some of
the staunchest nationalists upon returning to their respective countries after the wars.
They also had grievances for not receiving the same appreciation that their European
counterparts had received for their war service.

The rise of a native political elite, schooled in the European ideals of liberal democracy, provides the third explanation for the rise of nationalism in the colonized world.
The fact that a native elite arose out of colonialism points us to one of the inherent
contradictions of European imperialism. Most leaders of the nationalist movements in
the colonized areas were participants of the European educational systems, where they
learned about democracy, self-determination, and individual autonomy. Later on, the
native political elite was able to invoke such emancipatory ideas, alongside the native

people's deep sense of dignity and aversion to colonial oppression, to stir up nationalist fervor.

Immediately after gaining independence the political leaders promised both dignity and prosperity to their citizens – the two virtues that had been most desired under the oppressive weight of colonialism. However, for most of these countries, the post-colonial era was marked by a series of seemingly insurmountable challenges, further aggravated by the Cold War. During the Cold War era, developing post-colonial countries found themselves caught up in the middle of the East–West ideological divide. During this time, the political allegiance to one of the superpowers – either the United States or the USSR – meant obtaining financial support by the government of the newly independent country. For example, the battles for international supremacy between the United States and the USSR saw these superpowers supporting some of the most oppressive political regimes in the third world – like Mobutu's regime in Zaire (now the Democratic Republic of Congo), Mengistu's regime in Ethiopia, or Bokassa's regime in the Central African Republic. To date, decades after the end of the Cold War, the political mistakes committed during this period continue to hamper development efforts in many countries.

Notwithstanding the ghosts of the colonial past hovering over the developing world today, former colonies became independent sovereign states. New states were also created out of armed conflicts fought between and inside a number of post-colonial states. While, for some nations and ethnic groups, the formal recognition of statehood is still an aspiration. The international system of states operates on the principle of sovereignty, but it is helpful to remember the limits of such an arrangement. Because, as one scholar has noted:

> Two useful fictions are purported to lie at the basis of this system: the sovereign equality of states and the concept of anarchy as postulating the nature of relationships among them. These notions facilitate the management of conflict and cooperation among the units and the creation and functioning of international organizations. They obscure the perception that the actual relationships among the constituent units are hierarchical or clusters of hierarchically arranged units.[59]

Despite the official end of colonialism, the international system of sovereign states remains deeply asymmetrical. The divisions along economic lines are perhaps the most pronounced. It is impossible to separate the issues of poverty in the developing world from the colonial past. Past injustice cannot be remedied by gaining a membership in international organizations. The destabilizing processes of integrating former colonies into the global economy are viewed by Achille Mbembe as new modes of economic exploitation resulting in new technologies of domination taking over Africa.[60] This renowned scholar cautions against trying to define the needs of contemporary Africans as he exposes our limited, and guilt driven, depiction of Africa. He worries about the risks that threaten the coherence of African societies. There are threats of internal dissolution arising from external constraints and pressures associated with repayment of debts, structural problems, and internal wars. The ongoing disagreements and violence, sparked by worsening inequalities and corruption, create intense movements that "may well be the final defeat of the state in Africa as we have known it in recent years."[61] Mbembe still thinks the situation is far from hopeless because as "Africa is moving in several directions at once, this is a period that, at the same time, has been, is not yet, is no longer, is becoming – is a state of preliminary outline and possibility."[62] There are signs of vigilant optimism in his gaze over Africa, the continent of his birth, the land of people deeply harmed by colonization processes.

> In this sense, we must say that the postcolony, is a period of embedding, a space of proliferation that is not solely disorder, chance, and madness, but emerges from a sort of violent gust, with its languages, its beauty and ugliness, its ways of summing up the world.[63]

Notes

1 James Smoot Coleman (1994) "The Character and Viability of African Political Systems," in James Smoot Coleman (edited by Richard L. Sklar), *Nationalism and Development in Africa – Selected Essays*, Berkeley, CA: University of California Press, p. 93.
2 Heidi Holland (2008) *Dinner with Mugabe – The Untold Story of a Freedom Fighter who became a Tyrant*, Johannesburg, South Africa: Penguin Books, p. 212.
3 Achille Mbembe (2001) *On the Postcolony*, Berkeley, CA: University of California Press, p. 241.
4 Nathan Nunn (2009) "The Importance of History on Economic Development," *Annual Review of Economics*, Vol. 1, pp. 65–92.
5 Heidi Holland (2008) *Dinner with Mugabe*, op. cit., p. 215.
6 Nathan Nunn (2009) "The Importance of History on Economic Development," op. cit., pp. 236–237.
7 Jonathan Hart (2003) *Comparing Empires – European Colonialism from Portuguese Expansion to the Spanish-American War*, New York, NY: Palgrave Macmillan, pp. 13–19.
8 David K. Fieldhouse (1973) *Economics and Empire, 1830–1914*, Ithaca, NY: Cornell University Press, p. 490, map 9.
9 Henry Bernstein (2002) "Colonialism, Capitalism, Development," in Tim Allen and Alan Thomas (eds.), *Poverty and Development into the Twenty-First Century*, Oxford, UK: Oxford University Press, p. 250.
10 Ibid., p. 290.
11 James Smoot Coleman (1994) "The Character and Viability of African Political Systems," op. cit., p. 107.
12 Nayan Chanda (2007) *Bound Together – How Traders, Preachers, Adventurers, and Warriors Shaped Globalization*, New Haven, CT: Yale University Press, pp. 55–61.
13 Jared Diamond (1997) *Guns, Germs, and Steel – the Fates of Human Societies*, New York, NY: W.W. Norton & Company, pp. 78–81.
14 Jonathan Hart (2003) *Comparing Empires – European Colonialism from Portuguese Expansion to the Spanish-American War*, op. cit., pp. 15–16.
15 Adam Hochschild (1999) *King Leopold's Ghost*, Boston, MA and New York, NY: Houghton Mifflin Company, pp. 9–11.
16 Paul E. Lovejoy (2000) *Transformations in Slavery – A History of Slavery in Africa* (2nd edition), Cambridge, UK: Cambridge University Press, p. 68.
17 Nayan Chanda (2007) *Bound Together – How Traders, Preachers, Adventurers, and Warriors Shaped Globalization*. op. cit., p. 78.
18 John Isbister (2001) *Promises Not Kept: The Betrayed of Social Change in the Third World*, Bloomfield, CT: Kumarian Press, pp. 66–76.

19 David K. Leonard and Scott Straus (2003) *Africa's Stalled Development*, Boulder, CO: Lynne Rienner Publishers.

20 David K. Fieldhouse (1973) *Economics and Empire, 1830–1914*, op. cit., p. 173.

21 Niall Fergusson (2008) *The Ascent of Money – A Financial History of the World*, New York, NY: Penguin Press, pp. 143–148.

22 Ibid., p. 293.

23 Adam Hochschild (1999) *King Leopold's Ghost*, op. cit., pp. 81, 123–125.

24 James Smoot Coleman (1994) "Political Integration in Emergent Africa," op. cit., pp. 79–80.

25 William D. Grampp (1952) "The Liberal Elements in English Mercantilism," *Quarterly Journal of Economics*, Vol. LXVI, No. 4, November, pp. 465–501.

26 Ibid., p. 474.

27 Benjamin J. Cohen (1973) *The Question of Imperialism*, New York, NY: Basic Books, p. 21.

28 Curtis P. Nettels (1952) "British Mercantilism and the Economic Development of the Thirteen Colonies," *Journal of Economic History*, Vol. XII, No. 2, pp. 105–114.

29 Niall Fergusson (2008) *The Ascent of Money – A Financial History of the World*, New York, NY: The Penguin Press, p. 20.

30 Albert O. Hirschman (1981) *Essays in Trespassing – Economics to Politics and Beyond*, Cambridge, UK: Cambridge University Press, p. 171.

31 Vladimir Lenin (1938) "Imperialism, the Highest State of Capitalism," in Axel Hulsemeyer (2010) *International Political Economy – A Reader*, Oxford, UK: Oxford University Press, pp. 98–99.

32 Immanuel M. Wallerstein (2000) "The Rise and Future Demise of the World Capitalist System: Concepts for Comparative Analysis," in Axel Hulsemeyer (2010) *International Political Economy – A Reader*, Oxford, UK: Oxford University Press, pp. 116, 119–120.

33 John A. Hobson (1902) *Imperialism: A Study*, Excerpts Published by Panarchy, online, available at: www.panarchy.org/hobson/imperialism.1902.html.

34 David K. Fieldhouse (1973) *Economics and Empire, 1830–1914*, op. cit., p. 39.

35 Ibid.

36 Ibid., p. 40.

37 Ibid., p. 43.

38 Ibid., p. 476.

39 Ibid., p. 477.

40 Heather J. Sharkey (2013) "African Colonial States," in John Parker and Richard Reid (eds.), *The Oxford Handbook of Modern African History*, Oxford, UK: Oxford University Press, p. 161.

41 Duncan Ivison (2002) *Postcolonial Liberalism*, Cambridge, UK: Cambridge University Press, p. 40.

42 Gyan Prakash (1990) "Writing Post-Orientalist Histories of the Third World: Perspectives from Indian Historiography," *Comparative Studies in Society and History*, Vol. 32, p. 384, cited in Duncan Ivison (2002) *Postcolonial Liberalism*, op. cit., p. 40.

43 Duncan Ivison (2002) *Postcolonial Liberalism*, op. cit., p. 43.

44 Michael Ignatieff (2003) "The American Empire: The Burden," *New York Times*, January 5.

45 Jared Diamond (1997) *Guns, Germs, and Steel – The Fates of Human Societies*, New York, NY: W.W. Norton & Company, p. 358.

46 Ibid., pp. 354–359.

47 Daron Acemoglu and James A. Robinson (2012) *Why Nations Fail: The Origins of Power, Prosperity, and Poverty*, New York, NY: Crown Business, pp. 48–56.

48 Ibid., pp. 408–409.

49 Ibid., p. 112.

50 Daniel Philpott (2001) *Revolutions in Sovereignty – How Ideas Shaped Modern International Relations*, Princeton, NJ: Princeton University Press, pp. 30–33.

51 Ibid., p. 35.

52 Ibid., p. 33.

53 Ibid., p. 34.

54 David Potter (2002) "The Power of Colonial States," in Tim Allen and Alan Thomas, *Poverty and Development into the Twenty-First Century*, op. cit., pp. 272–284.
55 Daniel Philpott (2001) *Revolutions in Sovereignty*, op. cit., p. 36.
56 Quoted in William Easterly (2007) *The White Man's Burden – Why the West's Efforts to Aid the Rest Have Done So Much Ill and So Little Good*, New York, NY: Penguin Books, p. 23.
57 Ibid., p. 286.
58 John Isbister (2001) *Promises Not Kept: The Betrayed of Social Change in the Third World*, op. cit., pp. 101–144.
59 Robert W. Cox with Michael G. Schechter (2002) *The Political Economy of a Plural World – Critical Reflections on Power, Morals and Civilization*, London, UK: Routledge, p. 32.
60 Achille Mbembe (2001) *On the Postcolony*, op. cit., p. 67.
61 Ibid., p. 68.
62 Ibid., p. 241.
63 Ibid., p. 242.

Suggested further reading

M.E. Chamberlain (2013) *Scramble for Africa* (3rd edition), London, UK: Routledge.
John M. Hobson (2012) *The Eurocentric Conception of World Politics: Western International Theory, 1760–2010*, Cambridge, UK: Cambridge University Press.
Philip T. Hoffman (2015) *Why Did Europe Conquer the World?*, Princeton, NJ: Princeton University Press.
Duncan Ivison (2002) *Postcolonial Liberalism*, Cambridge, UK: Cambridge University Press.
Lawrence James (1996) *The Rise and Fall of the British Empire*, New York, NY: St. Martin's Press.
Jacob T. Levy and Iris Marion Young (eds.) (2011) *Colonialism and Its Legacies*, Lanham, MD: Lexington Books.
Edward Said (1979) *Orientalism*, New York, NY: Vintage Books.
Crawford Young (2012) *The Postcolonial State in Africa: Fifty Years of Independence, 1960–2010*, Madison, WI: University of Wisconsin Press.

Developing countries and the global economy

3

I Conceptualizing the global economy

There is little guidance what ordering principles should be used when studying the magnitude of the global economy. The key navigators come in different shapes and forms: private actors – multinational corporations, small businesses, individual entrepreneurs, private charity organizations; public actors – governments themselves, state-owned companies, regional blocs such as the EU; and finally, international organizations such as the World Bank, the IMF, and the WTO. These diverse actors do not necessary work well together but the idea of the interconnected world is particularly poorly understood in the context of economic development. Developmental policies are designed for and implemented by domestic actors. They impact the local populations in a very direct way. But it would be a mistake to treat such policies in isolation from the global environment. Economic globalization, understood as "the cluster of technological, economic and political processes that drastically reduce the barriers to economic exchange across borders,"[1] inevitably shapes options at the local level.

The framework for worldwide cooperation organized around a set of economic organizations is often used to explain a remarkable growth of transnational economic transactions in the decades after World War II. It all started with the momentous Bretton Woods agreement, which set the course towards formalized inter-state collaboration on economic matters. In this chapter, we examine the most prominent international organizations to probe how successful they have been in integrating developing countries into the operational architecture of the globalizing economy. Special attention is paid to trade, considered to be the primary force behind deepening economic internationalization. During the era of mercantilist trade, the desire to accumulate precious metals and exotic goods fueled the imperial greed and excused the conquest. In the time of global market integration, open trade has become the essential engine of growth in many parts of the world, driving forward previously poor economies such as China, South Korea, and India. In 2015, developing countries had a 42 percent share in world merchandise trade while the total merchandise exports of all WTO members (over 160 countries) totaled US$16.2 trillion in the same year.[2]

The unprecedented surge in the volume of international trade is the most visible aspect of the global economy, but other developments include the growth of foreign direct investment, international finance, and cross-border integration of production via globally spread value chains. These developments, however, have not created a more equal world. On the contrary, the opportunities offered by global markets are born out of fierce competition and are less homogenizing and less universalizing than initially thought. When discussing the social consequences of a globalizing economy, critics state that "the growth of transplanetary and supraterritorial connections empowers some people and disempowers others."[3] For example, least developed poor countries remain at the margin of global value chains (GVCs) of production and foreign direct investment flows.[4]

The globalizing momentum of the 1990s resulted in the establishment of the legally binding WTO, based on multilateral commitments made by over 100 countries. Such commitments can, to varying degrees, restrict the policy space of the WTO members. In the changing global environment, the realist view of the inter-state system is contested:

> Stop imagining the international system as a system of states – unitary entities like billiard balls or black boxes – subject to rules created by international institutions that are apart from, 'above' these states. Start thinking about a world of governments, with all the different institutions that perform the basic functions of governments – legislation, adjudication, implementation – interacting both with each other domestically and also with their foreign and supranational counterparts.[5]

The world of governments is still very relevant. When the 2008 global financial crisis began to ravage economies around the world, it was a concerted action by governments that prevented the crisis from becoming an even bigger catastrophe. The ensuing slow recovery put the brakes on the policies of economic globalization. Some scholars, in fact, blame the crisis on the global free-market ideology. They maintain such ideology pushed countries into needless deregulation and the privatization of trade and investment regimes, which contributed to credit expansion, slower growth, and rising inequality.[6] Another perspective sees the global economy as a single closed system in which every part is impacted by the flows of capital and trade. In this case, the global financial crisis was primarily driven by imbalances in global trade and capital, and "has unfolded in almost a textbook fashion."[7] This view takes the existence of the global integrated economy for granted, but it still asks the major state players – specifically Germany, China, the United States, and Japan – for their involvement to rebalance the system.

States continue to regulate important operations of the globalizing economy, although the extent of their interventions is the subject of intense debate. States also operate within the constraints of international agreements, conventions and treaties, guarded by different organizations and framed by international law. Understanding how such international arrangements manage collaboration, resolve conflicts, and facilitate exchanges, helps us clarify the relationship between pressures of the globalizing economy and the domestic developmental priorities of countries.

II Learning to cooperate: the Bretton Woods system

In the age of empires, economic agendas were motivated by the national interest and conceived in isolation from the rival states. The great powers occasionally cooperated on security matters, but seldom with respect to economic issues. Traditionally, international relations involved zero-sum game competition among different polities. This has changed since the end of World War II. A large volume of international treaties and organizations has been created by the cooperating countries. Some of these organizations have roots in the Bretton Woods conference held in New Hampshire in 1944, organized by the victorious allies.

Winding road to Bretton Woods

The Bretton Woods talks initiated a distinctive era of international institution building. The final agreement intended to balance a liberalizing world economic order with the domestic developmental priorities of nation-states. The treaties that existed before the two wars were mostly narrowly focused agreements between the European powers. During the nineteenth century, the massive industrialization pushed Western European societies on the brisk path of socio-economic development, altering the existing social and economic relations. From its beginning, the Industrial Revolution especially advanced those economies that were large enough to create formidable manufacturing outlets for diversified production lines. Thus, the United Kingdom, France and the Netherlands led the way, later followed by a newly united Germany and the northeast regions of the United States. The appetites of the European colonial powers for commodities and natural resources were satisfied by the exploitation of the overseas territories supplying many of the resources not available at home. This exploitation was possible since the Westphalian principle of sovereignty was not universally recognized under international law. Unlike the sovereign states of Europe, and a few other countries, including the United States, most polities were simply not eligible to be equal members of the international system.

One important institution that facilitated the commercial exchanges during the colonial era, and briefly during the interwar period, was the gold standard. Gold was the preferred unit of value since ancient times. During the golden years of the British colonial empire, many countries informally adopted the gold standard as it was managed by London. The convergence of different currencies around the gold–pound fixed exchange was the result of the strong position of the British pound fueled by the empire's monetary and trade policies. In 1819, the British formally institutionalized the practice of exchanging currency notes for gold on demand at a fixed rate.[8] The US Gold Standard Act of 1900 legitimized the dollar to gold exchange, although it had been informally in place since the time of the Civil War.

Officially, the period between 1870 and 1914 is known as the international gold coin standard, with some 40 countries participating.[9] By fixing currencies to gold, countries reduced uncertainty over exchange rates and achieved stability of their currencies.

The United Kingdom's commitment to the convertibility of the pound to gold was considered a type of collective good because it provided the stability for international commercial transactions. Other collective goods provided by the British system were: liquidity, and being a lender of last resort. It meant that in the time of crisis, the British banks could be approached for an emergency loan. The gold standard thus further bolstered the dominant position of the British empire by turning the pound into a tool of influence on the international stage.[10] There were still some costs for the British. The commitment to provide oversight to the system based on the gold standard exposed the United Kingdom to the possibility of depleting its own gold reserves. It also made the United Kingdom more sensitive to economic crises elsewhere. In the wake of World War I, the gold standard collapsed as countries renounced their gold standard obligations and allowed their currencies to float in the foreign exchange market. Given the expansionary fiscal policies introduced in anticipation of the war, huge imbalances in exchange rates ensued.[11]

After World War I, in 1918, the political geography of Europe and the Middle East was changed forever. Four great powers, Prussia, Austria-Hungary, Russia, and the Ottoman empire, were replaced by newly independent countries with governments of different sorts, including monarchies, sheikdoms, constitutional republics, and the Marxist socialist state of the USSR. The inter-war period thus proved to be particularly challenging for those who proposed multinational cooperative agreements. Renewed calls for such agreements culminated with President Wilson of the United States' project of the League of Nations, launched to prevent future wars and to encourage a concerted approach to world affairs. Under the league's mandate a financial commission was established to provide loans to countries in financial need. A series of conferences were organized between 1920 and 1922 to design a set of rules for monetary cooperation and the reduction of barriers to trade. All these attempts ultimately failed.[12]

Once the Great War was over, many countries decided to partially return to the gold standard under the leadership of Britain. Unsurprisingly, it became customary for governments to hold international reserves in the form of pound deposits in London. But when, during the 1920s, the British economy was stagnating, trust in the stability of the gold standard evaporated. London's gold reserves were indeed limited, and foreign holders of the pound were worried as to whether the United Kingdom could meet its obligations. The market crash of 1929 and the subsequent worldwide monetary contraction turned the situation for the worse, contributing to the 1930s Great Depression. The inflexibility of the gold standard required that countries protected their gold reserves. But by doing so banks refused, or were unable, to provide liquidity to their customers. Bank failures to meet exchange obligations, and capital outflows from London to countries with fixed and overvalued currencies, were the principal reasons why the United Kingdom eventually abandoned the gold standard in 1931.[13]

A fundamental risk under the "classic" coin gold standard was runaway inflation, caused by abandoning the regime either in response to an urgent situation at home or to meet foreign policy objectives. For example, at the outset of World War I, governments broke the rules and began to finance their colossal military expenditures by printing money. Once the war was over, several countries experienced debilitating inflation when their governments attempted to support the reconstruction process through massive public expenditures, which again meant printing more worthless money. The most famous case was Germany, where the annual rate of hyperinflation reached 182 billion by the end of 1923. Prices in Germany were then on average 1.26 trillion times higher than they had been in 1913.[14]

The short period between the two world wars was tarnished by the devastating Great Depression, triggered by the US stock market collapse in 1929. The long-term effects included massive unemployment, inflation, and spread of poverty in the industrialized world. In 1933, GNP in the United Stated was almost one-third less than in 1929, and nearly 13 million people (one in four of the labor force) were unemployed.[15] As the economic depression deepened, governments turned to protectionist policies with the hope of saving their own societies. Regardless of the consequences, political leaders on both sides of the ocean decided to ignore the problems of their neighbors. Instead of trying to coordinate a joint response to the international crisis, governments closed their economies to the rest of the world, and in doing so they prolonged the depression. This kind of destructive protectionist action became known as *beggar-thy-neighbor* policies.

The international economy basically came to a stop with the passing of the 1930 Smoot–Hawley Tariff Act in the United States. In a vain attempt to protect American manufacturing and agriculture, the Act raised the import tariff on some 21,000 items to prohibitive levels, exacerbating the already disastrous economic climate worldwide.[16] International trade collapsed around the world and countries' projectionist policies began to acquire a nationalistic flavor, concluding with the Nazi seizure of power in Germany in 1933. By then, the German economy, unable to pay the reparations imposed by the Versailles Treaty, was suffering from multiple ills after experiencing hyperinflation in the 1920s. The Nazis quickly enacted a new plan, which gave the state significant control over the economy, and Germany imposed a moratorium on all foreign debt.[17]

As the 1930s worldwide depression continued, countries struggled to maintain financial stability. For example, the United States left the gold standard in 1933 amidst panic caused by bank failures that led to the hoarding of gold by the public. In March of 1933, the newly elected President Roosevelt declared a four-day national moratorium that prohibited banks from paying out and exporting gold. A month later he ordered that all gold coins and saved gold certificates worth more than 100 dollars were exchanged for regular currency. By May 10, 1933, the US government absorbed US$300 million of gold coins and US$470 million in gold certificates. In another move, the United States devalued the gold–dollar exchange by issuing a permanent embargo on gold exports and by abrogating the gold clauses in private and public contracts. the United States then returned to a "qualified gold-bullion standard" under the Gold Reserve Act of January 30, 1934, having raised the dollar price of gold from US$20.67 (as mandated by the Gold Standard Act of 1900) to US$35 per ounce (as mandated by the new Act). The outcome was a partial gold standard that allowed only external convertibility for trade purposes. The partial gold standard had no internal convertibility, meaning that no US citizen could obtain gold for their money. The unintended consequence of this policy was the new gold rush, since gold was considered to be overvalued by the newly set US exchange rate. Overall, however, such moves were not fully successful in taming the depression.[18]

During the Great Depression, most countries engaged in actions aimed at maintaining confidence in their currencies. To counter the fear of gold and capital outflows, many central banks raised interest rates, triggering deflationary pressures on prices. Next, governments tried depreciating their currencies in the hope of assisting domestic producers. Internationally, these beggar-thy-neighbor policies were canceling each other. Major economic harm was done by protective tariffs placed on trade flows, which were replicated throughout the world as countries tried to discourage imports and keep aggregate demand protected at home. The foreign responses involved more retaliatory trade restrictions, prohibitions on private capital account transactions, and experiments with exchange rates. In short, the world economy disintegrated into an increasingly hostile inter-state disorder in the 1930s.[19]

The events surrounding the Great Depression inspired Keynes to write his famous book. The events also provided arguments about the necessity of international cooperation on economic matters. In 1934, the Roosevelt Administration repealed the misguided Smoot–Hawley Act and replaced it with the Reciprocal Trade Agreements Act, making the United States open to business again. It was an important legislation that authorized the president to negotiate tariff reductions with foreign states on a non-discriminatory basis. The Act symbolized the importance of trade for the US economy.[20]

Considerable turbulence ravaged the world markets until the end of World War II. In the late 1930s, the rise of political extremism in Germany, Italy, and Spain, and the perseverance of the Soviet Union as a communist replacement of tsarist Russia, further weakened attempts at international collaboration. Many countries simply curtailed their trading links with the rest of the world to eliminate the possibility of having significant external deficits. Protectionism imposed high costs on the world economy and prolonged the depression. All countries would have been better off in a world with open international trade and stable exchange rates, provided that there existed an international organization designed to help countries stabilize their balance-of-payments current and capital accounts without sacrificing domestic policy goals.[21] The need to create such an organization inspired the Bretton Woods agreement.

Bretton Woods agreement

Near the end of the war, the allies agreed that international institutions were necessary for states to collaborate to reduce the possibility of economic crises and more serious conflicts. During this period, the United Nations was created along with the Bretton Woods institutions: the IMF, International Bank for Reconstruction and Development (World Bank), and the GATT (General Agreement on Tariffs and Trade). The common international objectives of the Bretton Woods negotiations included post-war reconstruction and the long-term stable development of all countries.

The Bretton Woods conference took place in July 1944 in New Hampshire. It was inspired by the idea that sovereign states would have considerable freedom to pursue national economic objectives, while participating in a progressively liberalized world economy. The emphasis was on revitalizing international trade and building the international monetary order based on fixed exchange rates and currency convertibility for current account transactions. However, the negotiations at Bretton Woods were difficult at times, oscillating between clashing visions of the United Kingdom and the United States. The British delegation, led by J.M. Keynes, sought to avoid the threat of high unemployment and social instability caused, in his view, by austerity programs designed to prevent inflation and to restrict quantitative easing. Keynes insisted on a more flexible monetary system, where debtor countries could access the money reserves of wealthier nations when experiencing balance of payments problems.[22] With respect to trade, Keynes demanded exemptions allowing countries to regulate imports. The United Kingdom also wanted to maintain its reserve currency role, while keeping the system of preferential treatment of Commonwealth Countries. The US delegation under the leadership of Harry Dexter did not like the idea of exemptions, "escape clauses," or preferential treatments. The United States was focused on facilitating free trade around the world. To avoid exchange rate instability, the Americans wanted to get rid of all exchange controls.[23] The Bretton Woods negotiations did not embark on the trade agenda because separate trade talks between the United States and the United Kingdom were initiated in 1943 elsewhere. The trade talks would not be concluded until the end of 1945. The Bretton Woods conference mainly focused on monetary and banking issues and signing the charters of the IMF and the World Bank. Still, trade issues were firmly present at the negotiating table.

Some 44 countries took part in the Bretton Woods conference, but when the final deal went into effect in December 1945, only 30 of them signed the agreement. More countries joined later, with a notable exemption of the Soviet Bloc. To ensure the stability of the international monetary system, the Bretton Woods deal established a system of fixed exchange rates, with the US dollar pegged to gold at US$35 per ounce. In addition, the system was backed by a formal institution, the IMF, which committed countries to obey the agreed rules. Member countries who held their international reserves in the form of gold or dollars obtained the right to sell dollars for gold at the fixed price. This fixed exchange system was in effect a gold exchange standard, with the dollar as its principle reserve currency. Yet the new system was not as rigid as its classic predecessor from the colonial era. This post-war (alternative) gold standard was envisioned by Keynes in *A Treatise on Money*, published in 1930. Keynes wanted the new system to be more flexible and without actual gold coins in circulation but with an international clearing bank supporting the international monetary order. His proposal became a cornerstone of the Bretton Woods agreement.[24]

Under the IMF Articles of Agreement, Article I, the IMF was established to:

- Promote international monetary cooperation
- Facilitate the expansion and balanced growth of international trade, and to contribute to the promotion of employment and income
- Promote exchange stability and orderly exchange arrangements
- Assist in the establishment of a multilateral system of payments on current transactions between members
- Give confidence to members by providing temporary financial assistance to countries to help ease balance of payments adjustment.

IMF member-states were assigned quotas upon joining the organization and were obliged to pay a yearly membership fee proportionate to the quota. Each quota reflected the size of the member's economy and afforded them proportionally adjusted voting rights. The main purpose of quotas was to establish a permanent fund managed by the IMF and composed of contributions of member-states in gold and their own currencies. Under the original 1944 agreement, every quota was to be paid 25 percent in gold or currency convertible into gold (effectively the dollar, which was the only currency then still directly gold convertible for central banks) and 75 percent in the member's own money. Under the first calculated quota system in 1944, the United States was assigned US$2,750 million, which translated to almost 30 percent of voting power. The distant second quota belonged to the United Kingdom and was set at US$1,300 million.[25] Members could then borrow from this fund when experiencing current account deficits. To ensure additional flexibility, members could devalue their currencies against the dollar if the balance-of-payments was in *fundamental disequilibrium*. This term was never defined in the IMF Articles of Agreement, but *fundamental disequilibrium* was understood to mean either a sharp decrease in the value of the country's currency or its rapid increase, resulting in such a serious instability of the price level that the international demand for the country's products would be affected.[26]

The scope of the flexibility given to members with respect to their intended actions in the case of *fundamental disequilibrium* was somehow ambiguous. Schedule C, paragraph 6 stated:

> [the] member shall not propose a change in the par value of its currency except to correct, or prevent the emergence of, a fundamental disequilibrium. A change may be made only on the proposal of the member and only after consultation with the Fund.

It is well documented that some of the most important discussions during the Bretton Woods negotiations, and then during the early years of the IMF, concerned the meaning of the fund's consultation mandate and the extent of its authority. Especially what conditions should be applied when members wanted to use the financial resources of the fund. The executive board of the IMF issued a formal decision on this matter in October 1952, which allowed the fund to negotiate special arrangements with individual members.[27] This decision led to toughening of requirements placed on members in exchange for receiving financial help from the IMF. In time, the decision would have additional consequences for developing countries because of the structural reforms demanded as part of the negotiated arrangements with the IMF.

Under Article VIII section 2, the IMF urged signatories to make their national currencies convertible on international transactions to promote efficient multilateral trade: "no member shall, without the approval of the Fund, impose restrictions on the making of payments and transfers for current international transactions." The Article, in principle, allowed members to seek IMF approval via a consultation process in case they wanted to put such restrictions in place. The United States wanted to have strict rules behind any such consultation. The Americans pushed for giving the fund a broad authority to investigate and formally report on the factors that compelled a member to resort to particular exchange restrictions. The resistance from other IMF members, however, was fierce, causing the Americans to give up their hard-line position. Members pointed to Article XIV, which allowed them in the difficult post-war transitional period to maintain such restrictions. Overall, the key issue was the extent of the fund's authority over its members' domestic policies.[28] With time, the consultations with the IMF would resemble advisory sessions, only to translate into intrusive policy prescriptions in the 1990s.

The IMF Articles of Agreement have been amended several times since the Bretton Woods conference. The most recent change went into effect in January 2016. In the past, one of the most important amending resolutions was adopted in May 1968, concerning the overhaul of the quota system. The system is still in place today but under a different scheme. The increased instability of the Bretton Woods system prompted the resolution, which in 1969 created special drawing rights (SDRs), a type of international monetary reserve asset. Special drawing rights were intended to supplement the official reserves of the fund in response to worries concerning the inadequate supply of two key reserve assets: gold and US dollars, crucial for settling international accounts. In the early 1970s, the fundamental aspect of the Bretton Woods system collapsed, with the ending of the fixed exchange rate between the US dollar and gold. After that move, all major currencies switched to floating exchange rate regimes with the world's markets determining their values. As the IMF explains "the SDR is neither a currency, nor a claim on the IMF. Rather it is a potential claim on the freely usable currencies of IMF members."[29] SDRs can be exchanged for actual currencies of IMF members. Alternatively, SDRs can be bought by economically strong members to help members with weak external positions. The SDR also serves as an accounting unit of the IMF. The value of the SDR is based on a basket of four major currencies: the US dollar, the euro, the Japanese yen, and the pound sterling. The basket was expanded to include the Chinese renminbi as the fifth currency on October 1, 2016.

The IMF conducts its business based on the voting power of individual members via the operations of the executive board. Article XII (paragraphs a/b/c) makes it clear:

(a) The Executive Board shall be responsible for conducting the business of the Fund, and for this purpose shall exercise all the powers delegated to it by the Board of Governors.

(b) Subject to (c) below, the Executive Board shall consist of 20 Executive Directors elected by the members, with the Managing Director as chairman.

(c) For the purpose of each regular election of Executive Directors, the Board of Governors, by an 85 percent majority of the total voting power, may increase or decrease the number of Executive Directors specified in (b) above.

Below you can see quota and voting shares for certain IMF members. These numbers reflect changes introduced after the board reform amendment went into effect on January 26, 2016, all numbers shown are percentages.[30]

Countries with the largest quotas (top IMF contributors):

USA	16.58	Japan	6.17	China	6.11
Germany	5.34	France	4.05	UK	4.04

(Total: 42.29)

Other important IMF contributors:

Italy	3.03	Russia	2.60		
Canada	2.23	Brazil	2.23	Saudi Arabia	2.02

Countries with some of the lowest quotas (smallest IMF contributors):

Angola	0.18	Bolivia	0.06	Cambodia	0.06	Eritrea	0.03
Honduras	0.08	Samoa	0.03	Tanzania	0.11	Uganda	0.10

The voting scheme of the IMF means that the decision-making process is in the hands of the main IMF contributors. It is a formula based on the perception of fairness because those countries that pay more to maintain the system are given more influence in managing it. But given the inability of so many poor countries to become significant contributors, the IMF operating principles are criticized on the grounds that they reflect the developmental agendas of the world's dominant economies.

When a country joined the IMF, it also had to join the World Bank, which remained an obscure institution during its early years. When the bank opened for business in July 1946, it had a permanent staff of only 72 people and a budget of less than one billion dollars.[31] Its initial mandate was remarkably limited, with the intent of supporting postwar reconstruction and small infrastructure projects. The bank operated by collecting subscription fees from its members based on a similar formula to the IMF quota system.

Under the International Bank for Reconstruction of Development (IBRD) Articles of Agreement, Article I, the bank was established to:

- Assist in reconstruction and development
- Promote private foreign investment
- Promote balanced growth of international trade
- Arrange loans for useful and urgent projects
- Assist in a smooth transition to a peaceful economy.

According to Article V, "all the powers of the Bank" were vested in the board of governors. Every member of the bank was to nominate one country representative to the board. The voting formula for the board's meeting was again based on the voting power of individual members. Most importantly, the bank's operational decision-making processes were in the hands of the president and the group of 12 executive directors approved by the board of governors. Just as in the IMF, the World Bank was designed around the idea that the power within the organization should be awarded proportionally to the size of assigned contributions or shares. Article V guaranteed that five of the executive directors were from the five countries that contributed most to the bank's coffers. The rest were to be elected by the board of governors. The president would be selected by the executive directors to guide the general operations of the bank. The president of the World Bank acts as chairman of the executive directors. The text of the Article describing the position of the president, states that "he shall be responsible for the organization," which in time came to be interpreted as having considerable discretion in running and even reorganizing the organization. Over several decades, the bank evolved into the biggest multilateral lending institution for development.

Post-war developments

The IMF became very important in the first decade of the Bretton Woods system. Many countries experienced current account deficits as they reconstructed their war-torn economies. The new organization helped IMF members to overcome their financial difficulties and allowed them to trade with other nations. However, quite quickly it became obvious that the IMF was not sufficient to act as a pool of liquidity. Increasingly, the United States stepped in as a lender willing to risk its own deficits in order to maintain the stability of international monetary order and ensure smooth flows of international trade.

During the first decade of its existence, the World Bank remained in the shadows, especially after the announcement of the Marshall Plan for Europe. The formal name of the initiative was the European Recovery Program, but it was named after General George C. Marshall, who was secretary of state and secretary of defense under President Truman. Between 1947 and 1952, the United States provided financial assistance in the form of grants and loans – reaching at times nearly 2 percent of US GNP – towards the recovery of 16 western European countries.[32] It was an enormous amount of direct foreign aid and successfully worked as a catalyst for the economic revitalization of Europe. The Marshall Plan also had its own political objective crafted under the Truman Doctrine, which was to strengthen American allies in Western Europe and to contain the threat of the possible Soviet invasion from the East.[33]

The Bretton Woods agreement was primarily conceived for free market economies, although the Soviet Union sent its delegation to the conference and was assigned its IMF quota and the Bank subscription fee. Within a few years, however, the growing ideological discord between the two main superpowers became the underlying current of international relations. The significant political development of that period was the division of Europe and the deepening of Cold War animosities.

After the Soviet Union decided not to ratify the Bretton Woods agreement, the remaining countries under its sphere of influence were forced to withdraw their participation from the system and join a different set of institutions designed in Moscow. Under the label of state communism, such institutions were to advance the principles of state-run economies. Over time, the Soviets developed the Council for Mutual Economic Assistance (Comecon), which was the political framework for cooperation among the Soviet Bloc countries. The Comecon rejected the idea of free markets and had its own special rules, which included: currency inconvertibility, price controls, and state monopoly of foreign trade and payments. The idea was to achieve regional self-sufficiency in order to keep the interaction with the West at a minimum.[34] The world became divided into the Bretton Woods system, supported by the western free market economies and their allies in the developing world, and the state-run economies under the leadership of the Soviet Union. The Marshall Plan was initially intended for some of the Eastern European countries but it was rejected by them under the command of Moscow.

In the meantime, Western Europe was slowly recovering. The restoration of European convertibility in December 1958 marked an important date in the history of the global economy with a gradually expanding volume of international transactions, Japan followed in 1964. The British successfully pushed for making all the major European currencies fully convertible on current accounts after long and costly efforts by the United States to stabilize the monetary system. The United States achieved its goal at a very high price of seriously depleting its gold reserves. In response to the restored convertibility, the United States pushed for the expansion of the quota system in the IMF to provide the organization with additional capital.[35]

As international trade expanded, financial markets between countries became more closely integrated and moved towards the creation of today's global exchange market. With new opportunities to move private funds across borders, the linkages among countries' monetary policies tightened as the Bretton Woods system evolved. Given the increased capital mobility, countries' current account deficits and surpluses were closely monitored by central banks. Throughout the 1960s balance of payments problems and trade deficits became increasingly frequent in the West. A record British trade deficit in 1964 prompted the government to put in place an unprecedented 15 percent import tariff on manufacturing goods. This measure was abolished only two years later but it showed how desperate some policy-makers became in trying to juggle between the IMF rules, the requirements of the open trade system, and the weaknesses of the domestic economy. The short-term protectionist policies had adverse effects, leading to speculation against the pound – an unwelcome development that created additional liquidity problems for the United Kingdom. In 1967, the IMF provided the British government with a loan to help alleviate balance of payments problems and the pound was devalued.[36] The problems of the United Kingdom were not unique. Both France and Germany devalued their currencies in 1969. These financial problems shook the foundation of the Bretton Woods system of fixed exchange rates.

At the end of the 1960s, the United States' role as a main reserve currency country in the Bretton Woods system took on added significance. As the global economy expanded, suspicion grew that foreign holdings of US dollars had swelled to exceed US gold reserves. The problem had been foreseen by Robert Triffin almost a decade earlier. In his famous book, he signaled that the Bretton Woods arrangement was heading towards systemic instability because an international expansion of liquidity was financed by growing US deficits.[37] The fixed exchange system was facing a confidence problem, with the United States at the center of it. Eventually most financial institutions around the world came to believe that the United States was no longer able to fulfill its obligations and convert all dollar holdings around the world into gold.

Politics also played a significant role in unraveling the Bretton Woods system of fixed exchange rates. During the late 1960s, US government military purchases began rising as President Lyndon B. Johnson widened America's involvement in the Vietnam conflict.

At the same time, new domestic social programs were introduced, further accelerating the pace of the US expansionary fiscal policy. These policies caused a sharp fall in the US current account surplus, followed by a sudden inflation. Although the government under Johnson and then Nixon tried to introduce several stabilization programs, the economic situation was deteriorating quickly. The reality that the dollar would have to be depreciated was forcing itself on the minds of policy-makers. Foreign exchange markets experienced an unprecedented sellout of US dollars, threatening the US Treasury with massive gold losses. This impossible situation was unilaterally resolved by President Nixon, who announced on August 15, 1971 that the United States would no longer exchange gold to foreign banks for dollars. This action effectively ended the remaining link between the dollar and gold in place since the ratification of the Bretton Woods agreement.[38]

The 1970s turned out to be a volatile decade both in the North and in the South. Following the oil crisis of 1973–1974, the steady inflow of low-interest loans to the developing world resulted in the badly managed debt-led growth in many newly independent countries. The oil crisis happened in the context of the growing influence of the Organization of Petroleum Exporting Countries (OPEC). In September 1960, four Persian Gulf nations (Iran, Iraq, Kuwait, and Saudi Arabia) and Venezuela formed OPEC, with the purpose of coordinating the petroleum policies of its member countries. By 1973, eight other nations (Qatar, Indonesia, Libya, the United Arab Emirates, Algeria, Nigeria, Ecuador, and Gabon) had joined OPEC. In 1971, OPEC countries agreed to increase the price of oil from US$1.80 to US$2.29 per barrel. However, the turning point came with the October 1973 Arab–Israeli war. To punish the United States for supporting Israel during the conflict, OPEC decided to tighten the cartel by significantly reducing its oil production. Crude oil prices jumped dramatically in late 1973 and 1974 to almost US$30 per barrel.[39]

OPEC's actions destabilized the global economy for many years, while empowering oil producing countries. On the one hand, the increase in the price of oil caused OPEC countries to enjoy an unprecedented rate of growth and economic prosperity. This situation had largely negative consequences for other developing countries of the Global South. In coping with huge oil profits, the OPEC countries channeled the newly acquired petrodollars to the banks in the West.[40] To prevent the depreciation of the dollar, and fearing financial instability, the banks then extended significant loans to a number of developing countries on generous terms. Due to the growing economic instability at the end of the 1970s, the banks increased interest rates on such loans. This was the beginning of a serious debt crisis in the developing world.

III Attempting to help: debt, aid, and the cartel of good intentions

The petrodollar debt accumulated by developing countries in the 1970s was compounded by the loans negotiated with the World Bank and the IMF. Following decolonization, these organizations started to provide financial assistance to the newly independent governments. Money was intended towards infrastructure projects such as dams, schools, and roads. However, the principle of sovereignty complicated the process of monitoring how those loans were actually spent. While the debt was growing, the money also financed armed conflicts and secured authoritarian regimes in power. Inconsistent economic policies further aggravated the debt crisis. The critical moment happened when interest rates on borrowed money went up in the early 1980s.

In August 1979, a newly appointed chairman of the US Federal Reserve, Paul Volcker, started to raise interest rates domestically to curb inflation. His actions were followed worldwide, and loans previously given to various countries faced soaring interest rates. Borrowers in the developing world were confronted with a rapid rate increase, from approximately 7 percent to over 20 percent in one year. During the 1980s, dozens of countries in Africa and Latin America experienced severe financial and economic crises. The first sign of a major disaster happened in early 1982, when Argentina suspended payments on its US$37 billion debt. In August 1982, the government of Mexico shocked the financial world by announcing that it would be unable to meet its obligations on a US$85 billion debt. In December of that year, the contagion spread to Brazil, overwhelmed with US$91 billion debt obligations. By the end of 1983, most of Latin American and many African countries were unable to pay back their debts.[41] In several cases, the IMF had to step in and rescue countries from an economic collapse. Worried about the already huge level of debt experienced by some of the poorest countries in the world, the IMF introduced the reinforcement of the conditionality principle, which meant that countries would be required to undergo a process of structural adjustment reforms as a condition for obtaining a loan.

Structural adjustment often took the form of an economic shock therapy intended to overhaul the country's economy. In preparation for an annual meeting of the IMF and the World Bank in 1979, then president of the World Bank Robert McNamara and his deputy came out with the idea of structural adjustment loans.[42] The IMF typically attached some conditionality to its loans, and thus, given the instability of the international economy, the organization gladly embraced this new initiative. For the World Bank, however, a structural adjustment loan was a novelty, which became a norm when dealing with countries in need of developmental assistance during the 1980s and the 1990s. The rationale was simple. Instead of prolonging the state of economic uncertainty by introducing gradual changes, comprehensive structural reforms would quickly, and less painfully, mold a country into a perfectly operating free market system. Unfortunately, most structural adjustment loans failed to achieve their goals. Upon examining the record provided by 12 African countries and the then top ten ex-communist countries, which received most structural adjustment loans from the IMF and the World Bank between 1980 and 1999, it appears that in a vast majority of cases, these programs actually made the economic situation worse. For example, Côte d'Ivoire received 26 such loans over a 20-year period but continued to suffer a long and devastating economic depression.[43]

The convergence between the developmental policies advanced by the World Bank and the IMF in the era of structural adjustment loans came to be known as the Washington Consensus. The consensus promoted the liberalization of economies around the world along the neoclassical economic paradigm. By 1999, the term became synonymous with the failure of structural adjustment programs and the arrogance of the organizations that pushed for their implementation. The Washington Consensus will be discussed in greater detail later in the book. For now, it is important to mention that the idea was coined by John Williamson in 1989 to describe a number of common sense free market reforms to be introduced in Latin America.[44] However, the practices that emerged under the label of Washington Consensus routinely missed the point of the original concept.

The bulk of criticism was directed at the IMF and the World Bank after the failure of the Washington Consensus. Together with the UNDP, the USAID, the Inter-American Development Bank, these institutions became known for duplication of services and huge inefficient bureaucracies. Despite spending billions of dollars on programs designed to eradicate poverty, the aid organizations failed the very communities they tried to help. There were accusations of corruption and suggestions that political considerations often determined the terms of loans and financed projects.

Arguably, the appropriate name for all the major aid organizations is the Cartel of Good Intentions. Cartels have the ability to dictate terms of agreements to their customers, and similarly "in the foreign aid business, customers (i.e., poor citizens in developing countries) have few chances to express their needs, yet they cannot exit the system."[45] Another criticism points out how little these organizations have changed despite their attempts to introduce new themes and programs. Consider, for example, the previously examined 1970 Pearson Report named "A New Strategy for Global Development." Its focus on trade, investment, and economic growth, is expressed in the language that sounds remarkably similar to the policy prescriptions propagated during the reign of the Washington Consensus.

Both the World Bank and the IMF underwent a variety of reforms in order to become more effective. The institutional reforms conducted in 2009 streamlined IMF conditionality "in order to promote national ownership of strong and effective policies."[46] The reforms aimed at preventing a moral hazard, a situation when a borrower has incentives to take unnecessary risks knowing that losses can be forgiven or somebody else will pay for a borrower's mistakes. Moral hazard is a serious issue, but patterns of lending money that were established in accordance with structural adjustment programs pushed countries into adopting a blueprint of reforms, which in many cases exacerbated the problems instead of removing them.

As of August 2016, the IMF is an organization of 189 countries.[47] The organization maintains that "all conditionality under an IMF-supported program must be 'macro-critical' – that is, either critical to the achievement of macroeconomic program goals or necessary for the implementation of specific provisions under the IMF's Articles of Agreement."[48] However, in some geopolitically important cases, conditionality principles appear to be flexibly applied. An interesting case is Pakistan, which is the third largest recipient of developmental assistance in the world. Pakistan had negotiated numerous loans and grants since 1960, and some of them were written off even when such loans were worth billions of US dollars. Many loans were given to Pakistan without specific benchmarks or evaluative targets. A consistent tendency was to accept some IMF conditions while ignoring others. In September 2001, Pakistan negotiated another generous agreement with the IMF that doubled the loan that was just expiring. The new agreement addressed fears of political stability in the country. In fact, according to one expert: "Pakistan terms" has become a euphemism for "donor funding without accountability for policy objectives."[49] The IMF and the World Bank continue to offer Pakistan financial and technical assistance.[50]

The World Bank Group is now the largest multilateral aid organization, consisting of five interrelated units: the IBRD, the IDA, the International Finance Corporation (IFC), the Multilateral Investment Guarantee Agency (MIGA), and the International Center for Settlement of Investment Disputes (ICSID). The IBRD is the oldest part, with a membership of 189 countries in 2016. It offers a combination of financial and technical assistance programs to low-income and middle-income countries.

The IDA was established in 1960 to provide loans on preferential terms, including no interest, grace periods of more than a decade, and decades-long maturities for the poorest countries, defined as such by the organization. As of 2016, this branch of the World Bank was overseen by 173 members, who funded its operational programs for 77 countries. IDA programs hope to improve gender and income equality, economic growth, job creation, environmental damage, and living conditions. According to the IMF website, since 1960, the IDA has provided US$312 billion for investments in 112 countries.

Created in 1956, the IFC focuses exclusively on supporting the private sector in countries facing socio-economic, political, and post-conflict challenges. In 2016, the IFC was owned by 184 members of the World Bank, who governed it through their representatives to the board of governors. The IFC is the only one of its kind because it can provide direct financial assistance for private sector projects without a government guarantee – something that is prohibited in the mandates of the IBRD and IDA.[51]

The MIGA has a similar task in promoting the flow of foreign direct investment to the developing world by offering political risk insurance investors with equity participation. The MIGA offers guarantees against possible losses caused by political instability, armed conflicts, and unstable currencies. It expanded its mandate after reforms in 2009 that permitted it to extend the protection to banks and capital investors. The outline of its present strategy notes that such "changes have also allowed MIGA to introduce two new credit enhancement products: non-honoring of sovereign financial obligations in 2010 and non-honoring of financial obligations of state-owned enterprises in 2013." The agency was only created in 1988, but has been consistently growing and enlarging "its product line as well as its client base."[52]

The ICSID is a provider of international arbitration services for its 161 members, as counted in 2016. When created in 1966, the center was given a limited authority to adjudicate only those investment disputes where one of the parties was the host state. The number of arbitration cases dealt with by the ICSID has exploded since the 1990s – from two in 1992 to 50 in 2012 – mainly due to the proliferation of bilateral investment treaties around the world.[53]

In 2016, the World Bank had 7,000 employees and about 40 offices around the world. By contrast, the IMF had 2,300 staff members.[54] A major study into the institutional culture of the World Bank revealed systemic problems plaguing the organization. These problems are best encapsulated by using the term hypocrisy, which "reflect the conflicts between what the Bank as a collective actor says – its espoused goals, ideals, and policies – and what the Bank does."[55] According to the author of the study, organized hypocrisy is critical in permitting the World Bank to survive, especially when its external legitimacy is threaten by ongoing criticism. Blatant hypocrisy occurs when the bank openly violates its own mandates and policies. For example, when it funds infrastructure projects that go against its own environmental and social policies. There is also more subtle hypocrisy when the bank's proclaimed commitments to sustainable development or good govern-ance are not matched by its commitments in resource allocation.[56] Hypocrisy can be found in any international organization; however, when foreign aid institutions fall into the abyss of organized hypocrisy, they fail the most vulnerable people on the planet.

Another source of embedded hypocrisy in the way the World Bank conducts its activ-ities, is the "nonpolitical clause." Otherwise known as Article IV, section 10 of the 1944 Articles of Agreement, the clause states: "The Bank and its officers shall not interfere in the political affairs of any member; nor shall they be influenced in their decisions by the political character of the member or members concerned." The clause was put into the agreement to accommodate the Soviet Union among the growing uncertainties about the incompatibility of its state-run economic system with the liberal rationality of the Bretton Woods deal. The Soviets ultimately left the talks, but the clause stayed. By pro-hibiting political considerations to influence the decision-making processes in the World Bank, the organization gave itself a license to turn a blind eye to abuses committed by governments who requested its technical and financial assistance. The implications were far reaching. The clause solidified a technocratic approach to developmental assistance, and in the process it helped support governments run by tyrants and autocrats. Ironically then "the article against *politics* made it easier to use the Bank to pursue *politics*."[57]

Political leaders of recipient countries insist on nonintervention, and the World Bank and the IMF are said to be purely technical organizations. Still the same leaders quietly accept the conditionality terms that come with loans and investment guarantees. In the absence of a credible enforcement mechanism in the inter-state system international norms are not effective in constraining governments from selectively utilizing the sover-eignty principle. By the same token, nothing stops donor countries from demanding greater accountability and transparency when distributing financial assistance.

IV Expanding trade: from the GATT to the WTO and beyond

The issues of trade remained unresolved when the Bretton Woods agreement came into effect. Not until 1947 was the GATT inaugurated as a mechanism for managing cooperation on trade matters. The GATT was meant as a provisional arrangement in anticipation of the creation of the International Trade Organization (ITO). In November 1947, delegations from 56 countries met in Havana, Cuba, to consider the ITO draft as a whole. After long and difficult negotiations, the Final Act authenticating the text of the Havana Charter was approved in March 1948, but there were no commitments made by participating governments to ratify it. In the end, the Havana Charter was more or less abandoned by the key players, most notably the United States. One of the Havana Char-ter's chapters became the GATT, which stayed in place until 1995 as the only multilateral instrument governing world trade.

The idea behind the ITO was to design an effective multilateral regime with a broad mandate for the regulation of trade, including rules on restrictive business practices, primarily commodities, economic development, and foreign direct investment. The Havana Charter was so ambitious that it failed, possibly due to concerns over its scope. There were also objections over its conflicting provisions. Clair Wilcox, who headed the American negotiating team, tried to address the inconsistencies embedded in the charter. For example, by allowing countries to independently regulate and protect such commodity sectors as agriculture, mining, oil exploration, the Havana Charter was risking the creation of international tensions that could undermine the very principles of nondiscriminatory free trade. As a result, Wilcox suggested that in order to establish a free trading system "international agreement on commodity policy is required."[58] Wilcox assumed that it would be impossible for countries, including the United States, to completely abandon the idea of assisting domestic producers of primary commodities. He further recognized that some countries, especially underdeveloped economies, completely relied on the exports of a limited number of crops or minerals, and hence separate commodity agreements should be allowed. Special clauses permitting commodity agreements were included on the pages of the Havana Charter. Yet the charter was never ratified by the United States, and the ITO never materialized. Peter Kenen believes that the growing resistance against the ITO in the United States had to do with the fact that the charter allowed for too much government intervention. Other countries also had problems with accepting the charter, especially since it allowed excessive exemptions that in effect nullified the idea about having a fundamental set of multilateral rules.[59]

The fiasco of the Havana Charter could had been quite detrimental for the world trading system if it were not for the GATT, which was already in place at the time of the Havana conference. The fundamentals of the GATT agreement were negotiated early on when the British agreed with the American demands to restore, as soon as possible, their current account convertibility. Both sides also agreed to make international trade nondiscriminatory. There are two crucial rules expressing the principle of nondiscrimination: 1. Most Favored Nation rule (MFN) (Article I), and; 2. National Treatment rule (Article III).

The MFN rule is also known as the Favor One–Favor All rule, which means that any tariff concession made by one contracting party to the agreement must be extended to all other contracting parties. Unfortunately, the MFN principle was weakened by a number of allowable GATT exemptions. For example, by allowing countries to create customs unions and free trade areas, the GATT made the trade system vulnerable to fragmentation.

The National Treatment rule is meant to prevent the situation of a country introducing domestic regulations that give preferences to domestic producers and discriminate against foreign trading partners. In summary,

> what the principle of National Treatment dictates is that once border duties have been paid by foreign exporters, as provided for in a country's tariff schedules, no additional burdens may be imposed through internal sales taxes, differential forms of regulation, etc. on foreign exporters where domestic producers of the same product do not bear the same burden.[60]

There were also exemptions to this rule. One allowed a contracting party to maintain the preferences it had in place upon signing the GATT agreement. Furthermore, government agencies were allowed to discriminate when procuring goods.

The GATT was able to become operational as the 1946 trade negotiations were conducted in two streams. In parallel to the institutional talks concerning the charter of the ITO, a set of technical meetings took place between January and February 1947 in Lake Success, New York. During these talks, the GATT as a legal document was concluded. Later that year the talks moved to Geneva, where representatives from 23 countries conducted 123 bilateral tariff-cutting negotiations, resulting in about 50,000 tariff cuts, with the average tariff being cut by 35 percent.[61]

When the ITO failed, the trade negotiators agreed to protect the value of the tariff concessions by accepting the GATT by means of adopting the Protocol of Provisional Application. The protocol was completed on October 30, 1947, and came into force on January 1, 1948. The GATT consisted of 38 legal provisions, closely derived from chapter IV (*Commercial Policy*) of the Havana Charter. The GATT became a contractual arrangement to manage international trade flows because it represented the minimum acceptable set of provisions to fulfill that role. Apart from the 23 original contracting parties, 11 further countries signed the agreement in 1949 at the Annecy round of talks.

Since the GATT is based on chapter IV of the Havana Charter, it retains most of its general exceptions listed in GATT Article XX. One of them, however, is curiously taken from chapter VI (*Intergovernmental Commodity Agreements*) of the Havana Charter. The rest of the charter became dormant since the ITO was never born. It is somehow surprising that the framers of the GATT decided to keep the provision allowing the creation of commodity agreements, despite the fact that this issue was so divisive during the ITO negotiations. Consequently, the GATT offered an opportunity to establish such agreements. This opportunity still exists under the WTO.

The GATT also included the controversial Article XVII on state trading enterprises. It was a clause that permitted trading with the Soviet Bloc countries and facilitated the accession of non-market economies to the GATT. The Article is still an integral part of the world trading system under the WTO paradoxically tolerating the existence of state-owned transnational companies, although such companies in principle contradict the rationality of free trade. Article XXIV of the GATT was also problematic as it reinforced exemptions with respect to the application of the MFN principle. It did so by allowing contracting parties to negotiate preferential regional trade agreements. This would become a serious problem during the 1970s, and even more recently under the WTO when countries begin to favor regional deals over the multilateral rules.

As a provisional agreement, the GATT had several weaknesses. Its scope was very limited. The collapse of the ITO corresponded with the removal of some of its salient provisions from the multilateral trade agenda, such as those that addressed import controls, cartels, foreign investment, and development. The GATT was mainly about tariff cuts and the prohibition of quotas. The agreement was also legally awkward as it did not have binding interpretative legal powers. Without such powers, the GATT's dispute settlement process was in effect voluntary, and its proceedings could have been blocked by a participating party without any consequence. There was no formal institution underpinning the

GATT, so its signatories were not even called members, they were only contracting parties to the GATT. Viewed as a narrow technical agreement, it lacked guidance concerning decision-making procedures. With time, the relevant operating procedure would be slowly developed.

Although, the GATT was in its origin a club of developed nations, by the end of the Tokyo Round of multilateral trade negotiations (1973–1979) the number of GATT contracting parties rose to almost 90. Newcomers from the developing world hoped to participate on meaningful terms, but lacking a formal organization, the provisional GATT continued to be dominated by the top trading nations. The Tokyo Round enlarged the scope of the GATT by introducing six new agreements, called codes (on government procurement, on customs valuation, on technical barriers to trade, on antidumping, on subsidies and countervailing duties, and on an import licensing).[62] They were optional in the sense that individual countries could pick and choose among them. Most developed countries joined all the codes, but developing countries would only join one or two of them, if any, making the GATT system unmanageable. With growing frictions, contracting parties complained over its lack of predictability and transparency.

This is why the transformation from the GATT to the WTO during the Uruguay Round (1986–1993) of multilateral talks was very significant. The GATT and its original provisions are still at the core of the WTO. However, the WTO – officially born in January 1995 – is so much more than the GATT ever was. No more provisional arrangements. The WTO is a formal rules-based organization, which includes under its legal umbrella a set of old and new trade agreements, under the jurisdiction of the new dispute settlement mechanism. The Uruguay Round resulted in a further lowering of tariffs on industrial products and formulated first agreements on agriculture and textiles. The round created comprehensive new agreements on services and intellectual property, and attempted to deal with the issue of investment. Most importantly, by establishing the WTO as a formal rules-based organization, it strengthened the status of all WTO members (no longer contracting parties) by bestowing upon them the principle of judicial equality.

The Uruguay Round established a unified system of dispute settlement understanding (DSU) for all WTO agreements, with the capacity to interpret provisions and hence contribute to the building of international trade law. The new procedures include consultations, establishing of the panels, proceedings, issuing of the report, and a new appellate body. The crucial aspect of the DSU is the automatic acceptance of decisions reached by a dispute settlement panel and the appellate panel, which then become legally binding. No member of the WTO can arbitrarily block a trade dispute initiated in the WTO. The DSU permits sanctions against members who break the WTO trade rules. The WTO dispute settlement mechanism is the only system for the settlement of disputes between states with compulsory jurisdiction, appellate body and binding arbitration.[63]

For the first time, services have been brought within the multilateral framework of trade rules. The WTO General Agreement on Trade in Services (GATS) sets a general framework for the conduct and further multilateral liberalization of trade in services. The GATS is structured in two parts. The first part is the text of the agreement – its Articles and Annexes. The second part is composed of the schedules of country-specific commitments undertaken by WTO members. GATS rules are limited to sectors in which individual countries have accepted scheduled liberalization commitments. Many countries liberalized only some of their services sectors from the following, normally covered by GATS: telecommunication and audio-visual, financial, business (e.g., accounting, architecture), construction and engineering, transportation, distribution, tourism, recreational, health, environmental, and education services.[64]

Agriculture remains a problematic area within the multilateral framework of international trade.[65] The new WTO Agreement on Agriculture did not go far enough to open the borders for trade in agricultural products for developing countries. The agreement still accomplished three things. First, with a few temporary exceptions, it converted all quotas and unbound tariffs into bound tariffs. Second, it prohibited new export subsidies and cut existing ones. And third, it began to tackle domestic subsidies, which, in effect, protect farmers against foreign competition from the developing world in much the same way as tariffs. A point highly contested by the developing countries.

The WTO agreement on Trade-Related Aspects of Intellectual Property Rights (TRIPS) remains the most comprehensive international agreement on intellectual property ever negotiated.[66] The agreement was strongly advocated by a number of industries who believed that without a harmonized system of rules protecting intellectual property rights (IPR) around the world, foreign direct investment would suffer. In the knowledge economy, it was argued, goods and services increasingly depended on patents and copyrights, industrial designs, etc. which should be protected by adequate laws. TRIPS obliges all members of the WTO to have such laws in place and stipulates the necessary level of IPR protection and enforcement. Most developing countries, however, opposed the agreement, fearing the adverse impact of the monopolistic nature of patent rights, and worrying about the costs of creating new IPR laws, regulations, and sometimes the whole new administrative framework, necessary to comply with the TRIPS.[67]

The WTO offers each of its members an equal legal standing. There is no voting formula, as in the IMF. The WTO's decision-making process by consensus is greatly supported by developing countries because, if smaller countries disagree, the deal is off – decision making by consensus does not require unanimity. Any WTO member who is not present at a meeting when the decisions are made is considered to be in favor of consensus.[68] Another crucial principle guiding the WTO is the principle of single-undertaking. This means that every WTO member has to accept all WTO agreements as a single package. Single-undertaking stands in contrast to the GATT era, when countries could decide which of the newly negotiated agreements (codes) they wanted to sign.

The single-undertaking was an important factor in ensuring that the Uruguay Round could be concluded by rising stakes for those countries who were outside the organization. However, given the implementation problems faced by developing countries, the single-undertaking turned to be a deterrent against negotiating new WTO agreements. There have been at least three areas of concern for developing countries with respect to the WTO implementation process: new obligations that came with expanded scope of international trade rules; the necessity of creating new regulations, laws, enforcement procedures, and even administrative frameworks to address compliance; finally, insufficient progress in the negotiations on agriculture prevented the developing world from seeing the benefits of the WTO as a reciprocal deal.[69]

Table 3.1 Three pillars of the WTO – list of single-undertaking WTO agreements

WTO – List of single-undertaking agreements

I. Goods – GATT 1947 + agreements on:

agriculture, SPS (sanitary, phyto-sanitary) measures, textiles and clothing, technical barriers to trade, trade-related investment measures, anti-dumping, customs valuation, pre-shipment inspection, rules of origin, import licensing procedures, subsidies and countervailing measures, safeguards

II. GATS – General Agreement on Trade in Services

III. TRIPS – Agreement on Trade-Related Aspects of Intellectual Property Rights

Unified dispute settlement mechanism = Dispute Settlement Understanding (DSU) and the Appellate Body

Source: WTO website, online, available at: www.wto.org.

The WTO as an organization based on legal rules offered previously inaccessible opportunities to countries who traditionally lacked power and influence on the international stage. However, soon after the WTO was established, developing countries felt that the costs of implementing its new agreements exceeded their anticipated benefits.[70] Feeling overwhelmed by the existing WTO obligations, the developing countries have effectively resisted any new multilateral agreement from being negotiated in the WTO.

The WTO was created to be a universal organization, facilitating fairness and transparency in the global economy. New countries are welcome to join the organization. The WTO accession process aims to ensure the full compliance of a new member with the WTO agreements under the principle of single-undertaking. The candidates confront a set of comprehensive obligations, covering both border measures and domestic regulatory policies. WTO accession routinely emphasizes the internal reforms that the candidate state must undergo in order to behave in accordance with the legal principles of the WTO. As a result, the process can take over a decade to conclude, and sometimes stalls when a candidate refuses to meet the more intrusive demands for economic liberalization. All existing WTO members have to agree on the terms of individual accessions; hence, the process can be seen as a peculiar structural-adjustment process conducted under the watchful eyes of existing WTO members. Despite the financial crisis of 2008 and the diminished appetite for economic globalization, countries continue to apply for WTO membership. As of March 2017, the WTO had 164 members, with such diverse countries as Liberia, Samoa, Ukraine, Vietnam, Kazakhstan, Montenegro, and Russia joining over the last few years.

In 2001, the first round of multilateral negotiations under the WTO was launched in Doha, Qatar.[71] It was called the development round since it promised to address issues of particular importance to developing countries, such as the distorted rules of agricultural trade, the implementation of the existing WTO agreements, a problematic accession process, and special and differential treatment, just to name a few. The most problematic trade issues will be addressed later in the relevant parts of the book. The Doha talks are

still officially ongoing in 2017, although progress has been so slow that the round effectively undermined the role of the WTO as an organization capable of moving forward the
global economy. The weakening of the multilateral system resulted in the proliferation of
regional and bilateral agreements that threaten the coherence of the entire system,[72]
leading to the possible fragmentation of international trade law.[73] Scholars and practitioners suggest that institutional reforms may be necessary to transform the WTO into an
effective negotiating forum and a force capable of strengthening non-discriminatory
global trade practices.[74]

The two most cited problems with the WTO as an organization are the inflexibility of
the single-undertaking, and the decision making by consensus. Many developing countries continue to feel that they should not be forced into signing new trade agreements
that do not agree with their developmental priorities. At the end of the 2015 WTO ministerial meeting in Nairobi, Kenya, WTO members formally recognized that deep opposing
viewpoints on how to proceed with the Doha negotiations existed among them. Despite
reaching an important agreement on eliminating export subsidies in agriculture, the
Nairobi Declaration signaled trouble ahead for the organization, making it clear that the
WTO had not been able to reconcile existing divisions in order to strengthen the multilateral trading system.[75]

Given the ongoing tensions and divisions, some critics contest the existence of the
global economy. They claim that the present global dynamics are nothing exceptional,
and that we are merely witnessing a more vigorous intensification of economic interdependence among countries. To be specific: "the term 'international economy' has always
been a shorthand for what is actually the product of the complex interaction of economic
relations and politics, shaped and reshaped by the struggles of the great powers."[76] And
yet, while wrestling with the above statement, we observe that the contemporary global
economy differs from the past processes of internationalization in at least two critical
ways: the unprecedented importance of technology and the growing influence of developing countries.

The countless financial transactions that are taking place daily around the world would
not be possible without computers and satellites. Global market integration accelerated
when new technological inventions became widely used only about 25 years ago. The
connectivity factor is unique: "contemporary globalization has been marked by a large-
scale spread of supraterritoriality."[77] An enormous number of international economic
exchanges occur today, with little attention paid to time and distance. Advancements in
transportation technology further contribute to building more efficient transporting
vessels, capable of moving people, goods, components, and commodities faster and more
safely across the world.

Technological advancements have lowered coordination costs, allowing countries to
specialize in specific tasks or the production of specific components rather than entire
final products. The resulting fragmentation of global production has led to the growth of
GVCs, whose nature and scope has significantly changed over the last two decades. In
this context, new formidable actors from the developing world are having a transformative effect on the way the global economy operates. In 1996, the industrialized
countries accounted for almost two-thirds of the world imports of intermediate products,

but less than a half in 2012. China alone increased its share of trade in parts and components fivefold, from around 3 percent in 1996 to more than 15 percent in 2012.[78] However, China started to experience economic slowdown, and as a result, Asia's share in global import growth began to shrink from 73 percent in 2013 to 25 percent in 2015.[79] Negative trade growth was recorded in South and Central America, Africa, and the Middle East in 2015. Still, the global value of merchandise trade, and trade in commercial services, doubled between 2005 and 2015. During the same period, merchandise trade between developing economies increased from 41 percent to 52 percent of their global trade.[80] In 2016, world trade continued to experience sluggish growth at 1.3 percent, among renewed calls for global policy coherence. Overall, since 2005, both merchandise and services global trade flows have continued to grow, despite the financial crisis, disagreements among countries, and poor progress of multilateral negotiations at the WTO.

The global economy today presents a paradox. Despite the resistance of states to weaken the sovereignty principle, there is an ongoing need for multilateral organizations to solve collective problems. The challenge for countries today is not only to harness the possibilities offered by the global economy, but also to provide an enabling environment that promotes economic development.

Notes

1   Daniel W. Drezner (2007) *All Politics Is Global – Explaining International Regulatory Regimes*, Princeton, NJ: Princeton University Press, p. 10.
2   WTO (2016) *World Trade Statistical Review 2016*, Geneva, Switzerland: The WTO, p. 14.
3   Jan Aart Scholte (2005) *Globalization – A Critical Introduction* (2nd edition), New York, NY: Palgrave/ Macmillan, p. 83.
4   WTO (2014) *World Trade Report 2014 – Trade and Development: Recent Trends and the Role of the WTO*, Geneva, Switzerland: The WTO, pp. 82–88.
5   Anne-Marie Slaughter (2004) *A New World Order*, Princeton, NJ: Princeton University Press, p. 5.
6   Ha-Joon Chang (2012) *23 Things They Don't Tell You About Capitalism*, New York, NY: Bloomsbury Press, pp. 51–53.
7   Michael Pettis (2013) *The Great Rebalancing: Trade, Conflict, and the Perilous Road Ahead for the World Economy*, Princeton, NJ: Princeton University Press, pp. 2, 12.
8   The Resumption Act of 1819 was meant to stabilize the world order after the Napoleonic Wars.
9   Karl Gunnar Persson and Paul Sharp (2015) *An Economic History of Europe – Knowledge, Institutions and Growth, 600 to Present*, Cambridge, UK: Cambridge University Press, p. 196.
10 Andrew C. Sobel (2006) *Political Economy and Global Affairs*, Washington, DC: CQ Press, pp. 173–180.
11 Ibid., pp. 214–215.
12 Barry Eichengreen and Peter B. Kenen (1994) "Managing the World Economy under the Bretton Woods System: An Overview," in Peter B. Kenen (ed.), *Managing the World Economy – Fifty Years After Bretton Woods*, Washington, DC: Institute for International Economics, p. 9.
13 Ibid., pp. 216–217.
14 Niall Fergusson (2008) *The Ascent of Money – A Financial History of the World*, New York, NY: Penguin Press, p. 104.
15 John Kenneth Galbraith (1954) *The Great Crash 1929*, Boston, MA: Houghton Mifflin, p. 168.
16 Harold James (2001) *The End of Globalization: Lessons from the Great Depression*, Cambridge, MA: Harvard University Press. Electronic Edition, Location 418 of 3454.
17 Ibid., Location 1906–1920 of 3454.

18 Donald L. Kemmerer (2003) "Gold Standard," in *Dictionary of American History*, Farmington Hills, MI: The Gale Group Inc.

19 Andrew C. Sobel (2006) *Political Economy and Global Affairs*, op. cit., pp. 214–219.

20 Sylvia Ostry (1997) *The Post-Cold War Trading System: Who's on First?*, Chicago, IL: University of Chicago Press, pp. 57–58.

21 The current account measures a country's net exports of goods and services. The capital account comprises capital inflows and outflows related to domestic and foreign investment.

22 Karl Gunnar Persson and Paul Sharp (2015) *An Economic History of Europe – Knowledge, Institutions and Growth, 600 to Present*, op. cit., p. 202.

23 Sylvia Ostry (1997) *The Post-Cold War Trading System*, op. cit., pp. 57–58.

24 Nicholas Wapshott (2011) *Keynes and Hayek – The Clash That Defined Modern Economics*, New York, NY: W.W. Norton & Company, pp. 53–56.

25 IMF (1944) *Articles of Agreement of the International Monetary Fund*, Schedule A, July 22.

26 Karl Gunnar Persson and Paul Sharp (2015) *An Economic History of Europe – Knowledge, Institutions and Growth*, op. cit., p. 203.

27 Louis W. Pauly (1997) *Who Elected the Bankers? – Surveillance and Control in the World Economy*, Ithaca, NY: Cornell University Press, p. 89.

28 Ibid., pp. 89–90.

29 See IMF website. Online, available at: www.imf.org/external/np/exr/facts/sdr.htm.

30 IMF (2016) *IMF Annual Report 2016 – Finding Solutions Together*, Washington, DC: IMF.

31 Catherine Weaver (2008) *Hypocrisy Trap – The World Bank and the Poverty of Reform*, Princeton, NJ: Princeton University Press, p. 44.

32 Austria, Belgium (with Luxemburg), Denmark, France, West Germany, Greece, Iceland, Ireland, Italy, Netherlands, Norway, Portugal, Sweden Switzerland, Turkey, and the United Kingdom. On an important note, Spain, which was under Franco's military dictatorship, was omitted from the program.

33 Andrew C. Sobel (2006) *Political Economy and Global Affairs*, op. cit., pp. 273–274.

34 Ibid., pp. 328–329.

35 Fred L. Block (1977) *The Origins of International Economic Disorder: A Study of the US International Monetary Policy from WWII to the Present*, Berkley, CA: University of California Press, pp. 133–135.

36 Michael Collins (2012) *Money and Banking in the UK: A History*, London, UK: Routledge, pp. 498, 536.

37 Robert Triffin (1960) *Gold and the Dollar Crisis: The Future of Convertibility*, New Haven, CT: Yale University Press.

38 Andrew C. Sobel (2006) *Political Economy and Global Affairs*, op. cit., pp. 280–284.

39 Ibid., pp. 311–315.

40 A petrodollar is a dollar earned by a country through the sale of oil. The term was coined in 1973 by Ibrahim Oweiss, a professor of economics at Georgetown University, to describe the vast amounts of dollars obtained by the members of OPEC from the sale of oil.

41 Andrew C. Sobel (2006) *Political Economy and Global Affairs*, op. cit., p. 315.

42 William Easterly (2006) *The White Man's Burden: Why the West's Efforts to Aid the Rest Have Done So Much Ill and So Little Good*, New York, NY: Penguin Press, p. 65.

43 Ibid., pp. 66–67.

44 Narcis Serra and Joseph E. Stiglitz (eds.) (2008) *The Washington Consensus Reconsidered – Towards a New Global Governance*, Oxford, UK: Oxford University Press.

45 William Easterly (2002) "The Cartel of Good Intentions," *Foreign Policy*, No. 131 (July–August), p. 41.

46 See IMF website, online, available at: www.imf.org/en/About/Factsheets/Sheets/2016/08/02/21/28/IMF-Conditionality, accessed September 4, 2016.

47 IMF (2016) *IMF Annual Report 2016 – Finding Solutions Together*, op. cit.

48 Ibid.

49 Hilton L. Root (2006) *Capital and Collusion – The Political Logic of Global Economic Development*, Princeton, NJ: Princeton University Press, p. 183.

50 World Bank (2016) *The World Bank Annual Report 2016*, Washington, DC: The World Bank Group.

51 Catherine Weaver (2008) *Hypocrisy Trap – The World Bank and the Poverty of Reform*, op. cit., p. 46.
52 World Bank (2015) *MIGA Strategic Directions FY15–17*, Washington, DC: The World Bank Group.
53 Walter Mattli and Thomas Dietz (2014) "Rise of International Commercial Arbitration," in Walter Mattli and Thomas Dietz (eds.), *International Arbitration and Global Governance: Contending Theories and Evidence*, Oxford, UK: Oxford University Press, pp. 3–4.
54 See IMF website, online, available at: www.imf.org/external/pubs/ft/exrp/differ/differ.htm.
55 Catherine Weaver (2008) *Hypocrisy Trap – The World Bank and the Poverty of Reform*, op. cit., p. 19.
56 Ibid., p. 20.
57 William Easterly (2013) *The Tyranny of Experts: Economists, Dictators, and the Forgotten Rights of the Poor*, New York, NY: Basic Books, p. 118.
58 Clair Wilcox (1949) *A Charter for World Trade*, New York, NY: Macmillan, p. 115.
59 Peter B. Kenen (1994) *Managing the World Economy; Fifty Years After Bretton Woods*, op. cit., p. 14.
60 Michael J. Trebilcock and Robert Howse (1999) *The Regulation of International Trade* (2nd edition), London, UK: Routledge, p. 29.
61 John S. Odell (2000) *Negotiating the World Economy*, Ithaca, NY: Cornell University Press, pp. 162–164.
62 For a summary of the six Tokyo Round "code" agreements see: Gilbert R. Winham (1986) *International Trade and the Tokyo Round Negotiation*, Princeton, NJ: Princeton University Press, pp. 417–424.
63 Anna Lanoszka (2009) *The World Trade Organization: Changing Dynamics in the Global Political Economy*, Boulder, CO, and London, UK: Lynne Rienner Publishers, pp. 33–36.
64 Ibid., pp. 107–136.
65 Fiona Smith (2009) *Agriculture and the WTO – Towards a New Theory of International Agricultural Trade Regulation*, Cheltenham, UK: Edward Elgar.
66 Antony Touban, Hannu Wager, and Jayashree Watal (eds.) (2012) *A Handbook on the WTO TRIPS Agreement*, Cambridge, UK: Cambridge University Press.
67 Susan K. Sell (2003) *Private Power, Public Law – The Globalization of Intellectual Property Rights*, Cambridge, UK: Cambridge University Press, pp. 75–120.
68 Debra P. Steger (2009) "The Future of the WTO: The Case for Institutional Reform," *Journal of International Economic Law*, Vol. 12, No. 4, pp. 808–810.
69 Asoke Mukerji (2000) "Developing Countries and the WTO – Issues of Implementation," *Journal of World Trade*, Vol. 34, pp. 39–51.
70 Anna Lanoszka (2009) *The World Trade Organization: Changing Dynamics in the Global Political Economy*, op. cit., pp. 150–158, 231–235.
71 WTO (2001) *Doha Ministerial Declaration* (WT/MIN(01)DEC/1), November 20.
72 Ross Buckely, Vai Lo Lo, and Laurence Boulle (eds.) (2008) *Challenges to Multilateral Trade – The Impact of Bilateral, Preferential and Regional Agreements*, The Netherlands: Wolters Kluwer.
73 Adrian M. Johnston and Michael J. Trebilcock (2013) "Fragmentation in International Trade Law: Insights from the Global Investment Regime," *World Trade Review*, Vol. 12, No. 4, pp. 621–652.
74 Debra P. Steger (ed.) (2010) *WTO – Redesigning the World Trade Organization for the Twenty-First Century*, Waterloo, Canada: Wilfrid Laurier University Press.
75 WTO (2015) *Nairobi Ministerial Declaration* (WT/MIN/(15)/DEC), December 21.
76 Paul Hirst and Grahame Thompson (2002) *Globalization in Question – The International Economy and the Possibilities of Governance*, Cambridge, UK: Polity Press, pp. 13–14.
77 Jan Aart Scholte (2005) *Globalization – A Critical Introduction*, op. cit., p. 61.
78 WTO (2014) *World Trade Report 2014 – Trade and Development: Recent Trends and the role of the WTO*, op. cit., pp. 80–82.
79 WTO (2016) *World Trade Statistical Review 2016*, op. cit., p. 20.
80 Ibid., pp. 10–12.

Suggested further reading

Donatella Alessandrini (2010) *Developing Countries and the Multilateral Trade Regime: The Failure and Promise of the WTO's Development Mission*, Oxford, UK, and Portland, OR: Hart Publishing Ltd.

Eric Helleiner (2016) *Forgotten Foundations of Bretton Woods: International Development and the Making of the Postwar Order*, Ithaca, NY: Cornell University Press.

A.H.J. (Bert) Helmsing and Sietze Vellema (eds.) (2016) *Value Chains, Social Inclusion and Economic Development: Contrasting Theories and Realities*, London, UK: Routledge.

Craig N. Murphy (2005) *Global Institutions, Marginalization and Development*, London, UK: Routledge.

Thomas Piketty (2014) *Capital in the Twenty-First Century*, Translated by Arthur Goldhammer, Cambridge, MA: The Belknap Press of Harvard University Press.

Carmen M. Reinhart and Kenneth S. Rogoff (2009) *This Time Is Different: Eight Centuries of Financial Folly*, Princeton, NJ: Princeton University Press.

Ngaire Woods (2007) *The Globalizers: The IMF, the World Bank, and Their Borrowers*, Ithaca, NY: Cornell University Press.

Domestic strategies 4

Obstacles and opportunities

I Conceptualizing strategy

This chapter examines the role of the state in development, and focuses on aid-receiving countries and reforming economies. Policy making in the developing world had been frequently constrained in relation to the developmental priorities advanced by aid agencies and multilateral economic organizations. In this context, actions undertaken by aid-receiving governments were sometimes misguided, sometimes incomplete, or simply reflected the minimum acceptable level of compliance with the conditions stipulated by the donor agencies. The results varied, but top-down policy prescriptions seldom improved the lives of the affected communities. After decades of interventions and after trillions of dollars spent on foreign aid, poverty is still very much present in many parts of the world.

Perhaps one of the reasons behind the generally unimpressive record of poverty fighting practices by the international development community was the insistence on technical solutions or shock therapy plans instead of providing attentive support of long-term domestic strategies. The promotion of detailed policy instructions coincided with designing the blueprints for structural adjustment programs by the World Bank and the IMF at the height of the post-Cold War globalization era. As it turned out, many of these programs failed to place the targeted country on the path to prosperity. Bogged down reforms more often than not exacerbated pre-existing economic stagnation and remained unfinished. Structural reforms conducted in the name of democratization processes were based on a very cursory view of what democracy is. The plans collapsed when they could not adjust to changing circumstances.

Democracy cannot be instituted from the outside. Complex economies need more than an unyielding shock therapy plan. Technical solutions can only fill certain gaps. Successful economies are characterized by a pursuit of a long-term strategy that is oriented towards the future but capable of dealing with the systemic limitations of the present day. Lawrence Freedman observes that so many great plans failed because "pure control was always an illusion, at most a temporary sensation of success, which would soon pass as the new situation generated its own challenges."[1] A good strategy sets up priorities, outlining the anticipated path but remaining flexible enough to adapt to shifting realities on the ground. In short: "a productive approach to strategy requires recognizing its limits."[2]

It is hard to imagine that a developmental strategy can be effective without being grounded in indigenous political culture, anchored in local patterns of needs and limitations, and oriented towards consensus building. Developmental agencies can be supportive in assisting the implementation of self-directed developmental strategies of countries, but cannot orchestrate them. This kind of help begins with recognizing the pivotal role of state institutions – not only economic, but also social and political – because even the best developmental policy advice fails if there is no trustworthy institutional framework to make it work.

> Creating a political and social framework favorable to economic growth is often the greatest challenge these nations face. Yet developing countries are frequently advised to adopt models of economic institutions that can only succeed once appropriate social institutions exist. Enforceable property rights, for instance, often enjoy the status of being a necessary condition for economic development. The sustainability of the property rights requires social coherence and political accountability, since property rights can be overturned by the forces of social upheaval or confiscated by unchecked political discretion.... The security of property rights in highly unequal societies is often enforced by military might rather than the rule of law. The property-rights regime is only as stable as the social and political foundations on which it rests.[3]

Can individual states do more to improve the lives of their poverty-stricken citizens? The evidence says yes. Notwithstanding the legacy of colonialism, past tribulations and mismanagement, armed conflicts and natural disasters, accountable governments enable a domestic environment that reduces economic uncertainty and facilitates home-grown effective developmental strategies. On the other hand, regressive and corrupted governments assume responsibility for perpetuating socio-economic divisions and promoting economic insecurity.

However, while it is important to talk about the quality of government, the state has to be assessed in all its complex functions. The state remains a fundamental political unit of the international system. As such, the state cannot be simply ignored or treated as a government in a box by the advocates of economic globalization. The state also cannot be reduced to the rules that shape economic institutions, or to the incentive structures that constrain the choices of dominant political actors:

> the state is not just the political economic setting that structures the actions of private political economic actors, as posited by mainstream political economy. it also constructs the policies for reform, constitutes the political institutional setting that shapes the reform process, and can also be the political driver for reform.[4]

This chapter examines three historical views of the state in the developing world that evolved in the context of integration into the international economy. The first two views were part of the strategies constructed by the developmental agencies. Beginning with the post-colonization period of the twentieth century, the *relying on the state* approach meant that the state in the developing world was seen primarily as the friendly government receiving foreign aid. There were few questions asked regarding how the aid money was actually spent, the friendly government was relied on to distribute the money as promised. The second view was born during the period of debt crises and socio-economic meltdowns that overwhelmed many poor countries during the 1980s. At the time, the state in the developing world became viewed as structurally deficient and *in need to be fixed*. It was the time when the major developmental agencies began demanding a variation of pro-market structural reforms conducted under the label of democratization. The third view sees the state as the primary agent of domestic economic policy making. Here the government *utilizes the state* structures for state-directed development. This strategy is domestically conceived with little guidance from the development community. Some of the most successful economic transformations of the last century occurred in East Asia, among countries that engaged in state-directed development. This model still generates criticism from scholars who question its long-term sustainability. The connection is drawn between the former Soviet Bloc countries, whose economies eventually collapsed due to the inefficiencies caused by the authoritarian central planning. As a result, state-directed development cannot be considered fully successful without working towards creating democratically accountable state institutions. It appears that only under effective democracy, committed to ensuring rights, security, and freedom to participate, is long-term development sustainable.

II Relying on the state: the politics of aid in the Third World

The *relying on the state* view went hand in hand with the establishment of the foreign aid community. The view was based on the assumption that the newly independent governments were the driving force of socio-economic development and they had to be helped by the outsiders. In principle, the notion that rich industrialized countries should help poor countries to eradicate poverty by sharing wealth and providing assistance makes sense. But why, after over six decades of aid distribution, are there so many poor people around the world? A recent study measuring the level of absolute poverty (at approximately US$1 a day) in the developing world between 1981 and 2004 found that although there was a decline in the percentage of people living below this poverty line, less progress had been made in reducing the numbers of the poor. In fact, when a higher poverty line is used, the worldwide count of poor people rose over most of the period.[5] In this context, the period between 1993 and 2005 reveals a disturbing trend, as the authors of an important article explain:

> both the $1.25 and $1.45 [poverty] lines indicate a substantially higher poverty count in 2005 than obtained using our old $1.08 line in 1993 prices…. Focusing on the $1.25 line, we find that 25 percent of the developing world's population in 2005 is poor, versus 17 percent using the old line at 1993 PPP[6] – representing an extra 400 million people living in poverty.[7]

The most recent UNDP report reinforces these chilly statistics: in 2013 as many as 766 million people, 385 million of them children, lived on less than US$1.90 a day.[8]

Upon independence, most developing countries had fragile economies, tenuous social foundations, and weak political institutions. Their daily struggles were compounded by insufficient foreign reserves for purchases of necessary commodities and foreign goods. The foreign aid was to remedy some of those most urgent problems. Helping poor nations was heralded to the world as a noble initiative by President Harry Truman during his 1949 inaugural address:[9]

> we must embark on a bold new program for making the benefits of our scientific advances and industrial progress available for the improvement and growth of underdeveloped areas. More than half the people of the world are living in conditions approaching misery. Their food is inadequate. They are victims of disease. Their economic life is primitive and stagnant. Their poverty is a handicap and a threat both to them and to more prosperous areas.... Our aim should be to help the free peoples of the world, through their own efforts, to produce more food, more clothing, more materials for housing, and more mechanical power to lighten their burdens.... Such new economic developments must be devised and controlled to the benefit of the peoples of the areas in which they are established.

Not only did Truman usher in the first wave of foreign assistance programs, he also established a narrative that would become dominant in the post-war period. Specifically, he solidified the distinction between an underdeveloped and a developed world as he called on the world to help people in distant countries. Before long, foreign aid became a quasi-industry, worth billions of dollars, with the Cold War turning it into a political issue. Animosities between the United States and the Soviet Union triggered a fierce competition over the spheres of influence in the developing world. It is difficult to determine how much money was spent by both sides in the name of providing developmental assistance, but a rough estimate suggests trillions of dollars. The western agencies alone "spent $2.3 trillion on foreign aid over the last five decades and still had not managed to get 12-cent medicines to children to prevent half of all malaria deaths."[10] The African case has been particularly problematic because "by the early 1990s, Africa's relationship with the international economy was almost entirely mediated by public aid flows."[11]

Foreign aid comes in different forms. It could be money, various goods (including agricultural), services, or technical advice. Official development assistance (ODA) is defined as government aid designed to promote the economic development and welfare of developing countries. Consistent with this definition, the OECD reports that in 2014 there were 41 official donors, sending aid to over 140 recipient countries. The top five donor countries were: Japan (141 beneficiaries), United States (132 beneficiaries), closely followed by South Korea, France, and Australia.[12] There are no clear patterns when looking at the flows of foreign aid. For example, Luxemburg sends its modest contribution to 74 countries, dividing among recipients less money than the United States gave to Burkina Faso alone. China is still receiving developmental assistance from 35 countries. Canada's aid budget is the smallest among the top industrialized countries, but still divided among

48 recipients. Thailand gave to 55 countries and Turkey's budget was larger than Britain's. Many top developing economies were not included as donors in the report, such as Indonesia, India, Mexico, Brazil, and China. Indeed, these countries are documented to be donors as well as recipients. Especially China has been an influential donor, providing aid to over 120 countries.[13]

There are two main criticisms of foreign aid: 1. mismanagement and ineffectiveness of the foreign aid programs; 2. adverse consequences of aid programs for the development of recipient countries. A pointed reflection by Muhammad Yunus, illuminates the first criticism. He is the founder of the Grameen Bank and the winner of the 2006 Nobel Peace Prize for pioneering the concepts of micro-credit and micro-finance for the poor. Yunus remembers the situation experienced by one of the replicators of the Grameen Bank who attempted to work with the major provider of aid:

> I find multilateral donors' style of doing business with the poor very discomforting. I can cite one example of my experience in the island of Negros in the Philippines. Because hunger was so bad there, one of our replicators started Project Dunganon back in 1988. More than half the island's children were malnourished, and so in 1993 Dr. Cecile del Castillo, the replicator, still innocent about the nature and work habits of international consultants, asked the International Fund for Agricultural development (IFAD), a Rome-based UN agency created to assist the rural poor, for money to expand her successful program quickly. IFAD responded by sending four missions to investigate her proposal, spending thousands of dollars in airline tickets, per diems and professional fees. But the project never received a single penny.[14]

Three years later, the project eventually materialized, when an agreement was signed between the government of the Philippines, the Asian Development Bank and IFAD. The deal awarded the loan of US$37 million to the Philippines to support micro-credit programs in the country. When Yunus was writing his book in 1998, the money was still caught up in the layers of governmental bureaucracy. Yunus concludes: "had the Negros project simply received an amount equal to the cost of a single IFAD mission, it would have been able to assist several hundred poor families with micro-credit."[15]

There are three main assumptions made by the development community that have contributed to the problems of mismanagement and ineffectiveness of aid: certainty over prescriptive actions designed for the recipient countries, certainty that the right advice supported by money will lead to putting those actions to work, certainty that faceless experts can formulate the right advice under a broad mandate that represents everybody, from the multilateral agencies to celebrities.[16] Such assumptions stem from the state-centrist model, which relies on the recipient-state, and its friendly government, as an agent in charge of aid distribution. The model overlooks the individual agency of those people in the recipient country who are expected to be helped. The model also supports the view that development could be engineered from the outside. Over time, a growing number of failed developmental projects revealed that relying on the state for the distribution of aid had negative outcomes especially if the state was run by a corrupt autocrat.

The persistent problems with the *relying on the state* view do not mean, however, that all aid programs should be dismantled. One idea is to change the perspective with respect to how and for whom foreign assistance should be created.

> Once freed from the delusion that it can accomplish development, foreign aid could finance piecemeal steps aimed at accomplishing particular tasks for which there is clearly a huge demand — to reduce malaria deaths, to provide more clean water, to build and maintain roads, to provide scholarships for talented but poor students, and so on. It could seek to create more opportunities for poor individuals, rather than try to transform poor societies.[17]

Another criticism of the *relying on the state view* of foreign aid relates to the societal disruption caused by developmental assistance. It begins with the following question: who pays for and who benefits from foreign aid? By calling foreign aid "reverse Robin Hood," Mason Gaffney disagrees with its redistributive logic. According to him, the idea that resources from rich countries are used to support poor countries masks the hidden problems of this exchange. Gaffney points out that donor countries acquire a significant portion of their revenues by taxing the workers. This raises the issue of the accountability of western governments if the aid money is wasted. The recipient state should also be scrutinized by asking what happens on the receiving end of foreign aid. The money often goes to the wrong hands:

> In the recipient country, rich landowners are the primarily beneficiaries. Aid raises the wages of workers only if it increases productivity on marginal land. Otherwise, it just adds to the value of high-yielding land, which is owned by a small elite.[18]

Land ownership in the developing world is characteristically heavily concentrated and guarded by arbitrary property rights systems, often managed by the governmental elites. Consider Brazil, which still represents an extreme case of land concentration. According to Human Rights Watch, 2.8 percent of landowners own over 56 percent of arable land, while 1 percent of landowners own 45 percent of all land in Brazil. At the same time, nearly five million families are landless.[19]

There are also more direct forms of social disruption caused by foreign aid. Another scholar points to the problem of policy paralysis by examining a number of African countries. First, many African governments gave up on creating effective aid oversight institutions during the 1980s. As different governmental agencies repeatedly received large amounts of aid from multiple donors, sometimes for duplicate projects, the governments of the recipient countries had no incentives to invest in effective aid management systems. Second, although developmental assistance was administered by the state (government), over time it tended to weaken the state capacity because it encouraged public sector growth and consumption. Third, foreign aid encouraged corruption by increasing the resources that governmental elites could fight over. In some countries, foreign aid had contributed to a rise in ethnic violence by inflaming the existing societal divisions.[20] To summarize: "aid reinforced the biases and prejudices of the governments it assisted and provided extra resources to elites linked to the state, thus contributing to social stratification."[21]

While some economists still argue that foreign assistance is necessary, they completely reject the *relying on the state* view of foreign aid. In doing so they tend to underestimate the role of the state. Jeffery Sachs believes, for example, that the poorest societies are stuck in a self-perpetuating poverty trap, and only a sufficient infusion of aid can allow them to break through. Sachs identifies the Big Five developmental interventions that can lift developing countries from a poverty trap and place them on the ladder of economic development: 1. agricultural inputs; 2. investment in basic health; 3. investment in education; 4. power, transport, and communication services; 5. safe drinking water and sanitation.[22] Theoretically speaking, it is an interesting proposition, but implementation concerns raise practical questions. Sachs seems to be operating on the principle that developmental assistance can bypass the state because local NGOs can deliver the right aid to a local village. However, who can ensure that every village in the country gets what's required? Should every village obtain its own adequately equipped medical clinic? Villages cannot be thought as separate from the state. How significant is local education without connecting it to job opportunities throughout the country? How long can the local off-grid diesel generator remain a reliable source of electricity? Finally, how can transport between villages and the main airports be maintained without addressing country-wide infrastructure problems?

A village is part of the state and relies on state institutions. If these institutions are weak or arbitrary, run by people who are in collusion with the governmental elites, only selective villages get access to the foreign aid and to state resources. Village-based solutions can only work if they are considered in the context of the developmental strategy of the state where the village is located. As one practitioner observes, many issues that impact individuals "originate outside of people's own communities: most trade regimes, all epidemics, and just about anything to do with climate change. Should every community be manufacturing its own vaccines or pedagogic materials or shoes?"[23] Rethinking development and foreign aid means recognizing the need for local bottom-up solutions, but also recognizing the presence of the state where these local solutions take place. This means the prerogative to localize the aid, but also to provide incentives for improving state institutions.

III "Fixing the state": struggles over structural reforms

The view that the state in the developing world has to be fixed to facilitate its integration into the globalizing economy gained momentum after the collapse of the Soviet Bloc in the early 1990s. Francis Fukuyama wrote at the time that we had reached a moment in history when democracy had won and no other political order was possible. Liberal democracy was to be aspired to by all people freed from the shackles of authoritarian nightmares of the Cold War because it was the only system that held a promise of individual liberation. Furthermore, it was believed that freedoms promised by the liberal democracy also contained permission to test the limits of free markets: "Liberalism also made possible the modern economic world by liberating desire from all constraints and acquisitiveness, and allying it to reason in the form of modern natural science."[24]

Markets do not govern themselves. Markets operate in the policy space demarcated by the leading economies, trading states, and international agreements. Good functioning economies are supported by effective inclusive institutions that took generations to build. However, the state *needs to be fixed* reforms initiated under the IMF and the World Bank structural adjustment programs were expected to fix the state straight away. Reformist changes were supposed to produce immediate positive outcomes for societies urgently seeking an improvement in political and economic conditions. In the attempt to erase the inefficiencies caused by the past interventionist policies favored by oligarchic and autocratic governments, the changes necessitated establishing new electoral systems, reforming administrative and regulatory frameworks, redesigning constitutions, laws and courts, reforming the educational and health systems, reforming governmental bureaucracy and the military, just to mention a few crucial goals. Structural reforms took different forms in various parts of the world.[25] They all, however, proved to be much more challenging than ever anticipated.

The scope of structural reforms had been steadily enlarged since 1979, when the World Bank extended its first structural adjustment loan to Senegal. Initially, such reforms were meant to stabilize the struggling economies of Africa by demanding exchange rate devaluations. Soon the demands from the donors were more comprehensive: to reduce excessive government expenditures, to privatize the major industries, and to abandon domestic subsidies and liberalize trade. By 1989, more than half of the countries in Africa were subject to structural adjustment programs in exchange for loans that became known for the "explosive growth" in explicit and detailed conditions attached to them.[26] Given the changed global environment at the end of the Cold War, even more comprehensive structural adjustment programs were adopted in the 1990s. The reforms, ideologically informed by the Washington Consensus, were advocated throughout the former communist countries and the developing world as a unified prescription for transforming these states into free market economies.

The Washington Consensus was based on the understanding that the era of centrally run economies was over and the states could be better off if they liberalized their domestic economies, opened themselves to the world, and embraced macroeconomic discipline. It was "a consensus for liberalization and globalization rather than a consensus for equitable growth and sustainable development."[27] As a result, the policy prescriptions that framed many of the structural adjustments loans of the 1990s tended to neglect the developmental aspects of the reforms.

John Williamson, an American economist, coined the term "the Washington Consensus" when he formulated a set of ten policy instruments necessary to reform post-interventionist economies in Latin America. The following list later became known as the Washington consensus:[28]

1   Fiscal discipline – to combat huge balance of payments deficits experienced by the Latin American countries;

2   Re-ordering public expenditures priorities – especially subsidies, streaming public programs and reducing government bureaucracies;

3   Tax reforms – usually making conditions more favorable for business;
4   Liberalizing interest rates – financial liberalization, freeing money transactions;
5   Competitive exchange rates – to support foreign investment, it translated into "the two corner doctrine, which holds that a country must either fix firmly or else it must float 'cleanly.'"[29]
6   Trade liberalization – opening borders to foreign trade in goods and services;
7   Liberalization of inward foreign direct investment – lowering restrictions on foreign investment but without comprehensive capital account liberalization;
8   Privatization – intended to enhance competition but often resulted in switching from a state-owned to an elite-owned monopoly;
9   Deregulation – intended to enhance foreign trade and competition but often leading to the abolition of safety and environmental regulations;
10   Enhancing property rights – intended to strengthen the sense of ownership and hence boost the country's economic climate.

The philosophy of implementing the consensus favored the idea of unrestrained free markets, with the role for the state reduced to absolute minimum. As the Washington Consensus became a blueprint for 1990s structural adjustment processes, its basic assumptions were increasingly perverted in practice. For example, in the name of fiscal discipline and public expenditure reduction, governments were compelled to undertake severe austerity cuts in publicly funded programs aimed at helping the unemployed, supporting infrastructure, and assisting local farmers and small businesses. Fearing a reduction in the level of foreign aid granted by the IMF and the World Bank, many countries introduced massive public service layoffs that often aggravated the economic problems in the country.

Tax reforms frequently missed the point of the whole package of structural reforms when tax laws were rewritten in such a way as to please the IMF and the World Bank officials but not to upset the influential business groups at home. Routinely, tax reforms only benefited the elites, at the expense of the poor and small businesses. Financial liberalization was pushed on unprepared countries that lacked relevant prudential regulations, which opened them to speculative short-term investment flows. Eventually such unregulated financial speculations were the primary cause of a massive financial crisis of 1997 in Asia. The critical economic situation was exacerbated by a quick outflow of huge amounts of foreign currencies from seven Asian Tigers in a matter of several days. The crisis caused millions of people to lose their jobs and a descent into poverty.[30]

The Washington Consensus principle of trade liberalization also pushed many poor countries to become members of the WTO, only to face massive administrative costs associated with the implementation of new far-reaching WTO agreements. The benefits of many agreements for developing countries were in question when big trading nations continued to keep their agricultural subsidies.[31] Deregulation, which came with trade liberalization, often resulted in getting rid of, or not creating new, safety and environmental standards. The push to privatization led to the hasty sell-off of vital companies and utilities to a single, and thus monopolistic buyer, time and again connected to the oligarchy in charge of the country. The idea of creating a stable investment climate was frequently linked to a requirement that developing countries establish a strong regime for the protection of IPR and provided the rationality for including the TRIPS agreement in the newly established WTO. There was no doubt that a strong global regime for protecting IPR would increase the economic strength of the leading economies, who were leaders in innovation and intellectual property ownership, but there was no clear evidence that a strong IPR protection was a significant determinant of foreign direct investment flows. The technologically disadvantaged countries feared the adverse impact of the monopolistic nature of patent rights. In addition, the TRIPS requirement to create new laws and enforce the IPR protection on a level that routinely exceeded developing countries' financial and administrative capabilities.[32]

Overall, the structural adjustment processes included comprehensive but hasty implemented reforms, which often meant a deterioration in the living standards in the affected countries. The Washington Consensus did nothing to address the persistence of corrupted elites since it mainly focused on reforming those state institutions that were to ensure the protection of private property. The situation in many reforming countries was further aggravated by growing unemployment and deepening poverty, caused by massive public sector layoffs and significant reductions in public services (health, education) conducted in the name of fiscal discipline. As real per capita income stagnated, oligarchic privatization became a norm, the increase in unemployment was followed by low levels of economic growth.

Even if there were cases when the reforms did lead to some improvement in economic conditions, the Washington Consensus era of structural adjustment programs is considered a failure from the developmental point of view. The programs did not support developmentally sensitive policies that could enhance the capacity of individuals to advance and prosper. Despite a wave of structural adjustment programs on the African continent, too many countries in contemporary Africa continue to struggle with socio-economic problems. A large burden of external debt is another factor preventing countries from investing in their future. Consider Table 4.1, it demonstrates the extent of the

Table 4.1 IMF relationship with sub-Saharan Africa, 1991–1998 (in millions of US$)

Years	1991	1992	1993	1994	1995	1996	1997	1998
IMF purchases	579	527	1,146	918	2,994	652	524	837
IMF repurchases	614	530	455	467	2,372	596	1,065	1,139
IMF charges	228	186	138	170	559	124	101	88
Balance	−263	−189	553	281	63	−68	−642	−390

Source: World Bank, Global Development Finance 1999, in Jubilee 2000 Coalition, "IMF takes $1billion in two years from Africa," April 1999.

Notes: Balance shows the net transfer of funds from the IMF to sub-Saharan Africa; the negative sign indicates a net transfer from the countries to the fund. IMF purchases represent new resources (loans) taken out from the IMF, while IMF repurchases represent repayments of the principal of IMF loans. IMF charges represent repayments of the interest on IMF loans.

problem at the height of the structural adjustment period. Regardless of the sustained foreign aid support, most of the African continent had to pay back the money it received in the first place, making the region perpetually bankrupt.

Table 4.1 shows that repayments by African governments to the IMF more often than not outpaced the resources received. Just between 1997 and 1998, Africa's debt increased by 3 percent to US$226 billion. During the same period, African countries paid back US$3.5 billion more than they borrowed in 1998.[33] This trend continues even now. In 2015, the governments of 48 African countries in sub-Saharan Africa received US$32.8 billion in loans but paid US$18 billion in debt interest and due payments, with their overall debt level still rising. Collectively, in 2015, these countries were net creditors to the rest of the world in the amount of US$41.3 billion.[34]

The one-size-fits-all model of the Washington Consensus received massive criticism. The aid agencies continued to defend their policies, arguing that partial implementation of the reforms coupled with political interference were the reason for their failures. They turned a blind eye to the fact that structural adjustment programs provided rent-seeking opportunities, and for that reason reform processes were increasingly manipulated by opportunistic governments for political advantage.[35] As the extent of the capacity problems in implementing the reforms were identified, the attention shifted to the state and its institutions. Together with this shift came the argument for the "Augmented" Washington Consensus suggested by the economist Dani Rodrik. The augmented consensus added the following ten new principles to the previous list: legal/political reforms, regulatory institutions, anticorruption, labor market flexibility, WTO agreements, financial codes and standards, "prudent" capital-account opening, no-intermediate exchange rate regimes, social safety nets, and poverty reduction.[36] These principles constitute the building blocks of a liberal democratic state organized according to the rule of law and based on inclusive institutions with the government that retains autonomy over its monetary policy while it actively fights poverty.

The Augmented Washington Consensus has become the new standard in thinking about development. Under the augmented consensus the development community still prioritized fixing the state in the developing world but with the focus on democratization. Unfortunately, these new attempts fell short in understanding what democracy entailed. Policy advisors made mistakes in conflating the concept of state autonomy with state capacity when blaming the economic troubles of African states on their low level of autonomy. Nicolas van de Walle has demonstrated that African states combine a high level of autonomy with an extremely low capacity characterized by their limited institutionalization.[37] In other words, many developing countries lack institutional arrangements that allow the process of democratization to take hold.

However, autocrats are not interested in building democratic institutions unless they are provided with incentives to do so. Autocrats are not interested in delivering public goods based on sound policies as long as they feel secured to rule by winning the loyalty of critical business and military elites. This loyalty can be ensured if the foreign aid is sent to them with a few strings attached. The fundamental challenge for the development community is to create the right political incentives for corrupt rulers to work on institutional arrangements that pave a way towards democratic accountability.[38]

Therefore, the view that democratization can be engineered from the outside understates the scale of necessary changes that come with genuine democratic transformation. It also ignores the imperative of political survival that most politicians are operating under. Free elections can be difficult to organize in countries where the past oligarchic elites feel entitled to stay in power. Electoral campaigns can disintegrate into ideological struggles with numerous parties viciously competing with each other. And even if the old elites are purged and the new government is democratically elected and committed to democracy, the situation remains very fragile. Reformist governments face enormous pressures from multiple constituencies to deliver the promised improvement quickly before the disappointed voters turn against them. Free elections are thus a necessary – but insufficient – condition of democracy.

> A democratic transition is complete when sufficient agreement has been reached about political procedures to produce an elected government, when a government comes to power that is the direct result of a free and popular vote, when this government *de facto* has the authority to generate new policies, and the executive, legislative and judicial power generated by the new democracy does not have to share power with other bodies *de jure*.[39]

Working towards good democratic governance is a work in progress, but at the very minimum the country has to demonstrate its commitment to free and democratic elections. The absence of competitive elections signals outright autocratic decision making and the presence of a centralized and state-dependent economy. Perhaps the most egregious example of such a state is Zimbabwe, where periodic elections have taken place but they were neither free nor competitive. The country has been ruled by an autocratic ruler who refused to be contested. Robert Mugabe has stayed in power by promoting the sense of fear and by employing violence. Robert Mugabe exemplifies what many other autocratic regimes share, namely "an interest in promoting uncertainty because it compels people to accept outcomes that allow only the ruler and a small band of cronies to prosper through secrecy, endogamy, and violence – defeating open, mutual endeavors by unrelated stakeholders."[40]

Robert Mugabe was born in 1924 in Kutama, Zimbabwe, when it was part of Southern Rhodesia ruled by the white supremacist government. In 1975, he waged a guerilla war against the Rhodesian government for over a decade, unifying different partisan groups under the Zimbabwe African National Union and Patriotic Front alliance. Mugabe became the independent Zimbabwe's first prime minister in 1980 and then president. As of September 2016, he was the oldest and one of the longest serving rulers in the world. He is known for his human rights violations and his authoritarian rule, which have devastated the once quite promising economy of Zimbabwe. One researcher's quest to find an explanation as to why Mugabe, a freedom fighter, became the tyrant, provided several interesting perspectives. Ultimately:

> it was Robert Mugabe's own choices that destroyed Zimbabwe. Too weak to tolerate rejection, too angry to resist revenge, he succumbed to his power lust as well as

to retribution rather than serving Zimbabwe in the best interests of the people who once idolized him.[41]

Internal causes of poverty cannot be underestimated. The problem starts with the fact that citizens of Zimbabwe are denied the right to change a very bad ruler. Consistent with the findings of Nicolas van de Walle, Zimbabwe is characterized by a high level of autonomy but a very low capacity as it is constrained by derisory state institutions.

Despite Mugabe's dreadful reputation, in 2016 Zimbabwe received foreign aid from 24 countries, most notably from Australia, United States, and the European Union.[42] Zimbabwe is both one of the most repressive and one of the most economically shattered countries in the world. Its rate of inflation peaked at 500 billion percent in 2009 before the government was able to tame it. The inflation was halted by abandoning the local currency, the Zim dollar, in favor of foreign currencies and converting bank balances to US dollars at a rate of US$1 for every 35 quadrillion Zim dollars. Alas, when in 2016 the government ran out of foreign currencies, it announced the return of the "new" Zim dollar. This move stirred fears of another hyperinflation and disintegration of the economy.[43] For many years, Zimbabwe has been characterized by violence, corruption, and persistent poverty.

Important new research examines the link between violence, corruption, and persistent poverty, and probes how they are related. Corruption is cited as one of the most persistent factors behind the economic mismanagement of a country, and the internal cause of poverty. Defined by Transparency International as "the illegal use of public office for private gain," corruption is difficult to study, since by definition it happens out of sight.[44] Recent research has begun to assess the impact of corruption by looking for signs of collusion between governments who are authoritarian, or highly centralized, and business transactions conducted by the representatives of such governments, domestically and internationally. The institutional checks and balances imbedded within the frameworks of democratic states are intended to prevent the incidence of corruption. Many developing countries lack transparent institutionalized oversight systems. Donors of foreign aid can insist on transparency when it comes to the distribution of aid. A first step is the creation of randomized program evaluations.[45] Such randomized evaluations can serve to build program specific tools to combat corruption. As part of a long-term strategy, randomized evaluations can constitute important steps towards supporting the creation of a country-wide aid management system, which in turn, can facilitate the establishment of accountable state institutions.

In their seminal work, Stephen Haggard and Robert Kaufman observe that the persistence of authoritarian regimes depends both on the overall level of development and on economic conditions in the short run: "Authoritarian regimes vary according to which segments of the population are given preference, but all are responsive to the economic interests and demands of at least some sectors of their societies."[46] Some authoritarian regimes mainly cater to the military, some to certain strategic business interests, while some are based on personal ties with the traditional elites. During the post-Cold war era, the authors observed withdrawals of many "bureaucratic-authoritarian" regimes who were organized around a military–technocrat alliance and often were present in many middle-income countries.[47] These developments demonstrated that the more an authoritarian regime claimed legitimacy on the grounds of economic performance, the more vulnerable it became to economic crises: "Whatever the nature of the underlying 'authoritarian bargains,' poor economic performance – whether the result of external shocks, bad policy, or both – means a reduction in the resources available to political elites for sustaining bases of support."[48] The demand for democratic change often emerges as a result of an economic crisis of an authoritarian state.

In the past, once the developing country experienced a major economic crisis, the IMF and the Word Bank would step in with a set of *fix the state* reforms. In one particular case, the failure of such reforms led to a major independent study of what went wrong. The study was commissioned by the Global Development Network, and the Foundation for Development Cooperation was tasked with researching the internal causes of Fiji's continuing socio-economic problems following the implementation of democratization reforms in the 1990s. The conclusion pointed out seven problematic issue areas.[49] These areas are briefly examined here. The problems epitomize the experiences of many developing countries.

1   Incomplete complementary reforms. Under trade reforms, countries are asked to reduce tariff rates. This move means less revenue so it should be supported by the effective and fair restructuring of the domestic tax system but seldom does. Furthermore, trade and investment liberalization cannot produce expected results in a country without a system of property rights protection and without a transparent regulatory framework to provide predictability for investors and traders. Incomplete reforms lead to the loss of jobs and outflow of capital, which exacerbates the existing problems.

2   Political instability and investment uncertainty. Political instability generates a risky economic environment. An uncompromising reformist push for fiscal austerity can aggravate the sense of uncertainly over the country's future. Rapid trimming of the public sector leads to massive jobs cuts, with no corresponding alternatives for employment. Such outcomes fuel social resentment and prolong instability.

3   Bad governance and waste of public sector resources. Lack of attention to creating institutional checks and balances over the governability processes leads to a non-accountable government, which continues to make politically motivated decisions that waste already limited resources. As less and less funds are available for the improvement of infrastructure, and schools and hospitals become chronically underfunded. With no money for retraining or supporting workers who have already lost jobs, the economic situation deteriorates. When social safety nets disappear, the disillusioned population is less likely to accept new reforms.

4   Timing and sequencing of reforms. This is crucial for the actual implementation of the reforms and for maintaining the level of support for the reforms process. The multi-layer reform process must involve measures offsetting the short-term adjustment regarding employment losses and dislocation of people.

5   Political economy of the stakeholders. The existing elites must be subject to competition. Here donors can play a role by demanding transparency and rules in

distributing foreign aid. Corruption must be targeted not tolerated. Self-interest can be positively channeled by offering the current stakeholders opportunities for further advancement in a stable economic system, if they act according to the rule of law.

6   Institutional rigidities and constraints to reform. The problem stems from the persistence of extractive institutions that favor the elites to the detriment of the population at large. Government bureaucracy is routinely in need of overhauling in order to become a professional merit-based institution, and not a pool of corrupted rent-seeking administrators. In many countries, the institutionalized state-trading enterprises impede the reforms by maintaining the existing elites. Incentives must be created to invite democratic institution building that promotes long-term stability for all.

7   Lack of electoral competition and policy reform. Electoral reforms have to take into account the ethnic makeup of the society to avoid the deepening of ethnic divisions. Reforms must aim at promoting democratic competition and fairness.

It is fair to assume that the IMF and the World Bank will continue to be involved in helping developing countries to overcome their economic problems. The identified areas of concern remain useful for the purpose of preventing future failures.

IV Utilizing the state: state-directed development

State intervention is seen through two perspectives, based on the available evidence. In some countries, the state became the main engine of industrialization and innovation, leading to economic growth and poverty reduction. The classical case here is China, closely followed by South Korea, Taiwan, and Singapore. In other countries, state intervention has become an obstacle to development, producing policies that failed to stimulate socio-economic advancement, while political elites kept squandering country's resources. Here, countries like Nigeria, Zimbabwe, Tunisia, and Venezuela come to mind. On this subject, there is also the convincing historical example of the former Soviet Bloc. Remembering the spectacular and decades-long growth of such countries makes it hard to ignore Hayek's warning that centrally run economies degenerate into incompetent totalitarian systems.

The economic transformations of several East Asian countries, however, also make it difficult to dismiss the virtues of state-directed development. South Korea, for example, was one of the poorer countries in the world in the 1960s. By the mid-2000s, South Korea had achieved a status of a developed economy, with an estimated GDP per capita of US$39,900 in 2016. China is another example of a successful state-directed development. Its GDP per capita at around US$14,600 in 2016 is not as high, but taking into account the country's huge population of over 1.3 billion people, China's progress has been remarkable.[50] China is now considered the second-largest economy of the world, with a strong economic growth even during the fallout years of the 2008 global financial crisis.[51] China is now one of the world's largest trading countries and a manufacturing factory of the world. Just between 1981 and 2001, the country is said to have reduced the proportion of people living in poverty from 53 percent to 8 percent. China has also established growing relationships with Africa reaching to virtually every country on the continent. By 2006, nearly 900 Chinese companies had invested in factories, agricultural sector, and infrastructure projects in Africa, with the Chinese government promising billions in future investment: an amount that already exceeded the World Bank loan commitments to Africa.[52]

When looking for causes behind successful state strategies, Atul Kohli applauds cohesive–capitalist states like South Korea, who managed to use the political power of the state in a consistent way to promote economic growth. In contrast, the institutionally disorganized and fragmented-multiclass states, such as Nigeria, are characterized by unfocused and unpredictable policy-making. The worst cases, however, are the patrimonial states such as Zimbabwe where state power has been used destructively. The concept of a patrimonial state, was coined by Max Weber. A patrimonial state is unstable since it is arbitrarily governed by self-serving rulers using force, coercion, and personal connections over the state bureaucracy. State-directed development is short-lived in a patrimonial state. In summary, a successful state intervention happens in a cohesive–capitalist state with the following description:

> At a proximate level of causation, the variety of contextual variables that might have influenced the relative success of development efforts included the availability of experienced entrepreneurs, the competence and the work ethics of labor, the capacity of the society to absorb technology, and the general levels of health and education of the populace. A set of institutional factors, moreover, ... that often proved to be consequential in the analysis ... included the security of property rights, the ability to forge binding contracts, and the availability of banks and of other institutions to mobilize savings.....
>
> The state that is created is disciplined and disciplining, has a close working alliance with capitalists, and systematically incorporates and silences those who might detract from the state's narrow goals of industrialization and rapid growth. Authoritarian control and ideological mobilization are thus generally part of the ruling strategy of such states.... While such states may not persist beyond a few decades, ruling elites are often tempted to revert to such organized forms.[53]

China fits the description of being a cohesive–capitalist country. Over its long history, China has experienced periods of instability and economic collapse. Can China remain successful and continue to grow without becoming an effective democracy? Let's briefly examine China through the lens of its historical trajectory.

China's rich history is traced back to about 4000 BC. Over the millennia, China has been both a powerful empire and a country torn by internal conflicts. The Chinese people are credited with inventions such as the seismograph, wheelbarrow, compass, paper, and even gunpowder. The famous works of a Chinese teacher and philosopher by the name of Confucius (551–479 BC) infused Chinese political culture with the values of social harmony, loyalty, and high work ethics. For most of its past, China was skeptical about developing commercial links with the West. For a long time, China was an inward-looking country, building a wall along its borders, which in the fourteenth century reached 6,700 km. During the era of colonial conquest, China tried to remain outside the sphere of western influence. Repeated offers of commercial trade by the British empire with the hope of gaining access to tea failed. These offers were accompanied by large amounts of opium, quietly channeled to China from India. The resulting colonialism by addiction led to the Opium Wars, which ended in 1842 with the, humiliating for China, Treaty of Nanking.[54] The treaty marked the beginning of a particularly gloomy period in the country's history.

The decline of China carried on into the twentieth century. After World War I, Sun-Yat-Sen became the Chinese leader and supported the idea that the collective should take priority of the individual. H.D. Long, a Chinese economist educated in the United States, further elaborated a "conscious design" by the government approach to economic development. The policies of the ruling Nationalist Party of China under Chiang-Kai-Shek were informed by the ideas promoted by H.D. Long, who was actively supported by the Rockefeller Foundation. In 1937, Japan invaded parts of China, while western powers continued to use China as the extension of their territory. The co-opted government under Chiang-Kai-Shek was becoming increasingly unpopular.[55] The situation exploded in 1949, when communists under the leadership of Mao Tse-tung took over the reins of power. The nationalist government escaped the revolution by arriving on the island of Formosa (today's Taiwan) and declaring itself the only legitimate government of China.

On the mainland, Mao announced the establishment of the Peoples' Republic of China, for which his new country became widely ostracized. The United States pursued the policy of estrangement by recognizing Taiwan as the only real China, and moved its embassy there. Mao ruled China in the name of communist principles, which in practice meant a state-run economy where no private property was allowed, farms were collectivized, and individual rights were suppressed. The *Great Leap Forward* began in 1958, as an attempt to accelerate the country's industrialization, as resources were shifted from agriculture into heavy industries. This led to one of the worst famines in history, estimated to cause the death of 45 million people.[56] In 1966, the beginning of the Cultural Revolution grew out of Mao's idea to clean China's past and create a socialist society. Supporters of the Cultural Revolution and the Red Guard for years terrorized the country by destroying all cultural artifacts, "rooting" out all capitalists and killing scores of people.

Considered illegitimate, communist China was isolated from the world until a momentous visit by America's President Nixon in 1972.[57] The visit once again changed the path of China's relationship with the West towards engagement. It also paved the way for the economic reforms under Deng Xiao Ping. In 1979, the Carter Administration enacted the Taiwan Relations Act and transferred diplomatic recognition from Taiwan to mainland China. The Act reaffirmed the One-China policy, leaving Taiwan in a legal-limbo. As the United States moved its embassy from Taipei to Beijing, the United States vowed to maintain unofficial ties with Taiwan.

The current reformation period, which began with the rise of Deng Xiao Ping to power in the late 1970s, has attempted to remedy the damage of the communist past. The key year was 1979, when Deng Xiao Ping initiated the *Four Modernizations*, comprising agriculture, industry, education, and science. The first major reforms dealt with the agricultural sector, but soon an "open door" policy was initiated with the opening of special economic zones in the coastal provinces aimed at export-oriented production. These reforms were extraordinarily successful, and have helped to transform China from one of the poorest countries in the world, to one of the most economically powerful. In 2001, China joined the WTO. The following statistics provide a glimpse into the present state of the Chinese economy:

Box 4.1 WTO statistics regarding China's economy

During the period under review (2014–2016), China's economic growth slowed down … with GDP growth rates of around 6.5%–7% a year.

China's current account surplus saw an upward trend during the review period. It totaled US$330.6 billion in 2015, and was equivalent to 3% of GDP.

Although China remains the world's largest trader (excluding intra-EU trade), trade, particularly imports, lost considerable momentum during the review period…. In 2015, both exports and imports of goods declined, with exports totaling US$2.28 trillion, down from US$2.34 trillion in 2014, and imports amounting to US$1.68 trillion, down from US$1.96 trillion in 2014. Import contraction in value terms reflects, to a considerable extent, lower oil and other commodity prices.

Manufactured products remained the dominant component of exports, accounting for slightly over 94% of the total. Among them, office machines and telecommunication equipment; and textiles and clothing continued to be China's main exports. Manufactured products accounted for 64.4% of imports in 2015. The main categories include office machines and telecommunications equipment; and chemicals. Fuels and other mining products accounted for some 21% of China's imports in 2015, while agricultural products accounted for 9.5%.

In 2015, services represented 12.3% of China's total exports and 22.9% of its imports. Exports of travel, construction, telecommunications, financial and business services were the most dynamic in the review period, while, among imports, travel services continued to gain a considerable market share, accounting for 62.3% of the total in 2015.

China remains one of the largest recipients of foreign direct investment (FDI) in the world. In 2014, FDI inflows reached US$119.6 billion, 1.7% more than those registered in 2013. The main sectors attracting FDI in 2014 were manufacturing (33.4% of the total), real estate (29%), leasing and business services (10.4%), and wholesale and retail trade (7.9%).[58]

Despite China's economic growth, numerous problems have emerged. Inequality is growing across the country and within its cities. China is also known for having one of the laxest environmental standards in the world. The environmental damage has the possibility to render much of the country uninhabitable, and health concerns are beginning to become major problems: "Beneath the surface of China's enormous accomplishments is a complicated story: The country's astonishing and exploding domestic market promise huge opportunities. But that growth also masks systemic weaknesses."[59] The main questions concerning the long-term viability of China's state-directed development tend to focus on the continuation of the uncontested one-party system of governance, which includes corruption, patronage, oppression, and an inability to deal with inefficient SOEs, environmental problems, and growing inequality.

The social consequences of the Chinese economic success have largely been masked over the past 25 years; however, they are quickly coming to the forefront. The inequality that currently exists is at its worst point in the recent history of the country.

> China's level of income inequality, as measured by the Gini coefficient, has increased sharply, from 0.28 in the early 1980s to 0.49 in 2007. By this measure, China is more unequal than Japan (0.31), India (0.34), the United States (0.36), and Russia (0.44). Estimates vary, but it is likely that around 100 million Chinese still live in absolute poverty. Meanwhile, more than one million Chinese are millionaires (even measured in dollars), flaunting their palatial mansions, private jets, and foreign luxury cars.[60]

When the country was poor, equality prevailed, but this kind of equality is not what the people desire. Coastal bias and inland poverty have emerged as major problems. The biggest problem with the inequality gap between rich and poor, however, has to do with the fact that rich are the powerful and the poor are the powerless. Since the 1989 military crackdown on pro-democracy protesters in Beijing, the government has continued to purge labor activists and critics of the regime. Yet protests are multiplying as anger spreads over a political system that allows unchecked injustice.[61] During the eighteenth congress of the Chinese Communist Party (CCP) in November 2012, Xi Jinping, for the first time, was given all three roles as head of the state, the party, and the military. Chinese President Xi Jinping oversees the main decision-making body, which is the Politburo Standing Committee, consisting of seven men. The influential party leaders are well known for their wealth. At the same time, the Hurun Report provided the following data: in 2011, the net worth of the 70 richest delegates in China's National People's Congress rose to US$89.8 billion, a gain of US$11.5 billion from 2010.[62] There is, however, some reason for optimism, given the new anticorruption campaign initiated by Xi Jinping. Yet the size of the problem is quite alarming. Within just two years of taking office, the party has punished some 270,000 members for corrupt activities that penetrated every level of the government and its administration.[63]

In terms of environmental concerns, the issue is equally as severe. Many of the problems are the result of the economic progress promoted at the expense of the environment.

> Chinese citizens are swimming in a sea of toxic pollutants. Many rivers and lakes are unusable even for irrigation; air quality is dangerously unhealthy; and desertification, erosion, and dust storms are depleting arable lands. Public costs can be measured in early death, chronic disease, lost productivity, and diminished quality of life.[64]

Such environmental issues are a serious threat to China's long-term socio-economic sustainability. In terms of economic concerns, many problems are structural and they exist because
of the role the central government plays within the country's economy. Issues of corruption
continue to frustrate many foreign investors, and what makes this even more significant is
that the Chinese judiciary is not known for its independence. The four state-owned banks
have dominated the Chinese financial system, since the government severely limits any
international competition. There is very little risk-assessment, and the state-owned banks
are known to freely lend money to a party member, or a state run business, with few strings
attached. Furthermore, to cope with the impact of the 2008 global financial crisis, the government encouraged banks to lend. This has resulted in a massive accumulation of debt:
"China's total debt in 2007, counting that of private households, independent firms, and
government institutions, equaled 158 percent of the country's GDP. In 2014, it reached 282
percent of GDP – among the world's highest levels for a major economy."[65]

Another issue is China's corporate debt, which now stands at around 170 percent of
GDP. It is the highest corporate debt in the world, and as such it threatens global financial
stability. The situation is particularly problematic since three-quarters of China's corporate debt belongs to inefficient SOEs.[66] The Chinese state has subsidized and shielded
these firms from foreign competition to ensure that they continue to employ a large
section of the population.

> On the demand side, with the old players (SOEs) and the old economy (heavy indus-
> try) in trouble, Beijing has opted for an old instrument (credit growth) to rescue them.
> Over the last couple of years, the People's Bank of China has injected a huge amount
> of liquidity into the economy, with credit growing at around 13 percent annually.
> This led to a housing boom that has, in turn, boosted demand for commodities, con-
> struction equipment, and materials – sectors dominated by state-controlled actors.
> According to the IMF, it could take over three years to work off housing inventory in
> China's smallest cities, even absent new residential construction.[67]

There are approximately 155,000 SOEs in China and they continue to be a drain on the
economy. The closure of unproductive firms would lead to significant job losses, so the
government is reluctant to address the problem.

The future can still belong to China. After all, China has accumulated huge foreign
exchange reserves, partially managed by the China Investment Corporation. This sovereign wealth fund has been making some spectacular purchases around the world, to the
tune of billions of dollars. Most recently it made the US$13.7 billion acquisition of
Logicor, a European warehouse and logistics company.[68] Critics, however, worry that,
without democratization reforms, the Chinese style state-directed development is not sustainable. It has been only three decades since China started experiencing spectacular economic growth, and already the inadequate provisions of social services, severity of
environmental problems, and the government patronage promotion schemes are threatening the country's stability. Frustrated with procedural injustices, arbitrary decision
making, and the corruption of government officials, people find it difficult to tolerate
growing inequality. In the words of one observer:

Since 1989, the CCP has not adopted any genuine political reforms, relying on high growth rates to maintain its rule. This strategy can work only when the economy is booming – something Beijing cannot take for granted.... It would be far better for the political system to change gradually and in a controlled manner, rather than through a violent revolution.[69]

Eric Li, a Shanghai-based entrepreneur, maintains, however, that the debate over the democratization of China is over. The CCP possesses self-correcting qualities, which will allow it to become even more creative and progressive in advancing China's development.

The country's leaders will consolidate the one-party model and, in the process, challenge the West's conventional wisdom about political development and the inevitable march toward electoral democracy. In the capital of the Middle Kingdom, the world might witness the birth of a post-democratic future.[70]

If China manages to adapt its socio-economic system to the changing domestic demands, address the existing problems, and withstand global pressures, this would be a very successful strategy indeed. The new President, Xi Jinping, has promised national revival, the eradication of government corruption, and a great future for all citizens. He has also put in motion a massive infrastructure program called the One Belt, One Road initiative in the attempt to create a new Silk Trade Road connecting China with multiple countries through the Asian continent to Europe.

If successful, the ambitious program would make China a principal economic and diplomatic force in Eurasian integration. One Belt, One Road calls for increased diplomatic coordination, standardized and linked trade facilities, free trade zones and other trade facilitation policies, financial integration promoting the renminbi, and people-to-people cultural education programs throughout nations in Asia, Europe, the Middle East, and Africa.[71]

The future will show whether this will be sufficient to generate a sustainable economic growth, reduce inequality, and make people content, despite having limited rights and curtailed freedoms.

So far, a number of countries that have pursued state-directed development without complementary democratic reforms have turned increasingly authoritarian, with limited successes to stimulate egalitarian economic growth throughout the country. Russia is a good example here. Although Russia has reasserted itself on the global stage and became an influential petrostate,[72] its model of economic development has yet to produce a prosperous and open society. The promised democratic reforms were never consolidated in Russia, instead the political power became increasingly consolidated vertically.

Centralized economies with unfinished democratic reforms can fall into severe crisis under the weight of mismanagement and fraud. Take Brazil, for example, where the government never fully implemented promised electoral and land reforms. Given the awkward sequencing of reforms – failure to openly privatize the SOEs first and conduct an early land reform – the elites retained a considerable amount of power, forcing the government to be politically motivated in its decisions in order to survive.[73] Mismanagement and corruption led to a growing waste of public sector resources. For example, billions were spent on major sports events (the World Cup in 2014 and Summer Olympics in 2016), while there was no money spent on essential infrastructure projects. As the economic situation deteriorated, the political instability damaged Brazil's investment climate. The policy space stagnated, political instability increased, the extractive institutions persisted, and the reforms stalled. As a result, in 2009, the Brazilian government still controlled 115 firms, with assets worth around US$756.8 billion, which included Petrobras, a leading oil company, and a number of powerful banks. Apart from having over a 58 percent share of votes, the government officials occupied seven out of nine seats on the Petrobras board of directors, and every major investment project had be approved by the federal government.[74] In 2014, Petrobras became engulfed in a huge corruption scandal that involved governmental interference, tainted contracts, and vast mismanagement of its revenues.[75] In 2016, Brazil's problems exploded in a national crisis, with the president forced to resign and the economy further deteriorating.

Both approaches to the state in the developing world previously discussed in this chapter – *relying on the state* and *fixing the state* – were rooted in the strategies conceived by the development community. These strategies, in turn, were informed by the prevailing theoretical paradigms of the time, with a focus on modernization theory. The fundamental idea was to help the poor countries of the Global South develop and modernize in the image projected by the industrialized economies. Another imperative was to accelerate the integration of developing countries within the international economic networks. The policy prescriptions of the IMF and the World Bank became increasingly intrusive and demanding over the years, culminating in vastly misguided reform programs ideologically driven by the Washington Consensus.

State-directed development, on the other hand, emphasizes an indigenous effort by the government to stimulate and provide rules for a growth oriented economy. The East Asian states, especially China, South Korea, Taiwan, Singapore, Hong Kong, and Japan, have done exceptionally well by pursuing this model. Some of them, however, achieved success at a high cost. China's past experiments during the Mao years left many societal scars. The galloping progress of the South Korean economy also came with high costs:

> Rapid industrialization occurred within the framework of a highly authoritarian state characterized by overtones of fascism, a state that I have labeled cohesive–capitalist. Political dissent was not tolerated. many opposing the regime were repressed, and labor was corporatized and state-controlled. Income, wealth, and power inequalities in the society also became more skewed, especially starting in the 1970s, as the government deliberately encouraged economic concentration in the hands of big chaebols.[76]

South Korea continued to have a weak banking system into the 1990s, making it very dependent on foreign financial flows. These problems were exposed during the 1997

Asian financial crisis, including high debt to equity ratios. Following the crisis, South Korea introduced a series of reforms aimed at assuring greater transparency of its investment regime. The chaebols are highly concentrated corporate business groups, such as Samsung or Hyundai, habitually controlled by the funding families, and still dominate the economic landscape of South Korea. The country managed to overcome the financial crisis to resume a modest but steady rate of economic growth.

Most importantly, South Korea offers a refreshing example of a previously highly authoritarian country slowly transforming into an effective democracy. The peaceful removal of the former president from office on corruption charges normally would be seen as a sign of trouble. In the case of South Korea, however, such an event should be viewed as the evidence that the country's democratic institutions provide the necessary checks and balances. In March of 2017, South Korea's Constitutional Court determined that President Park Guen-hye had violated the constitution by releasing state secrets while colluding with private corporations. The court unanimously upheld Park Guen-hye's impeachment, declaring that she had "seriously impaired the spirit of democracy and the rules of law." This was an important moment in South Korea's history. New democratic laws and institutions put in place in the late 1980s worked as they were designed.

> Observers around the world have hailed South Koreans' impressive display of peaceful resistance as millions of citizens consistently gathered to demand Park's removal and the restoration of the rule of law. It is certainly true that the country, often referred to as the "Republic of Protests," has mastered the art and politics of peaceful mobilization.[77]

It was true that Park's departure had divided the country. Younger Koreans and pro-democracy activists wanted her to go. She also had many supporters, mostly older conservatives, who felt loyalty to her, possibly because of her father, former President Park Chung-hee, who ruled the country from 1961 until his assassination in 1979.[78]

The late Park Chung-hee is credited with pushing through the most critical phase of Korean industrialization reforms. He consolidated the cohesive–capitalist state and "engineered a growth-oriented alliance of state and capital, recorporatized labor, and used economic nationalism to exhort the entire society into the service of economic advancement."[79] His state-directed model continued to operate into the 1980s to become the consistent political driver for reform. As the Korean economy expanded, more and more liberalizing reforms took place. Ironically, those reforms eventually ended the political career of Park's daughter.

In a spirit of renewal, only two months after the impeachment, in May 2017, South Koreans decisively elected progressive liberal candidate Moon Jae-in. He will be closely watched by those who believe that genuine efforts to democratize the country may be the only option capable of sustaining a long-term thriving path of state-directed development. If Moon Jae-in is successful in uniting the country, respecting the democratic constitution, and continuing to steer South Korea on the path of sustainable economic progress, the county will prove that the authoritarian state can transform itself. It will also show that the state can be a powerful political driver of socio-economic progress leading to democratization.

Notes

1 Lawrence Freedman (2013) *Strategy – A History*, Oxford, UK: Oxford University Press, p. 242.
2 Ibid.
3 Hilton L. Root (2006) *Capital and Collusion – The Political Logic of Global Economic Development*, Princeton, NJ: Princeton University Press, p. 6.
4 Vivien A. Schmidt (2009) "Putting the Political Back into Political Economy by Bringing the State Back in yet Again," *World Politics*, Vol. 61, No. 3, p. 516.
5 Shaohua Chen and Martin Ravallion (2007) *Absolute Poverty Measures for the Developing World, 1981–2004*, Washington, DC: The PNAS Organization, October 23.
6 PPP means Purchasing Power Parity, which is an economic concept that equalizes countries' purchasing powers by calculating it in US dollars.
7 Shaohua Chen and Martin Ravallion (2010) "The Developing World is Poorer than we Thought, But No Less Successful in the Fight against Poverty," *Quarterly Journal of Economics*, November, p. 1611.
8 UNDP (2016) *Human Development Report 2016 – Human Development for Everyone*, New York, NY: UNDP, p. 29.
9 Harry S. Truman (1949) *Inaugural Address*, January 20, 1949. Online, available at: www.trumanlibrary.org.
10 William Easterly (2006) *The White Man's Burden: Why the West's Efforts to Aid the Rest Have Done So Much Ill and So Little Good*, New York, NY: Penguin Press p. 4.
11 Nicolas van de Walle (2001) *African Economies and the Politics of Permanent Crisis, 1979–1999*, Cambridge, UK: Cambridge University Press, p. 189.
12 OECD (2014) *2014 Global Outlook on Aid*, Paris, France: OECD.
13 *The Economist* (2016) "Where Does Foreign Aid Go?" August 10.
14 Muhammad Yunus with Alan Jolis (2001) *Banker to the Poor – The Autobiography*, Dhaka, Bangladesh: The University Press Limited, p. 14.
15 Ibid., p. 15.
16 William Easterly (2007) "Was Development Assistance a Mistake?," *American Economic Review*, Vol. 97, No. 2, pp. 328–332.
17 Ibid., p. 331.
18 Mason Gaffney (2009) "Foreign Aid: Reverse Robin Hood," in Clifford W. Cobb and Philippe Diaz (eds.), *Why Global Poverty? – A Companion Guide to the Film "The End of Poverty,"* New York, NY: Robert Schalkenbach Foundation and Cinema Libre Studio, p. 170.
19 USAID (2010) *Brazil – Country Profile* (Property Rights and Resources Governance).
20 Nicolas van de Walle (2001) *African Economies and the Politics of Permanent Crisis*, op. cit., pp. 188–210.
21 Ibid., p. 210.
22 Jeffrey Sachs (2005) *The End of Poverty – Economic Possibilities for Our Time*, New York, NY: Penguin Press, pp. 233–234.
23 Paul Farmer (2013) "Rethinking Foreign Aid – Five Ways to Improve Development Assistance," *Foreign Affairs*, December 12. Online, available at: www.foreignaffairs.com/.
24 Francis Fukuyama (1992) *The End of History and the Last Man?*, New York, NY: The Free Press, p. 333.
25 William J. Baumol, Robert E. Litan, and Carl J. Schramm (2007) *Good Capitalism, Bad Capitalism and the Economics of Growth and Prosperity*, New Haven, CT: Yale University Press, pp. 153–174.
26 Nicolas van de Walle (2001) *African Economies and the Politics of Permanent Crisis*, op. cit., pp. 214–215.
27 Narcis Serra and Joseph E. Stiglitz (eds.) (2008) *The Washington Consensus Reconsidered – Towards a New Global Governance*. Electronic Edition.
28 John Williamson (2008) "A Short Story of the Washington Consensus," in Narcis Serra and Joseph E. Stiglitz (eds.) (2008) *The Washington Consensus Reconsidered – Towards a New Global Governance*, op. cit., pp. 16–17.
29 Ibid., p. 17.

30 Jan Aart Scholte (2005) *Globalization – A Critical Introduction* (2nd edition), New York, NY: Palgrave/Macmillan.

31 Nayan Chanda (2007) *Bound Together – How Traders, Preachers, Adventurers, and Warriors Shaped Globalization*, New Haven, CT: Yale University Press, pp. 272–273.

32 Keith E. Maskus (2000) *Intellectual Property Rights in the Global Economy*, Washington, DC: Institute for International Economics.

33 Robert Naiman and Neil Watkins (1999) *A Survey of the Impacts of IMF Structural Adjustment in Africa: Growth, Social Spending, and Debt Relief*, Washington, DC: Center for Economic and Policy Research, April. Online, available at: www.cepr.net.

34 Global Justice Now (2017) *Honest Accounts 2017 – How the World Profits from Africa's Wealth*, Report, Health-Poverty-Action, May. Online, available at: www.healthpovertyaction.org/.

35 Nicolas van de Walle (2001) *African Economies and the Politics of Permanent Crisis*, op. cit., pp. 11–14.

36 Dani Rodrik (2001) *The Global Governance of Trade As if Development Really Mattered*, New York, NY: UNDP, October.

37 Nicolas van de Walle (2001) *African Economies and the Politics of Permanent Crisis*, op. cit., pp. 20–63.

38 Hilton L. Root (2006) *Capital and Collusion*, op. cit., p. 47.

39 Juan J. Linz and Alfred Stepan (1996) *Problems of Democratic Transitions and Consolidation – Southern Europe, South America, and Post-Communist Europe*, Baltimore, MD: The Johns Hopkins University Press, p. 4.

40 Hilton L. Root (2006) *Capital and Collusion*, op. cit., p. 5.

41 Heidi Holland (2008) *Dinner with Mugabe – The Untold Story of a Freedom Fighter who became a Tyrant*, New York, NY: The Penguin Group, p. 216.

42 *The Economist* (2016) "Where does Foreign Aid Go?," August 10.

43 *The Economist* (2016) "Zimbabwe's New Currency – Who Wants to be a Trillionaire?," May 14.

44 Raymond Fisman and Edward Miguel (2008) *Economic Gangsters – Corruption, Violence, and the Poverty of Nations*, Princeton, NJ: Princeton University Press, p. 18.

45 Ibid., pp. 192–202.

46 Stephen Haggard and Robert Kaufman (1996) *The Political Economy of Democratic Transitions*, Princeton, NJ: Princeton University Press, p. 28.

47 Ibid., p. 29.

48 Ibid., p. 29.

49 Mahendra Reddy, Biman C. Prasad, Pramendra Sharma, Sunia Vosikata, Ron Duncan (2004) "Understanding Reform in Fiji," Study Commissioned by the Global Development Network (GDN) (Washington, DC) and the Foundation for Development Cooperation (Brisbane), May.

50 Central Intelligence Agency (2017) *World Factbook – South Korea*, Washington, DC: US Government – CIA. Online, available at: www.cia.gov/library/publications/the-world-factbook/.

51 WTO (2012) *China Trade Policy Review* (WT/TPR/S/264), May 8.

52 Deborah Brautigam (2009) *The Dragon's Gift – The Real Story of China in Africa*, Oxford, UK: Oxford University Press, pp. 2–9.

53 Atul Kohli (2005) *State-Directed Development – Political Power and Industrialization in the Global Periphery*, New York, NY: Cambridge University Press, pp. 418–419.

54 W. Travis Hanes III and Frank Sanello (2002) *The Opium Wars – The Addiction of One Empire and the Corruption of Another*, Naperville, IL: Source Books.

55 William Easterly (2013) *The Tyranny of Experts: Economists, Dictators, and the Forgotten Rights of the Poor*, New York, NY: Basic Books, pp. 47–70.

56 Frank Dikötter (2011) *Mao's Great Famine: The History of China's Most Devastating Catastrophe, 1958–1962*, London, UK: Bloomsbury.

57 Margaret MacMillan (2006) *Nixon in China – The Week that Changed the World*, Toronto, Canada: Viking Canada.

58 WTO (2016) *China Trade Policy Review* (WT/TPR/S/342), June 15.

59 Kenneth Lieberthal and Geoffrey Lieberthal (2003) "China Tomorrow: The Great Transition," *Harvard Business Review*, Vol. 81, No. 12, p. 70.

60 Martin King Whyte (2013) "China Needs Justice, Not Equality – How to Calm the Middle Kingdom," *Foreign Affairs – Snapshot*, May 5.
61 Ibid.
62 Bloomberg (2013) "Report – China," *Bloomberg News*. Online, available at: www.bloomberg.com, November 8.
63 James Leung (2015) "Xi's Corruption Crackdown – How Bribery and Graft Threaten the Chinese Dream," *Foreign Affairs*, Vol. 94, No. 3, pp. 32–38.
64 Judith Shapiro (2016) *China's Environmental Challenges*, Cambridge, UK: Polity Press. Electronic Edition. Location 3530 of 4411.
65 Zhiwu Chen (2015) "China's Dangerous Debt – Why the Economy Could Be Headed for Trouble," *Foreign Affairs*, Vol. 94, No. 3, p. 13.
66 Edoardo Campanella (2017) "Beijing's Debt Dilemma: Why China's Bubble is a Threat to the Global Economy," *Foreign Affairs – Snapshot*, June 29.
67 Ibid.
68 Claire Milhench (2017) "Sovereign Funds' Corporate Acquisitions Triple in Second Quarter of 2017," *Reuters*, July 3.
69 Yasheng Huang (2013) "Democratize or Die – Why China's Communists Face Reform or Revolution," *Foreign Affairs*, Vol. 92, No. 1. Electronic Edition.
70 Eric X. Li (2013) "The Life of the Party – The Post-Democratic Future Begins in China," *Foreign Affairs*, Vol. 92, No. 1. Electronic Edition.
71 Jacob Stokes (2015) "China's Road Rules – Beijing Looks West toward Eurasian Integration," *Foreign Affairs – Snapshot*, April 19.
72 Marshall I. Goldman (2008) *Petrostate – Putin, Power, and the New Russia*, New York, NY: Oxford University Press.
73 Hernan F. Gomez Bruera (2013) *Lula, the Workers' Party and the Governability Dilemma in Brazil*, London, UK: Routledge, pp. 157–160.
74 Aldo Musacchio and Sergio G. Lazzarini (2014) *Reinventing State Capitalism – Leviathan in Business, Brazil and Beyond*, Cambridge, MA: Harvard University Press. Electronic Edition, Location 1515 of 5528.
75 Bello (2015) "Whose Oil in Brazil?," *The Economist*, February 14.
76 Atul Kohli (2005) *State-Directed Development*, op. cit., p. 122.
77 Katharine H.S. Moon (2017) "Park Leaves Behind a Divided South Korea," *Foreign Affairs – Snapshot*, March 17.
78 Carter J. Eckert (2017) "South Korea's Break with the Past – The End of the Long Park Chung-hee Era," *Foreign Affairs – Snapshot*, May 11.
79 Atul Kohli (2005) *State-Directed Development*, op. cit., p. 123.

Suggested further reading

Daron Acemoglu and James A. Robinson (2006) *Economic Origins of Dictatorship and Democracy*, New York, NY: Cambridge University Press.
Eva Bellin (2002) *Stalled Democracy: Capital, Labor, and the Paradox of State-Sponsored Development*, Ithaca, NY: Cornell University Press.
Domingo Felipe Cavallo and Sonia Cavallo Runde (2016) *Argentina's Economic Reforms of the 1990s in Contemporary and Historical Perspective*, London, UK: Routledge.
Alexander Cooley and John Heathershaw (2017) *Dictators Without Borders: Power and Money in Central Asia*, New Haven, CT: Yale University Press.
Hernando De Soto (2000) *The Mystery of Capital – Why Capitalism Triumphs in the West and Fails Everywhere Else*, New York, NY: Basic Books.

Kwadwo Konadu-Agyemang (2017) *IMF and World Bank Sponsored Structural Adjustment Programs in Africa: Ghana's Experience, 1983–1999*, London, UK: Routledge.

Mariana Mazzucato (2013) *The Entrepreneurial State: Debunking Public vs. Private Sector Myths*, New York, NY: Public Affairs.

Aldo Musacchio and Sergio G. Lazzarini (2014) *Reinventing State Capitalism – Leviathan in Business, Brazil and Beyond*, Cambridge, MA: Harvard University Press.

Armed conflicts, violence, and development

5

I Conceptualizing armed conflict

In contrast to the past, when armed conflicts were dominated by large-scale wars between countries and empires, over the last few decades growing numbers of people have been killed in violent struggles taking place within states. Internal armed conflicts tend to persist over a prolonged period of time. Such enduring warfare is exacerbating poverty and consuming the developmental potential of societies. Often considered civil wars or ethnically motivated conflicts, internal wars routinely escape the scrutiny of major media outlets. In actuality, violent internal conflicts have been overshadowing the international landscape for decades. Between 1989 and 2014, as many as 135 of all 144 active armed conflicts around the world were waged within states. With time, internal wars frequently spill beyond the state, fueling regional animosities. As a result, the involvement of external actors has been rising, with 27 percent of armed conflicts displaying this tendency in 2013.[1]

The renowned Correlates of War Project defines civil war as:

> any armed conflict that involved: (1) military action internal to the metropole of the state system member; (2) the active participation of the national government; (3) effective resistance by both sides; and (4) a total of at least 1,000 battle-deaths during each year of the war.[2]

John Burton distinguishes between armed conflicts (such as civil and ethnic wars) and management problems or disputes that can be resolved by compromise or settlement and by arbitration or negotiation. Armed conflicts are considered as:

> conflicts that occur through the pursuit and denial of human needs – recognition, identity, and developmental needs generally – in relation to which there cannot be compromise. Conflicts are situations or processes in the course of which persons are prepared to sacrifice themselves as martyrs, become "terrorists" or even starve themselves to death.[3]

Despite these definitions, some military operations around the world are difficult to classify. The destruction that follows civil wars has a lasting impact on societies, resulting from the spread of devastation and loss of human lives.

Most conflicts display unique local characteristics, which elude broader analytical patterns. Civil wars are fought over at least one of the following reasons: resources, contested territory, or past grievances. Statistically speaking prolonged civil wars are more likely to occur in low-income countries with a chronically troubled economy, and which is dependent on primary commodities.[4] Many present-day conflicts revisit the wounds that persist from the time of colonial violence or years of totalitarian occupation. The multilayered grievances and competing agendas expressed in most conflicts make it impossible, however, to isolate clear causes of today's internal wars. Factors that originated in the past, such as ethnic polarization, structural inequalities, and manipulated indigenous political processes matter, but the present pursuit of economic and political objectives by rebel groups and their leaders are also important. In addition, geo-political pressures and external actors can influence the local conflict to the extent that it is difficult to desegregate "the global-local nexus, and capture the exact nature of external–internal connections and how they relate to each other."[5] In summary, each conflict compels a careful examination of historical legacies and current developments from the three levels of analysis: individual, state, and international, but no single factor determines the outcomes.

With the spread of terrorism, the refugee crisis, and uncertainties about the future course of foreign policy in the United States and the EU, the world appears to be entering its most dangerous chapter in decades.[6] In 2017, all 10 dangerous conflicts in the world are considered internal, although several of them have already attracted international involvement. Armed conflicts in Syria, Iraq, Afghanistan, and Ukraine have divided the international community over alliances and the preferred type of response. Conflicts in Turkey, Yemen, Greater Sahel and Lake Chad basin, Congo, and Myanmar are having a hugely destabilizing regional effect. Regional instability is also a growing problem in South Sudan and Mexico. All these conflicts are consuming lives and destroying communities. The violence has resulted in waves of international migrants, who are targeted for recruitment or emigration purposes by armed gangs of traffickers and corrupt officials. While failed states continue to be exploited by militant groups and terrorist organizations, nationalistic populism is on the rise worldwide. We also observe the consolidation of authoritarian regimes in the Middle East and elsewhere. In this perilous environment, the Geneva Conventions are progressively more ineffective in their role to protect civilians and regulate the conduct of war. International law leaves many questions unanswered with respect to how some of the armed operations around the world should be classified and what kind of response from the international community they warrant.[7] Attempts to secure cooperation are viewed with skepticism, although this is the time when meaningful international collaboration is particularly wanted.

> Many world leaders claim that the way out of deepening divisions is to unite around the shared goal of fighting terrorism. But that is an illusion: Terrorism is just a tactic, and fighting a tactic cannot define a strategy. Jihadi groups exploit wars and

state collapse to consolidate power, and they thrive on chaos. In the end, what the international system really needs is a strategy of conflict prevention that shores up, in an inclusive way, the states that are its building blocks.[8]

II Conflicts in theory and practice: resisting simplification

The realist school of international relations maintains that war is inevitable simply because violence is a fundamental characteristic of nature into which all human beings were born. The classical political thinker, Thomas Hobbes, believed this to be a natural human condition. Hobbes experienced the turmoil preceding the Civil War in England and fled to France in 1640, where he remained until 1651. In his most famous work, *Leviathan*, he argued that people were naturally wicked and should not be trusted to govern. According to Hobbes, all people are naturally in a state of war since they are opportunistic and distrustful and crave power.

Not everybody agreed with Hobbes. On the other side of intellectual debate, liberal thinkers maintained that people could create peaceful societal order and avoid violent conflict through deliberate action. In his 1795 essay entitled "Perpetual Peace," Immanuel Kant, a German philosopher from Prussia, argued that it is our destiny, as humans, to achieve peace because nature through giving us reason has willed it. Kant claimed that by creating a system of "republican states" – what we would call democratic states today – humanity could eliminate wars. Such a system of states would enter into a federation governed by a constitution where the equal rights of all members would be secured.[9] The United Nations can be thought of as an embryonic form of Kant's vision.

Those interested in conflict prevention support the "democratic peace theory," which maintains that democratic states do not engage in wars with each other and that democracies display the least amounts of internal and external violence.[10] The theory is based on two important assumptions. First, democracies require the consent of the majority of a state's population before entering into violent conflicts abroad, which is often very difficult to obtain. At the very least the democratic government needs to take into account its citizens' wishes if it wants to be reelected. Whereas in authoritarian countries, a single ruler or a handful of elites can either force the population into a war, if the society has no interest in fighting, or can convince individuals through the control of information and the media that going to war is necessary. The second assumption is that while disagreements persist in all societies, such discords are much more likely to be channeled into peaceful political competition within democracies.

Democratic peace theory is consistent with the human needs approach to conflict prevention and conflict resolution:

> there are some human needs, such as those of individual and group identity and recognition that will be pursued regardless of costs and consequences. They are ontological and not within the control of the individual or identity group. The only effective control of the pursuit of such needs is another need, that of valued relationships. Only if there are valued relationships with authorities – that is to say

legitimized relationships – can there be social harmony at any societal level from the family to the international.[11]

Democratic peace theory extends its reach towards promoting democratization processes in the name of peacebuilding. It is believed that only in a well-functioning democracy can social harmony happen and primal human needs be fulfilled, hence removing any need for violence. Proper democratic governance would include the processes of ongoing societal cooperation:

> the ultimate challenge is the establishment of social and political institutions that are problem solving, and not adversarial or confrontational.... What are sought are leaders who do not have a defined political program which they seek to promote, but capabilities not unlike those of a facilitator whose prime function is to bring different view-points and interests together, and to help an analysis that can suggest constructive outcomes.[12]

Democratization continues to be viewed as a preferable option for both conflict prevention and conflict resolution. Democratic processes have to emerge organically within the country. It would be difficult to find a convincing case where intervention by external forces successfully instituted democratization aimed at peacebuilding. Author Roland Paris notes that the 1990s were especially hospitable to international missions to restore peace to war-torn countries that

> were guided by a generally unstated but widely accepted theory of conflict management: the notion that promoting "liberalization" in countries that had recently experienced civil war would help to create the conditions for a stable and lasting peace. In the political realm liberalization means democratization.[13]

These peacebuilding efforts turned to be largely problematic in trying to create working liberal-market democracies at an accelerated rate. There were 14 such post-conflict peacebuilding missions: Namibia (1989), Nicaragua (1989), Angola (1991), Cambodia (1991), El Salvador (1991), Mozambique (1992), Liberia (1993), Rwanda (1993), Bosnia (1995), Croatia (1995), Guatemala (1997) East Timor (1999), Kosovo (1999), and Sierra Leone (1999). The results were mixed in terms of economic progress, but the missions were hardly successful in resolving internal conflicts. Unfortunately, "international efforts to transform war-shattered states have, in a number of cases, inadvertently exacerbated societal tensions or reproduced conditions that historically fueled violence in these countries."[14] The most miserably failed missions were in Angola, Liberia, and Rwanda.

Democracy is particularly difficult to adopt in a country whose history shows signs of deep societal divisions. The case in point here is Rwanda, where hostilities between two ethnic groups festered over several generations. The calamity of Rwanda, however, should not be hastily explained as an ethnic conflict. Upon closer examination, multiple causes emerge. Tensions intensified as the country started to experience massive economic problems, which, in turn, were manipulated by political leaders and aggravated by external geopolitical factors. A divided society became overwhelmed by multiple challenges and eventually exploded in the tragedy of genocide.

The tragedy of Rwanda – a case study

From early April to mid-July 1994, the central African country of Rwanda witnessed a genocide, initiated by one of its indigenous ethnic groups, the Hutu, against another, the Tutsi. In just 100 days, approximately one million ethnic Tutsi were killed in massacres committed by Hutu militias and elements of the army with the participation of the civilian population. Genocide, a term that borrows from the Latin words *genus*, meaning "race" or "kind," and *cid*, meaning "kill" or "cut," involves the "intent to destroy, in whole or in part, a national, ethnical, racial or religious group."[15] The killing, although organized with the extensive help of the state, was largely carried out by the average Rwandan with machetes or other unsophisticated weapons. Ethnic fears and hatreds, already simmering below the surface from years of civil war, were played upon by elites, primarily via radio broadcasts to induce the population to violence.[16]

A look into the country's history offers a partial explanation behind the ethnic animosities that persisted over generations. The story begins 400 years ago with the migration of Tutsi cattle herders to the area that is now Rwanda. Although Hutu farmers had been already living off the land there for some time and outnumbered their new neighbors, the Tutsis' wealth allowed them to establish a monarchy to rule over the region. For hundreds of years Hutu and Tutsi shared a single language, religion, and culture, and intermarriage between groups had been very common. Despite the lack of primordial differences, however, the socially constructed class differences between the groups were hierarchical and very real. The Tutsi retained greater political and economic power than their Hutu counterparts throughout this period.[17]

The colonial powers reinforced the existing ethnic hierarchy. In 1895, Rwanda became a German colony, but following the German defeat in World War I, Belgium took over. Both German and Belgian colonizers favored the Tutsis, believing they were racially superior, and used "modern scientific methods" to determine who was a member of which group, including preposterous tests such as head and nose measurements. Tutsis were given the opportunity of education, and they received the best jobs in the colonial bureaucracy. In 1926, the Belgians introduced administrative reforms that proved to be decisive in shaping Rwanda's socio-ethnic relations. Traditional Tutsi chiefs were used to implement forced labor policies and collect high taxes. Tutsi officials were also tasked with administering corporal punishment on behalf of the colonial masters. The objective was to instill a climate of fear and uncertainty and to keep the society ethically divided as a means of maintaining political control.[18] Until 1959, while under Belgian colonial rule, virtually all the chiefs and community leaders that dominated Rwandan political and economic life were Tutsi.[19] The Belgian colonizers ordered that all Rwandans must carry identification indicating their ethnicity – a project that further served to encourage ethnic polarization.

In the late 1950s, as in many parts of the world, Rwandans started to rise up against their colonial masters. Striving to maintain their dominance, the Belgians once again exploited the ethnic divisions and supported a Hutu intelligentsia in their successful 1959 attempt to overthrow the Tutsi monarchy to replace it with a presidential republic. This power shift, also called the Hutu or Social Revolution, was accompanied by the killing of hundreds of Tutsis and the eviction of many more from their homes. Some tens of thousands of Tutsis fled to neighboring countries.[20] Subsequently, a radical Hutu government was elected to rule Rwanda. Its members tended to blame the Tutsi for most of the country's problems and started to institutionalize Tutsi subordination. Following a few years of growing marginalization, the Tutsi tried to organize resistance against the Hutu ruling regime from Burundi and Uganda, but their guerilla tactics continued to be defeated. The Hutu government retaliated with waves of country-wide repressions, resulting in the killing of up to 30,000 Tutsi and more than 100,000 escaping across the borders.[21] Rwanda became an independent country in 1962, with the government firmly in Hutu hands.

For several decades, ethnic hostility was kept relatively subdued, but in the late 1980s tensions rose when Rwanda was hit by a combination of financial and agricultural crises. Rwanda's main agricultural crop, responsible for the majority of farmers' incomes and a significant portion of government revenues, was coffee. According to a 1993 study, the government of Rwanda earned between 60 and 80 percent of its foreign currency from export taxes on coffee.[22] From 1969 to 1981, the government introduced policies to push coffee production everywhere at all costs. For example, it mandated coffee production on government sponsored ranches, while it prohibited the removal of coffee plants to replace them with another crop. As a result, coffee production grew on average 4.4 percent, but it also made Rwanda's economy vulnerable to any change in price or demand. At the time, however, world coffee prices were fairly predictable and the Rwandan government felt secure.

The International Coffee Agreement (ICA) helped stabilize world prices through an established system of quotas allocated to each producer. In 1987, growing pressure from major international coffee traders intensified differences of opinion among the signatories of the ICA. The news that the ICA was about to collapse depressed world prices for coffee. Immediately, Fonds d'égalisation (Rwanda's stabilization fund), which purchased coffee from domestic farmers at a fixed price, started to accumulate a considerable debt. A serious blow to Rwanda's economy came in June 1989, when the United States indeed suspended its support for the ICA and talks reached a deadlock. The ICA system of quotas was abolished and world coffee prices plunged in a matter of months by more than 50 percent.

The falling coffee prices reduced incomes across the country and threatened the economic security of a large portion of the population when famines erupted throughout the Rwandan countryside. The government faced serious financial problems as export earnings declined by 50 percent between 1987 and 1991. According to the World Bank, Rwanda's GDP per capita declined from 0.4 percent in 1981–1986 to −5.5 percent in the period immediately following the slump of the coffee market (1987–1991).[23] The World Bank had been assisting Rwanda for decades before this crisis. Its officials consistently concluded that the country was making successful socio-economic progress. One report noted that the government was rightly concerned about the rural population and hence pursued policies of prudent, sound, and realistic management.[24] The World Bank 1986 report on Rwanda even applauds the "social and cultural cohesion of its people."[25] This is particularly puzzling because even during the best of times the country was not unified or socially cohesive. The manufacture of such positive images of Rwanda continued until the genocide. It is possible that it helped to mask the growing animosities between Hutu and Tutsis.

The Rwandan government was repeatedly praised in the World Bank reports, even when the bank's own evaluations of its Rwandan projects identified problems with corruption and mismanagement. This positive image helped Rwanda to receive a large amount of foreign aid. In the 1989–1990 period, official development aid accounted for 11.4 percent of Rwanda's GNP, one of the highest in Africa. According to the World Bank, foreign assistance financed over 70 percent of public investment between 1982 and 1987. Overall, the size of foreign aid increased from an average of US$45 per person in the 1980s to US$80 per person in the 1990s. In summary, Rwanda was one of the most externally supported countries in the world.[26]

Nevertheless, the coffee economy provided vital revenues. When it collapsed and famines became widespread in the country, a World Bank mission was dispatched to Rwanda in 1988 to deal with the crisis. The government resisted introducing the World Bank/IMF reforms package for years, fearing their impact, but finally agreed in exchange for financial assistance.[27] The mission prescribed a set of reforms, namely: "trade liberalization and currency devaluation, alongside the lifting of all subsidies to agriculture, the phasing out of the Fonds d'égalisation (the State coffee Stabilization Fund), the privatization of state enterprises and the dismissal of civil servants."[28] Some of these reforms were implemented in November 1990, only weeks after the incursion from Uganda of the rebel army known as the Rwandan Patriotic Front (in French, the Front Patriotique Rwandais, FPR). The timing of the imposed structural reforms could not have been worse. The devaluation of the currency eroded salaries, jobs in the public sector became exposed, and the government continued to accumulate debt as it increased military spending on fighting the FPR. At the same time, severe food shortages again hit Rwanda. The shortages were aggravated by the deepening agricultural crisis resulting from a combination of factors such as climatic fluctuation, land degradation, and lack of crop diversification. The incoming civil war compounded the agricultural crisis.[29]

The civil war formally started with the invasion by the FPR. This rebel army was officially set up in 1987, but its origins date to 1979 when Tutsi intellectuals in exile in Uganda formed a party bent on returning to Rwanda and regaining control of the state. In October 1990, driven by past and ongoing grievances, the military wing of the FPR entered Rwanda, leading to a civil war that lasted three-and-a-half-years. The rebel army was small but experienced, and included many of the former Tutsi refugees. The army was largely funded by the western nations and claimed to be on an "eight-point plan" mission that "spoke of 'national unity' and the elimination of corruption, placing the return of refugees as only a fifth objective."[30] The corruption charge targeted Habyarimana, who had ruled Rwanda since 1973, easily winning subsequent "elections" by a 99 percent margin, since he was running unopposed.

The invasion by the FPR was pushed back by the Rwandan army, but the rebels managed to gain control of the northeast part of the country, presenting a permanent threat to the government. When this happened, the ruling clique under the leadership of President Habyarimana decided to radicalize racist prejudice. The FPR was described as the primary threat, but all Tutsis living inside the country were labelled as dangerous. On October 4, 1990, the army staged a series of shooting incidents in the capital Kigali, which resulted in the imprisonment of some 10,000 Tutsis. Financial and moral support started to pour from the government to extremist newspapers and radio stations, where Tutsis were constantly the subject of hateful propaganda. As the civil war unfolded, thousands of Tutsis were killed in targeted massacres organized by local and national officials in collaboration with the army and the police.[31] In radicalizing the population, the government kept openly talking about the killing of between 100,000 and 200,000 Hutu in Burundi in 1972 by the Tutsi-dominated army.[32] On the other side of the conflict, the FPR rebels continued to engage in guerilla attacks.

The research shows that both sides in this civil war were financed by the very same donor institutions. Although most of the arms purchases were negotiated privately and outside the aid framework agreement, the record shows that arms became part of government expenditures, and as such part of the state budget monitored by the World Bank. The files collected from the National Bank of Rwanda demonstrate that Habyarimana's ruling party (MRND) used money obtained from the donors to import approximately one million machetes, later used during the genocide. Machetes were imported through various channels, including Radio Mille Collines, an organization linked to the radical anti-Tutsi Interhamwe militia, known for fomenting ethnic hatred.[33]

The fighting officially ended in August of 1993, and a United Nations peacekeeping force was established by October later that year. The pressure came from the international community, which having embraced a paradigm of post-conflict liberalization, demanded that the Habyarimana government enter into a power-sharing agreement with the FPR. The complicated process, known as the Arusha peace negotiations, proceeded slowly, as it threatened to deprive the Habyarimana clique of their total control of the Rwandan state. In the background of the peace negotiations, the hateful propaganda targeting Tutsis continued. During this time, a series of critical decisions were made by senior members of the ruling party, Éduard Karemera and Matthieu Ngirumpatse, to turn young people affiliated with the Interahamwe into proficient soldiers. In its origins, the Interahamwe were a paramilitary group of young men, loosely supported by the Hutu-governing party of Rwanda. In 1993, the nature of this support changed, when the Interahamwe started receiving direct financing and a large number of weapons. Dallaire, in his book, describes how the decisions made by individuals at the top level of the party prepared the ground for the genocide:

> the Interahamwe were being trained at army bases and by army instructors in several locations around the country, and on a weekly basis a number of young men would be collected and transported for a three-week weapons and paramilitary training course that placed special emphasis on killing techniques. Then the young men were returned to their communes and ordered to make lists of Tutsi and await the call to arms.[34]

When, on April 6, 1994 Habyarimana was killed in a plane crash on the way back from talks in Arusha, the situation quickly deteriorated. Violence first erupted in Kigali, instigated by the army, the presidential guards, and the militia, but it soon spread to the rest of the country, in the orgy of killing that consumed hundreds of thousands of defenseless people.[35]

The UN's peacekeeping mission turned out to be a fiasco. The UN's articulated goal was to monitor the peaceful transition to democracy, believed to be conducted in parallel to the processes of ongoing economic liberalization pushed by the IMF and the World Bank. From the perspective of these external actors, the dual reforms were expected to promote socio-economic stability in the country, but in the end they turned out to be tragically misplaced.[36]

The UN mission was not adequately equipped to prevent the resurgence of war, which led to the 1994 genocide, and most of its personnel quickly withdrew from the country when hostilities broke out. A small group of peacekeepers under the command of General Roméo Dallaire remained in Rwanda and hopelessly attempted to stop the genocide as the UN mandate prohibited them from employing force to protect civilians. When the UN Security Council finally approved a second peacekeeping mission to Rwanda, the governments were sadly reluctant to act. A frustrated witness to the tragedy, Dallaire later condemned the red tape politics of the UN and the penny-pinching resources of the mission, concluding that the root of the problem was: "the fundamental indifference of the world community to the plight of seven to eight million black Africans in a tiny country that had no strategic or resource value to any of the world power."[37] In June 1994, France launched a UN sanctioned mission to the region – Operation Turquoise – to protect civilians. It was limited and controversial. The civil war ended later that year, with an FPR victory paving the way to a Tutsi government.[38]

In 1995 Rwanda's external creditors entered talks with the Tutsi dominated new government regarding the debts of the former regime. The talks also concerned the money used to finance the massacres. The new government recognized that these "odious debts" were legitimate. In 1998, at a special donors' meeting in Stockholm, a multilateral trust fund of US$55.2 million was set up under the banner of post-war reconstruction. The money was destined to service Rwanda's "odious debts" with the World Bank (i.e., IDA debt), the African Development Bank, and the International Fund for Agricultural Development.[39]

Violent armed conflicts escape neat theoretical explanations. Every conflict displays a distinct cobweb of factors deemed to be responsible for its outbreak. Though there is one poignant connection: all armed conflicts cause long-lasting psychological, social, and economic damage. Developmental policy grounded in the human-centered approach can make a positive difference if it works to dismantle the state structures that preserve a climate of intolerance and sustain social polarization. In summarizing his findings about the causes of the 1994 tragedy in Rwanda, Peter Uvin concluded:[40]

> Rwanda's genocide was the extreme outcome of the failure of a development model that was based on ethnic, regional, and social exclusion, that increased deprivation, humiliation, and vulnerability of the poor; that allowed state-instigated

racism and discrimination to continue unabated; that was top-down and authoritarian; and that left the masses uninformed, uneducated, and unable to resist orders
and slogans. It was also the failure of a practice of development cooperation based
on ethnic amnesia, technocracy, and political blindness.

We cannot forget, however, that there were individuals on both sides whose decisions
played an important role in the conflict. These people made plans, they trained the
fighters, and they organized the operations. They also intensified ethnic and cultural differences. As of December 2015, the International Criminal Tribunal for Rwanda (ICTR)
indicted 93 of them for crimes against humanity. The Rwanda genocide occurred over
two decades ago, but its ghosts were never put to rest. No theory predicted the tragedy
and no theory can fully explain it. In trying to understand the events that took place in
Rwanda, we hope to gain insights into other civil wars whose causes resist simplification.

III Questions about the importance of culture

Going against the tide of the post-Cold War globalization momentum, Samuel Huntington published an article entitled "The Clash of Civilizations," which quickly became the
basis for a new fairly influential theory of conflict. It argues that geo-political transformations, the reaction to economic globalization, and competition for resources are vital in
explaining the growing importance of cultural identity. Under Huntington's line of reasoning, many contemporary armed conflicts have been given cultural dimensions. In fact,
the argument is made that culture will become the most significant factor as a source of
future wars. He declares that:[41]

> [The] fundamental source of conflict in this new world will not be primarily ideo-
> logical or primarily economic. The great divisions among humankind and the dom-
> inating source of conflict will be cultural. Nation states will remain the most
> powerful actors in world affairs, but the principal conflicts of global politics will
> occur between nations and groups of different civilizations.

Huntington anticipated that countries with similar cultural features will group together
with other "kin" countries to confront other cultural groupings. Huntington breaks down
these cultural groupings into eight separate entities: Sinic (Chinese/other common communities in Southeast Asia, including Vietnam and Korea), Japanese, Hindu, Islamic,
Orthodox (largely Russian), Western (Europe, North America, Australia, and New
Zealand), Latin American, and (possibly) African. He suggests that it is difficult to talk
about a distinct African civilization, given the multiplicity of cultures that exist there.[42]
Huntington sees the West currently as the most dominant of all civilizations; however,
he predicts that the emergence of new powerful countries will cause the power of the
West relative to that of other civilizations to decline. The two main challengers to
western dominance are believed to be the Islamic and Asian cultures. The Islamic challenge stems from its "pervasive cultural, social, and political resurgence of Islam in the

Muslim world and the accompanying rejection of western values and institutions."[43] The strength of this resurgence is based upon social mobilization within Islamic societies and their sustained population growth. The Asian challenge takes its strength from the successful economic performance of Asian countries, which emphasize the superiority of their cultural values. These two cultures are expected to be on a collision course with western civilization for years to come, competing over economic resources and global influence.[44]

Huntington presents a very interesting argument, although many scholars disagree with his predictions. The first major controversy and rebuttal to his theory can be found through a rejection of the primacy of culture. To many scholars, the world is still dominated by states, not cultures or civilizations. This contra-argument rests upon realist's assumptions that states are rational and self-interested entities that are concerned with their own progress and not with uncertain and unpredictable cultural relationships. Fouad Ajami, for example, criticizes the importance of cultural argument by stating that: "Huntington would have nations battle for civilizational ties and fidelities when they would rather scramble for their market shares, learn how to compete in a merciless world economy, provide jobs, move out of poverty."[45] This critique is based on the understanding that it is not the civilizations that influence, or even control, the state, yet rather it is the state that influences and controls the civilizations. Consequently, the governments have a big role to play in unifying their populations, despite the ethnic differences among their people.

The second criticism of the "clash of civilizations" thesis was expressed by a group of scholars who conducted quantitative analysis of armed conflicts fought during a given period of time. The results appeared to contradict the thesis, since: "Of the 50 most serious ethno-political conflicts being fought in 1993–1994, only 18 fell across Huntington's civilizational divides."[46] Sierra Leone, a country that suffered a civil war between 1991 and 2002 is also not on a fault line between civilizations. The conflict in Sierra Leone was believed to have largely agrarian roots, and the issues that led to violence related to "coercive forms of labor exploitation once associated with domestic slavery."[47]

However, one can argue the thesis has more to do with predictions for the future than with an analysis of a number of past conflicts over a limited period of time. On this point, the events of September 11, 2001 come to mind. It was on this day that a coordinated terrorist attack in the United States organized by a fundamentalist Muslim group took place. These developments led many scholars to re-think Huntington's argument. While for some, the 9/11 events provided terrifying evidence that the world has indeed entered a period of conflicts between civilizations, others were quick to point out the economic and political dimension of terrorism, on the one hand, and the purely irrational dimension, on the other.

The final critique of Huntington accuses him of promoting cultural essentialism. In all the cases of civil wars involving distinct religious groups, one can identify a number of other factors that contributed to the eruptions and continuation of violence. In addition, it appears that the critical conflict is taking place within Islamic civilization itself. Since 2014, the self-declared Islamic State has attempted to conquer a territory across parts of Libya, western Iraq and Syria in the name of civilizational mission. In essence, Islamic

State is a fundamental militant Sunni movement that rejects not only western values but also more liberal forms of Islam and other branches of Islam, most notably Shia Muslims. Islamic State has tried to establish a caliphate with exclusive political and theological authority over the world's Muslims. In reality, its actions have been characterized by unending violence and little institution building.[48] The movement grew on the ruins of a failed state in Iraq in the years following the US invasion of 2003. Islamic State continues to mobilize young people around the world and to pose threats worldwide, as recent attacks in the United Kingdom and France have demonstrated.[49] On June 7, 2017, suicide bombers and gunmen attacked the Iranian parliament and the holy shrine of Ayatollah Khomeini in Tehran, killing scores of people. It was an unprecedented attack by Islamic State on the predominantly Shia Muslim country.[50] To a certain degree, the culturally based violence instigated by Islamic State fits well within Huntington's theory. However, in rejecting the legitimacy of Shia Muslims and violently targeting them, Islamic State escapes an easy theoretical fit within Huntington's thesis.

To sum up, no matter if we draw the connection between religiously motivated terrorist attacks and the ideas first formulated by Huntington, or rather if we reject the explanatory value of his theory, his work remains important. Huntington's significance stems from recognizing that cultural identity is becoming one of the defining characteristics of the modern era. Despite the criticism that he ignores the prominent role of the state as a rational, self-interested entity, he compels us to think beyond clearly delineated borders inside murky international politics increasingly influenced by complex cultural beliefs.

What sets the West apart in this new international reality is the separation between the state and the religion that relegated the secret realm into the private sphere. In contrast, the ascendency of civilizations based on cultural values justifies the central role of religion in a country's political life. Huntington predicted that the cultural nature of conflict would define the way societies interpret relations among nations. One can argue that consistent with this prediction several newly elected governments – from Poland to India – used religious overtones to secure power. President Putin of Russia is a great supporter of the Russian Orthodox Church. The Church and its affiliated organizations was the biggest recipient of presidential grants between 2013 and 2015, totaling more than 256 million rubles. In 2008, as many as 72 percent of Russians identified themselves as Christian Orthodox, compared to 31 percent in 1991. Putin's imperial ambitions are widening.[51] While the Kremlin keeps supporting rebels in the ongoing civil war in Ukraine and defending the Assad regime in Syria, its rhetoric against the West has acquired a cultural flavor. Strengthened by the new legislative developments (the 2012 Foreign Agents Law, the 2013 Law Protecting Religious Feelings, and the 2015 Undesirable Organizations Law), the Kremlin strives to expose western civilization as permissive, corrupted, and too individualistic. In short, the attitudes of western civilization are considered incompatible with Russian values.[52] Critics worry how the Russian Orthodox Church mobilizes religion for political purposes, while promoting conformity and patriotism in coordination with government propaganda.[53]

The misunderstanding regarding cultural factors relating to sectarian and ethnic identities weighed heavily on the course of operations in Iraq after the 2003 US invasion.

Cultural factors played a major role in the disintegration of Iraq. The internal culture of armed forces is increasingly recognized as influencing the military strategy of both sides of conflicts. The need for exploring culture's relationship to armed conflict is clear. Two recent books on the subject argue that policy makers and military strategists must understand culture's impact on warfare, or be faced with dire consequences of their ignorance.[54]

To heal broken societies and consolidate peace, several countries introduced Truth and Reconciliation Commissions. Perhaps the best known was in South Africa. Launched in 1995, it attempted to bring about a national reconciliation by revealing the truth about human rights violations during the apartheid period. Since then, there have been more than 20 such commissions around the world, for example, in Sierra Leone, Sri Lanka, Colombia, and Rwanda. These commissions are viewed as a means to reclaim a sense of community in the post-conflict societies. The work done by them has been mostly well received, but the commissions have also been criticized for having limited mandates and for not implementing suggested recommendations. Most importantly, cultural psychologists raised doubts about the commissions' practices being based on western and Christian concepts. The commissions operate on the assumption that there is an objective truth, while stressing openness, and confession. These practices often clash with indigenous cultural values and with the subjective personal needs that appreciate privacy. Furthermore, when a commission asks participants to re-live the conflict by confronting evidence and listening to testimonies, the process can prolong the experience of trauma. Collective memory practices that take clues from liberation and cultural psychology may be more helpful in healing damaged communities.[55]

IV Igniting conflicts: multiple causes of violence

Huntington's hypothesis is somehow similar to the idea that *ethnicity* is the basis for most contemporary conflicts. The argument is generally stated as follows: different ethnic groups, who are differentiated by customs and culture, and who usually share the same geographical region, often develop an exclusive identity based on an intense dislike for one another, which sometimes leads to violence.

> Markers of ethnic identity in terms of language and networks of "home" also figure prominently, both in the self-perception of groups of young people and in their identification of "others" within the community. This can sometimes find expression in quite stereotyped prejudice.[56]

When coupled with religious and linguistic differences, this mutual prejudice can be relentless and long lasting. It has been suggested that the tightly enforced measures of the Cold War era prevented such ethnic animosities from leading to bloodshed. This can explain why, once it was over, ethnic differences long kept under the lid of Cold War politics were allowed to manifest themselves, leading to increased ethnic violence. An example often cited in support of this argument is that of the former Yugoslavia.[57]

This southern European country remained peaceful and fairly prosperous following World War II. Its political leader, Josif Broz Tito, ruled the country peacefully until his death in 1980, despite the fact that the country consisted of an ethnically diverse society. Six respective nations in particular had distinct histories that pre-dated the creation of the Yugoslavian state: Serbia, Croatia, Bosnia, Slovenia, Montenegro, and Macedonia. Many of which were comprised of a combination of Roman Catholic, Eastern Orthodox, and Muslim populations. The unifying factor was a threat of Soviet invasion. The argument goes that once the Cold War was over and the Soviet Union collapsed, Yugoslavia descended into devastating and ethnically divided civil war.

Compounded by a severe economic crisis in the late 1980s, which opposing groups blamed on each other, ethnically different nations began to declare themselves independent in 1991. Rising tensions and attempts by the central government to keep the country together eventually led to an explosion of violence.[58] The devastating conflict engulfed the whole region for almost a decade. In one particularly appalling battle in July 1995, described as an outburst of genocide, Serb forces overran the predominantly Muslim city and UN "safe-area" of Srebrenica. Some 30 Dutch peacekeepers were captured and immobilized while soldiers systematically killed 8,000 local men over a three-day period.[59] Although the consensus seems to be that most of the violence in Bosnia was carried out by Serbs, it is important to note that members of all ethnic groups in the region committed atrocities. Former Yugoslav president and Serbian politician, Slobodan Milosevic, who ruled the remnants of Yugoslavia between 1997 and 2000, was arrested by the newly elected government in 2001 and flown to The Hague to face charges of genocide and war crimes in front of the International Criminal Tribunal for the Former Yugoslavia. With over 300,000 people killed, it was alleged that Milosevic and his cronies aggravated the Yugoslavian conflicts by exaggerating cultural and ethnic differences and prejudices.[60]

Ethnicity as an explanatory variable behind armed conflicts was used when previously colonized nations attempted to gain their independence.[61] The impulse towards national self-determination justified the rejection of "threatening strangers" that deepened the ethnic fragmentation of many nations. The initial trigger could had been the need for retribution stemming from long festered grievances. In addition, competing claims over territory, resulting from the artificiality of state borders drawn both by colonizers and Cold War regimes – especially in Africa, Eastern, Central Europe, and Central Asia – made the situation worse.

The ethnic warfare hypothesis is accused of having several flawed assumptions. The first takes for granted that there is something unchangeable and primordial about ethnicity. The research from political psychology contradicts this view. Ethnicity is said to be socially constructed rather than biologically rooted, and the "symbols, myths and memories" shared by an ethnic group can and do change over time.[62] Political psychology insists that ethnicity is a human invention, which, in turn, can be manipulated by sinister humans for political purposes.

This is not to say that ethnicity is not an important motivator for violence; however, it seems that the idea of ethnicity is quite adaptable and matters much less because ethnicity is not based on tangible differences but rather on the fact that people *believe* in these differences. Another mistake is to assume that the ethnic hatreds of today have ancient roots. In fact, most contemporary ethnic antagonisms are surprisingly modern. In most cases, warring ethnic groups had been living among each other in peace for centuries, until governmental elites have created the conditions for violence by psychologically manipulating ethnic myths and fears.[63]

A growing number of scholars have embraced this idea that ethnicity is a socially constructed category, and have taken the concept one step further. Ann Hironaka, for example, believes that the vast majority of wars since 1945, despite frequent claims that they were initiated by ethnic grievances, actually have no objective ethnic basis. She asserts that attachments to and identification with "ethnic group," "nationality," and even "state" in some cases have simply been *imagined*, or in other words, socially manufactured and exploited by cynical military and political leaders. Rwanda serves as a fairly good illustration of this phenomenon. More often than not, the post-colonial state did not allow its population to form attachments at the national level, rather than imagining membership in a sub-national ethnic group. Hironaka thus believes that the lack of strong institutions is one of the most important contributors to civil war, claiming it to be a factor much more important than ethnic rifts. A weak state, she argues, would be unable to create the conditions necessary for economic development, will not provide an outlet for the peaceful management of conflict, and will not protect minority rights, all of which can lead to violence. She notes that institutionally strong states, such as Canada and the United States, host many ethnic groups yet do not suffer from internal ethnic conflicts.[64]

Another explanation talks about economic motivations behind armed conflicts. Poor countries are by definition characterized by great difficulties for the average person to meet basic economic needs. Inefficient farming practices, scarcity of usable land, underdeveloped transportation infrastructure, high population density, social divisions, corruption, and mismanagement can all lead to shortages of the resources that people need. Oftentimes the natural resources are available in abundance, yet certain systemic conditions such as those mentioned above, prevent the population at large from benefiting from them. In such environments, the desire to acquire wealth by exercising control over profitable natural resources prompts "violent entrepreneurs" to initiate conflicts. Despite the appearance to the contrary, many internal conflicts are not stimulated by ethnic hatred or reflect an ongoing battle between government and rebels. Rather they are stimulated by the promise of economic benefits.

> What is usually considered to be the most basic of military objectives in war – that is, defeating the enemy in battle – has been replaced by economically driven interests in continued fighting ... the image of war as a contest has sometimes come to serve as a smokescreen for the emergence of a wartime political economy from which rebels and even the government (and government-affiliated groups) may be benefiting. As a result of these benefits, some parties may be more anxious to prolong a war than to win it.[65]

Past grievances are exploited and other reasons are commonly manufactured to attract new recruits. By skillfully exploiting past and manufactured injustices, rebel groups are able to enlarge their membership more quickly. Violence can expedite the predatory goals of a rebel army or a militia group by serving several useful economic functions.

David Keen has identified seven such short-term economic functions of violence, which are worth reproducing here:[66]

1   *Pillage and looting*, which are probably the easiest ways of obtaining wealth in weak states;

2   *Extortion*, specifically collecting "protection money" from those who these groups have the power to hurt;

3   *Control of trade*, especially manipulating the supply of resources to increase their price;

4   *The exploitation of labor*, usually involving forcing people to accept low wages or no wages (slavery) with the threat of violence;

5   *Controlling land directly*, in order to exploit its natural resources (which often involves depopulation of the area);

6   *Benefiting from foreign aid*, not in the traditional sense that it is sent to the neediest to meet basic needs, but rather that it is interdicted and then sold to local populations;

7   Wars also may be initiated or prolonged in hopes that the central government or an outside power will *increase funding to military services*.

The presence of any number of these motivators, in combination or singly, may trigger violence and will certainly prolong any violent conflict once it has begun. The devastating effect of such economically motivated violence on the development process is unambiguous.

Economic greed and conflict in Angola – a case study

During the 1990s, the southern African country of Angola found itself by all measures in severe crisis. Four decades of internal war placed the state near the bottom of the 2002 United Nations' HDI, being ranked in position 161 out of 173 nations. The statistics spoke for themselves, since almost 30 percent of children died before they reached the age of six, nearly half were underweight, and a third had no formal educational opportunities whatsoever. Two-thirds of adults lived on less than US$1 per day, 42 percent were illiterate, drinking water was unsafe, health services were non-existent, and food shortages were a fact of daily life. Accordingly, average life expectancy in the country was a meager 47 years. In addition, nearly half a century of civil war had turned four million people, or almost one-third of Angola's population, into refugees, with one million of these relying on foreign food aid to survive.[67]

This extreme development fiasco would not be so perplexing were it not for the state's immense wealth of natural resources. Angola is a large country, boasting more than twice the geographical size of France, but most importantly it is endowed with significant oil, diamond, and gold reserves. Its production of oil quadrupled to about 800,000 barrels per day in the 1980s and 1990s, making it the second largest oil producer in sub-Saharan

Africa. During the same period, Angola became the world's fifth largest producer of non-industrial diamonds.[68]

Angola's war began in 1961, with its struggle to gain independence from Portugal, which it officially won in 1975. Almost from the time the first shot was fired, the country became a Cold War battleground, with the superpowers and their allies supporting one or two out of the three pro-independence parties as the war went on. From the beginning, the Soviet Union, Cuba, and China supported the MPLA (Popular Movement for the Liberation of Angola) with weapons, troops, and training. MPLA was a leftist party, which came to form the country's first official (albeit weak) government in 1975. The United States and South Africa, fearful of the spread of communism on the continent, began supporting rebels, who eventually united under the banner of the National Union for the Total Independence of Angola (UNITA). Early skirmishes between the opposing groups included a South African assault on MPLA forces in 1965. Assistance from both sides of the Cold War divide continued to pour to both sides of the conflict for decades until a 1988 agreement, partially brokered by the UN, allowed for the withdrawal of Cuban and South African troops from the country. By this time, the ideological battle between western democracy and communism had long since ceased to function as motivation for perpetuating the violence in Angola. The desire to control Angolan oil, diamond, and gold reserves became the main rationale for MPLA and UNITA actions in the late 1980s. In fact, it seems likely that economic greed functioned as these rival groups' driving force for much of their post-independence war.

In addition to its lengthy duration, the Angolan war stands out because the UN sent three peacekeeping missions to the country, all of which were unsuccessful. The first United Nations Angola Verification Mission (UNAVEM I) was created in 1988 as a complement to that year's ceasefire, and was tasked with verifying the withdrawal of Cuban forces. It was hoped that with the removal of external support the war would end, but the war continued. Mission UNAVEM II, sent to Angola in 1991, was to monitor a ceasefire and to aid the peace process, but because it was inadequately supported, full-scale war resumed, causing UN personnel to withdraw. The third UNAVEM of 1995 also ended in failure, despite being given a more comprehensive mandate and greater support.[69]

The persistence of war even in the face of repeated UN peacekeeping missions can only be explained by the fact that both the MPLA government and UNITA rebels were benefiting from the war and had no desire to bring it to a close. Most researchers of the Angolan war agree that the leaders of both parties were using the country's natural resources primarily for their own personal gain. The military elites on both sides would channel significant portions of revenues acquired from the sale of diamonds, gold, and oil to their families and loyal friends. The rest of the income was used to purchase even more military hardware, with the goal of capturing a greater share of productive land. The process in essence became a self-reinforcing cycle. It must be noted that the incentives to loot the country were great: between 1992 and 1998 UNITA received US$3.7 billion from its diamond sales, while the MPLA squeezed between US$2 and US$3 billion per year out of its oil sales during the same period.[70]

It appears that neither ethnicity nor culture were the dominant factors behind the long war in Angola. The main motive for prolonging the war was the personal economic gain of the Angolan elites. It was a rational objective of individuals whose already high stakes in winning the war were raised by the external actors. Neither government nor rebels would have had the means to continue the violence if the international market for their goods had not existed. Foreign investors played a large part in exploiting the country's diamond and oil reserves. Even after a 1998 embargo on the export of UNITA diamonds, they continued to be sold on international markets. Simultaneously, arms continued to flow from abroad to the rebels, despite a 1993 prohibition on these sales. Arguably, some of the world's largest oil firms, including Chevron, Elf Aquitaine, BP, and ExxonMobil, operating in Angola were complicit in perpetuating the war because they provided the necessary revenues. Moreover, it was alleged that much of the nearly US$900 million in payments made to Angola's coffers by these companies was used to buy arms. In the late 1990s the companies paid to secure exploration and production rights in offshore regions of Angola.[71] The war ended in March 2002, after the UNITA's leader was killed in a battle. His name was Jonas Savimbi and it was reported that by the early 1990s he had accumulated a fortune of approximately US$4 billion from UNITA's control of Angola's diamonds. He used the money to re-launch the civil war.[72] The 2002 peace treaty saw UNITA giving up arms and assuming the role of an official opposition.

The war left Angola's economy in disarray. To finally take advantage of its rich natural resources – gold, diamonds, extensive forests, Atlantic fisheries, and large oil deposits – Angola introduced limited political and economic reforms. Over the last decade, the economy started to grow, and the political situation stabilized after the 2008 elections and the introduction of a new constitution in 2010. Despite these developments, Angola remains one of the poorest countries in the world. The country is ruled by the MPLA under the leadership of Jose Eduardo dos Santos, who has been in power since 1979. *Forbes* magazine ranked his daughter Isabel as the richest woman in the African continent, with a fortune worth around US$3 billion. In 2016, Isabel was appointed by her father to become the chief executive of the state-owned oil company Sonangol. According to Africa and the World News, more than half of its population lives in poverty. The country's life expectancy is also one of the lowest in the world.[73]

Economic reasons behind armed conflicts tend to emphasize structural conditions as crucial in shaping the decisions undertaken by political leaders. The abundance of strategic resources or precious metals and stones in countries otherwise known for systemic poverty led to the term "resource curse," which Richard Auty coined when he observed that, paradoxically, many resource-rich countries are not doing well and are poorer than less endowed countries.[74] The curse hypothesis is rooted in the Dutch Disease phenomenon, first observed in early seventeenth-century Netherlands during the Dutch Tulip Bulb Market Bubble. The disease happens when a sudden rise in the value of a particular product leads to massive currency appreciation and a shift in the domestic priorities towards production of this product. The massive flows of revenues from oil, for example, can make a government reliant, and ultimately dependent, on oil revenues to the detriment of other industries. Apart from distorting the economic growth of the country, the "resource curse" can have significant political consequences:

One of the greatest risks concerns the emergence of what political scientists call "rent-seeking behavior." Especially in the case of natural resource, a gap – commonly referred to as an economic *rent* – exists between the value of that resource and the costs of extracting it. In such cases, individuals, be they private sector actors or politicians, have incentives to use political mechanisms to capture these rents.[75]

In poor countries, opportunities for rent-seeking by state-owned corporations, government officials, and rebel armies intensify existing social tensions. Hostilities between groups competing for access to critical resources often explode in violent conflicts. Despite being generously endowed, Angola represents an extreme case of squandered opportunities, which invites the "resource curse" hypothesis. Civil wars are more likely to erupt in oil-endowed states because at stake is the control of the whole apparatus of the state in charge of the oil revenues.[76] Geo-economic considerations also play a role. In an institutionally weak but resource-rich country, internal conflict can be triggered by surging international demand for its resources. As the global demand increases the inflow of revenues, the conflict escalates and more weapons are bought to fight over the control of the resources.[77]

There is another explanation behind internal wars. It centers on the nexus connecting development, aid, and conflict. The scholar who coined the argument examined the processes behind the genocidal conflict in Rwanda. In conclusion, he implicated the aid agencies in shaping multiple processes responsible for the eruption of violent conflicts. It is argued that a narrow economic-technical view of development promoted by the aid agencies ignores human rights violations, persistent income inequality, social exclusion, and targeted humiliation of selective ethnic groups, and by doing so it perpetuates the conditions that foster resentment and facilitate structural violence. Furthermore, by functioning along top-down externally defined parameters, the development aid system hampers peoples' own capacities, overlooks their needs, and deprives them of the sense of value. Consequently, "it should not be surprising that people will resist these reductionist schemes – whether by being passive, obnoxious, fundamentalist, cynical, racist, or violent."[78] In this context, the price of misguided projects sponsored by the development community takes on an added significance. Economic crisis raises stakes for autocrats who are supported by foreign aid to stay in power, even if the price of staying in power is a violent war.

V Escaping the conflict trap

The conflict trap signifies structural factors, which underlay a country's proneness to civil war. Empirical research by Paul Collier shows that:

> certain economic conditions make a country prone to civil war, and how, once conflict has started, the cycle of violence becomes a trap from which it is difficult to escape.[79]

These economic conditions are: poverty, economic stagnation, dependence on primary commodities. The risk of conflict is different in every case, depending on specific economic characteristics, which in turn are affected by specificity of every conflict.[80] Collier downplays the significance of other factors such as historical grievances and ethnic differences, but he recognizes that opportunistic politicians and disgruntled rebels like to capitalize on existing divisions to inflame the violence.

The list of the most troublesome conflicts of 2017 includes at least four countries that appear to be stuck in the conflict trap: Afghanistan, South Sudan, Congo, and Chad. We talked about the chronically disordered Afghanistan before. The 2001 peacebuilding Bonn Conference, after the removal of the Taliban government with the help of western intervention, did not deliver stability. As early as 2003, the Taliban reemerged by launching an unending insurgency against the government and its western allies. As of 2016, the country continues to fall apart, with civilian casualties rising, the Taliban regaining influence, and Islamic State targeting Shia Muslims to incite sectarian violence.[81]

The war in Southern Sudan was among the cited examples of cross-civilizational conflicts over two decades ago because it was fought between a largely Muslim north and Christian/indigenous south.[82] The present violence in South Sudan officially began in 2013 with a collapse of a fragile coalition between the president and other party leaders, who initially helped him to consolidate power when the country became independent in 2011. The current conflict displays familiar complexity along three dimensions: a political dispute within the ruling party, the Sudan People's Liberation Movement; a regional and ethnic war; and a crisis within the army itself.[83] Given the long-running secessionist ambitions of the South all three dimensions are rooted in the past, as conflicts and disagreements have marked the region for decades.[84] The August 2015 peace agreement was short lived and the country soon saw the terrible escalation of violence. There is a substantial UN peacekeeping presence on the ground in South Sudan, but it is mostly tasked with protecting civilians at UN bases. The country is saturated with weapons, which is a major obstacle when it comes to containing the war. In 2011, it was estimated that there were between 1.9 and three million small weapons in the country, translating to one gun per every third person. The weapons have continued to flow in from China, Canada, Ukraine, South Africa, and other countries, prompting the call for an immediate arms embargo.[85] What makes the South Sudanese conflict particularly tragic is the wasted developmental potential of its economy. Just before the outbreak of current violence, South Sudan was expected to be one of the fastest growing world's economies. Shortly after independence, South Sudan was collecting about US$400 million in oil revenues per month, representing 82 percent of its GDP. When the country seceded from Sudan, it obtained 75 percent of the existing oil production, but it had to arrange a deal with the North who retained control over pipelines and infrastructure. Past grievances made the dealmaking extremely difficult.[86] Competing views about how to reach such a deal became the primary reason behind the political discord within the ruling party that eventually resulted in the current war.

Next on the list are overlapping conflicts, extending over the Greater Sahel and Lake Chad basin, involving several countries including Nigeria, Niger, Burkina Faso, Côte d'Ivoire, Cameroon, and Chad. These regions have seen internal wars for generations.

The current conflict in the Democratic Republic of Congo, at a minimum, can be traced to the 1996 war when similar protagonists were involved. The war started as "the second independence" revolt against the former president of Zaire/Congo, Mobutu Sese Seko, who had ruled the country since 1965. The rebels succeeded in overthrowing the Mobutu regime but the country remained in a precarious condition with unresolved tensions and unchanged state institutions. Rebel leader Laurent Kabila managed to secure US$500 million worth of deals with extraction companies for the region's oil and diamonds, but was not willing to share them.[87] There are multiple reasons for the renewed violence, but competing interests between the former "liberation" leaders and allies cannot be underestimated.[88]

Chad represents another society penetrated by violence. Oil rents and diplomatic rents have fueled armed conflicts for generations. The societal saturation with violence started in colonial times, and for this reason history matters. But the state plays a role in perpetuating the violence by being a "government by arms," as Marielle Debos calls it. A dysfunctional state that uses the threat of violence as a tool of governing creates the sense of ongoing insecurity and normalizes violence, even in peaceful times. In her book about Chad, Debos makes the argument that the institutionalized acceptance of violence can become so widespread in a society that it becomes part of the daily functioning of a country.[89]

> In Chad, the mode of government rests in part on armed men, even if no one is in a position to identify and quantify them. Disorder is a way of governing the army and the state. Men in arms who are not all fully integrated into its bureaucratic apparatus are still linked to the state.... Ending war is not enough. The issue is to escape from the inter-war situation maintained and reproduced by the state.[90]

Once the country has fallen into the conflict trap, people's propensity to survive forces them to rely on those who are in the position of power. Having no rights and seeing no prospects of life outside the existing conditions, people tend to attach themselves to anybody who offers protection and possibility of employment. Those in the positions of power will use the structures of the state to solidify their alliances with the threat of violence. The constant potential for violence degenerates the governability processes of the state.

While most conflicts involve rational decisions of military and political leaders pursuing their objectives, certain leaders are guided by delusions of divine providence. The most notorious case in point is Joseph Kony, who, by claiming to be a modern-day prophet, managed to institutionalize a sense of chosen community among his followers. For almost 30 years, Kony has been savaging villages in the region north of Uganda, and later Congo and South Sudan, as the leader of the Lord's Resistance Army. Any atrocity was excused in the name of violating the code of conduct invented by this brutal self-proclaimed prophet, who nevertheless was successful in attracting thousands of loyal disciples. He is responsible for the death of more than 100,000 people and the kidnapping of 50,000 children.[91] In 2017, Kony continues to be Africa's most wanted man, pursued on the charges of genocide and crimes against humanity for which in 2005 he was indicted, in absentia, by the International Criminal Court.

Armed factions in civil wars have often used children under the age of 18 as fighters in violation of the UN's Convention on the Rights of the Child. The documented case is Sierra Leone.[92] Several factors predispose children to becoming soldiers. The most obvious reason is living in the midst of a war. Direct exposure to violent conflict presents children with the perverse opportunity to participate as soldiers, perhaps voluntarily, with the idea that war is heroic or to avenge a family member's death, or more troublingly, by being kidnapped and forced to fight. Often children are tricked into joining the military with promises of gaining education or receiving a nonviolent job. Children are also forced to join by being threatened with prison or with harm to themselves or a family member. Insecurity is also a major factor. In many countries, being in an armed group and carrying a gun means having less of a chance of being killed. Belonging to an army may also mean the only reliable access to food and a decent pay.[93] Although exact data on the number of children used in armed conflicts is difficult to confirm because of the illegal nature of child recruitment, UNICEF estimates that tens of thousands of children are participating in armed conflicts worldwide. It is reported that:[94]

Box 5.1 UNICEF data regarding children and conflict

- Since 2013, an estimated 17,000 children have been recruited in South Sudan and up to 10,000 have been recruited in the Central African Republic.
- In Nigeria and neighboring countries, data verified by the United Nations and its partners indicate that nearly 2,000 children were recruited by Boko Haram in 2016 alone.
- In Yemen, the UN has documented nearly 1,500 cases of child recruitment since the conflict escalated in March 2015.

The same UNICEF report says that over the last ten years at least 65,000 children have been released from armed groups. The reintegration of children scarred by violence is very challenging. When a demobilized solder who served as both a rebel and a member of the army during armed conflicts in Chad was asked what happened to the young who were demobilized like him, the answer was quick and heartbreaking. He said that: "Half are in the garrisons, half are in the rebellions, half are road bandits."[95] In a country with frequent conflicts, experience of being a soldier can be used as work in the army, the rebellion, or road banditry. There are fluid loyalties at work, and as rebel leaders sign international agreements and make deals with new allies, their followers have to orient themselves in this factional system. However, the author observes:

> fluidity is not a symptom for anarchy and that armed violence as a practical occupa-
> tion does not emerge from a collapse of the state or from a society that has become
> bogged down in a culture of violence. The careers of men in arms are part of a
> political context; they are linked to armed factionalism and to the post-war repro-
> duction of the conditions that led to war.[96]

The conflict trap is treacherous and difficult to escape from because it has multiple dimensions: political, psychological, cultural, and social. The primary goal in any post-conflict society is to prevent the resurgence of violence, and to this end, an organized attempt to create *order* is a necessary step to allow war-devastated societies to embark on the path to peaceful socio-economic development. To consolidate peace during this period, a democratic electoral system can be designed in such a way as to reward moderation and punish those who attempt to incite violence (e.g., outlawing hate speech). Once there is peace and order in the country, democratic and market reforms can be implemented gradually in order not to create great social uncertainty.[97] Democratization is a complex process and it is not enough to introduce a series of open market reforms to successfully instill political and economic liberalization. To this end, Roland Paris proposes a new peacebuilding strategy he calls "Institutionalization before Liberalization." He explains:

> What is needed in the immediate postconflict period is not quick elections, democratic ferment, or economic "shock therapy" but a more controlled and gradual approach to liberalization, combined with the immediate building of government institutions that can manage these political and economic reforms.[98]

Notes

1 Lotta Themnér and Peter Wallensteen (2014) "Armed Conflicts, 1946–2013," *Journal of Peace Research*, Vol. 51, No. 4, pp. 541–554.
2 Meredith Reid Sarkees (2017) *The Correlates of War Typology of War: Defining and Categorizing Wars* (Version 4 of the Data), p. 5, The Correlates of War Project at The Pennsylvania State University. Online, available at: www.correlatesofwar.org.
3 John W. Burton (2001) "Peace Begins at Home – International Conflict: A Domestic Responsibility," *International Journal of Peace Studies*, Vol. 6, No. 1, p. 5.
4 Paul Collier (2008) *The Bottom Billion: Why the Poorest Countries are Failing and What Can Be Done About It*, Oxford, UK: Oxford University Press, pp. 34–35.
5 Paul Tiyambe Zeleza (2008) "The Causes and Costs of War in Africa," in Alfred Nhema and Paul Tiyambe Zeleza (eds.), *The Roots of African Conflicts – The Causes and Costs*, Oxford, UK, Athens, OH, Pretoria, SA: James Currey Ltd., Ohio University Press, and Unisa Press, p. 22.
6 Jean-Marie Guéhenno (2017) "10 Conflicts to Watch in 2017 – From Turkey to Mexico, the List of the World's Most Volatile Flashpoints got a Lot More Unpredictable This Year," *Foreign Policy*, January 5. Online Edition.
7 Laurie R. Blank and Benjamin R. Farley (2015) "Identifying the Start of Conflict: Conflict Recognition, Operational Realities and Accountability in the Post-9/11 World," *Michigan Journal of International Law*, Vol. 36, No. 3, pp. 466–539.
8 Jean-Marie Guéhenno (2017) "10 Conflicts to Watch in 2017, op. cit.
9 Immanuel Kant (2010) *Kant's Principles of Politics: Including His Essay on Perpetual Peace (1891)*, translated by W. Hastie, Whitefish, MT: Kessinger Publishing.
10 Michael Doyle (1983) "Kant, Liberal Legacies, and Foreign Affairs II," *Philosophy and Public Affairs*, Vol. 12, Fall, pp. 323–353.
11 John W. Burton (2001) "Peace Begins at Home – International Conflict: A Domestic Responsibility" op. cit., p. 6.
12 John W. Burton (2001) "Conflict Prevention as a Political System," *International Journal of Peace Studies*, Vol. 6, No. 1, p. 28.

13 Roland Paris (2004) *At War's End: Building Peace After Civil Conflict*, Cambridge, UK: Cambridge University Press, p. 5.

14 Ibid.

15 UN (1948) *Convention on the Prevention and Punishment of the Crime of Genocide, Article 2, Adopted by Resolution 260 (III)A of the General Assembly of the United Nations*, December 9.

16 Alison Des Forges and Timothy Longman (2004) "Legal Responses to Genocide in Rwanda," in Eric Stover and Harvey Weinstein (eds.), *My Neighbor, My Enemy: Justice and Community in the Aftermath of Mass Atrocity*, Cambridge, UK: Cambridge University Press, p. 50.

17 J. Matthew Vaccaro (1996) "The Politics of Genocide: Peacekeeping and Disaster Relief in Rwanda," in William J. Durch (ed.), *UN Peacekeeping, American Policy, and the Uncivil Wars of the 1990s*, New York, NY: St. Martin's Press, p. 369.

18 Michel Chossudovsky (2003) *The Globalization of Poverty and the New World Order*, Montreal, Canada: Global Research Publishers, Electronic Edition, Location 2988–3295 of 11018.

19 Roland Paris (2004) *At War's End: Building Peace After Civil Conflict*. op. cit., p. 70.

20 Nigel Eltringham (2006) "Debating the Rwandan Genocide," in Preben Kaarsholm (ed.), *Violence, Political Culture and Development in Africa*, Oxford, UK, Athens, OH, Pretoria, SA: James Currey Ltd., Ohio University Press, and Unisa Press, pp. 78–79.

21 Peter Uvin (1996) *Development, Aid and Conflict: Reflections from the Case of Rwanda*, Working Paper 24, Helsinki, Finland: United Nations University World Institute for Development Economics Research, pp. 6–9.

22 Marysse, S., E. Ndayambaje, *et al. (1992) Revenues ruraux au Rwanda avant Vajustement sturcturel. Cas de Kirarambogo*. Louvain La Neuve: CIDEP, cited in Peter Uvin (1996) *Development, Aid and Conflict: Reflections from the Case of Rwanda*, op. cit., p. 26.

23 Michel Chossudovsky (2003) *The Globalization of Poverty and the New World Order*, op. cit.

24 Peter Uvin (1996) *Development, Aid and Conflict*, op. cit., pp. 13–15.

25 Ibid., p. 14.

26 Ibid., p. 15.

27 Roland Paris (2004) *At War's End: Building Peace After Civil Conflict*, op. cit., p. 71.

28 Michel Chossudovsky (2003) *The Globalization of Poverty and the New World Order*, op. cit.

29 Peter Uvin (1996) *Development, Aid and Conflict*, op. cit., p. 29.

30 Nigel Eltringham (2006) "Debating the Rwandan Genocide," op. cit., p. 85.

31 Peter Uvin (1996) *Development, Aid and Conflict*, op. cit., pp. 31–32.

32 Nigel Eltringham (2006) "Debating the Rwandan Genocide," op. cit., p. 74.

33 Michel Chossudovsky (2003) *The Globalization of Poverty and the New World Order*, op. cit., Location 3407 of 11018.

34 Roméo Dallaire (2003) *Shake Hands with the Devil: The Failure of Humanity in Rwanda*, Toronto, Canada: Random House, p. 142.

35 Peter Uvin (1996) *Development, Aid and Conflict*, op. cit., p. 32.

36 Roland Paris (2004) *At War's End: Building Peace After Civil Conflict*, op. cit., pp. 71–72.

37 Roméo Dallaire (2003) *Shake Hands with The Devil*, op. cit., p. 6.

38 J. Matthew Vaccaro, "The Politics of Genocide: Peacekeeping and Disaster Relief in Rwanda," op. cit., pp. 367–369.

39 Michel Chossudovsky (2003) *The Globalization of Poverty and the New World Order*, op. cit., Location 3476 of 11018.

40 Peter Uvin (1996) *Development, Aid and Conflict*, op. cit., p. 34.

41 Samuel P. Huntington (1993) "The Clash of Civilizations?," *Foreign Affairs*, Vol. 72, No. 3 (Summer).

42 Samuel P. Huntington (2011) *The Clash of Civilizations and the Remaking of World Order*, New York, NY: Simon & Schuster, pp. 45–47.

43 Ibid., p. 102.

44 Ibid.

45 Fouad Ajami (1993) "The Summoning," *Foreign Affairs*, Vol. 72, No. 5, p. 2.

46 Bruce M. Russett, John R. Oneal, and Michaelene Cox (2000) "Clash of Civilizations, or Realism and Liberalism Déjà Vu?," *Journal of Peace Research*, Vol. 37, No. 5, p. 588.
47 Paul Richards (2006) "Forced Labour and Civil War – Agrarian Underpinnings of the Sierra Leone Conflict," in Preben Kaarsholm (ed.), *Violence, Political Culture and Development in Africa*, Oxford, UK, Athens, OH, Pretoria, SA: James Currey Ltd., Ohio University Press, and Unisa Press, p. 181.
48 Zachary Laub (2016) *The Islamic State – Backgrounder*, Washington, DC: Council of Foreign Relations, August 10.
49 Jean-Marie Guéhenno (2017) "10 Conflicts to Watch in 2017," op. cit.
50 Russell Goldman (2017) "How the Iran Terror Attacks Unfolded at Two Sites," *New York Times*, June 7.
51 Marvin Kalb (2015) *Imperial Gamble: Putin, Ukraine, and the New Cold War*, Washington, DC: Brookings Institution Press.
52 George Soroka (2016) "Putin's Patriarch – Does the Kremlin Control the Church?," *Foreign Affairs – Snapshot*, February 11.
53 Sergei Chapnin (2015) "A Church of Empire – Why the Russian Church Chose to Bless Empire," *First Things*, November.
54 Peter R. Mansoor (2011) "The Softer Side of War – Exploring the Influence of Culture on Military Doctrine," *Foreign Affairs – Review Essay*, Vol. 90, No. 1.
55 Glenn Adams and Tuğçe Kurtiş (2012) "Collective Memory Practices as Tools for Reconciliation – Perspectives from Liberation and Cultural Psychology," *African Conflict and Peacebuilding Review*, Vol. 2, Issue 2, pp. 5–28.
56 Preben Kaarsholm (2006) "Politics and Generational Struggle in KwaZulu-Natal," in Preben Kaarsholm (ed.), *Violence, Political Culture and Development in Africa*, Oxford, UK, Athens, OH, Pretoria, SA: James Currey Ltd., Ohio University Press, and Unisa Press, p. 153.
57 Read the excerpts from the speech by John Major, former British prime minister, cited in David Patrick Houghton (2009) *Political Psychology – Situations, Individuals, and Cases*, London, UK: Routledge, pp. 180–181.
58 Michael Mann (2005) *The Dark Side of Democracy: Explaining Ethnic Cleansing*, Cambridge, UK: Cambridge University Press, pp. 254–366.
59 Michael Mann (2005) *The Dark Side of Democracy: Explaining Ethnic Cleansing*, op. cit., pp. 256–258, 363–365, and 395–397.
60 Ibid.
61 Yahya Sadowski (2000) "Ethnic Violence: Fact or Fiction?," in Myra H. Immell (ed.), *Ethnic Violence*, San Diego, CA: Greenhaven Press, p. 32.
62 Joseph Nye (2005) *Understanding International Conflicts: An Introduction to Theory and History* (5th edition), New York, NY: Pearson Longman, p. 154.
63 Patrick Houghton (2009) *Political Psychology*, op. cit., pp. 168–183.
64 Ann Hironaka (2005) *Neverending Wars: The International Community, Weak States, and the Perpetuation of Civil War*, Cambridge, MA: Harvard University Press, pp. 53–60.
65 Mats Berdal and David M. Malone (eds.) (2000) *Greed and Grievance: Economic Agendas in Civil Wars*, Boulder, CO: Lynne Rienner Publishers, pp. 2, 26–27.
66 David Keen (2000) "Incentives and Disincentives for Violence," in Mats Berdal and David M. Malone (eds.), *Greed and Grievance: Economic Agendas in Civil Wars*, op. cit., pp. 29–31.
67 Michael Renner (2002) *The Anatomy of Resource Wars*, Washington, DC: World Watch Institute, pp. 5–6.
68 Ibid., p. 32.
69 Yvonne C. Lodico (1996) "A Peace That Fell Apart: The United Nations and the War in Angola," in William J. Durch (ed.), *UN Peacekeeping, American Policy, and the Uncivil Wars of the 1990s*, New York, NY: St. Martin's Press, pp. 104–122.
70 Michael Renner (2002) *The Anatomy of Resource Wars*, op. cit., pp. 32–34.
71 Ibid., p. 34.
72 Paul Collier (2008) *The Bottom Billion*, op. cit., p. 28.

73 The AfricaW (2017). Online, available at: www.africaw.com/major-problems-facing-angola-today, accessed January 30, 2017.
74 Richard M. Auty (1993) *Sustaining Development in Mineral Economies: The Resource Curse Thesis*, London, UK: Routledge.
75 Macartan Humphreys, Jeffrey D. Sachs, and Joseph E. Stiglitz (eds.) (2007) *Escaping the Resource Curse*, New York, NY: Columbia University Press, p. 5.
76 Ibid., p. 13.
77 Paul Collier and Anke Hoeffler (2004) "Greed and Grievance in Civil War," *Oxford Economic Papers*, Vol. 56, No. 44, pp. 563–595.
78 Peter Uvin (1996) *Development, Aid and Conflict*, op. cit., pp. 34–35.
79 Paul Collier (2008) *The Bottom Billion*, op. cit., p. ii.
80 Ibid., p. 34.
81 Scott Worden (2017) "How to Stabilize Afghanistan – What Russia, Iran, and the United States Can Do," *Foreign Affairs – Snapshot*, April 26.
82 Samuel P. Huntington (2011) *The Clash of Civilizations*, op. cit., p. 137.
83 Alex de Waal and Abdul Mohammed (2014) "Breakdown in South Sudan – What Went Wrong and How to Fix it," *Foreign Affairs – Snapshot*, January 1.
84 Richard Cockett (2010) *Sudan: Darfur and the Failure of an African State*, New Haven, CT: Yale University Press.
85 Aditi Gorur and Rachel Stohl (2016) "Two Options for South Sudan – And Neither of Them is Good," *Foreign Affairs – Snapshot*, July 27.
86 Alex de Waal (2013) "Sizzling South Sudan – Why Oil is Not the Whole Story," *Foreign Affairs – Snapshot*, February 7.
87 Paul Collier (2008) *The Bottom Billion*, op. cit., pp. 21–22.
88 Philip Roessler and Harry Verhoeven (2016) *Why Comrades Go to War: Liberation Politics and the Outbreak of Africa's Deadliest Conflict*, Oxford, UK: Oxford University Press.
89 Marielle Debos (2016) *Living by the Gun in Chad: Combatants, Impunity and State Formation*, London, UK: Zed Books. Electronic Edition.
90 Ibid., Locations 3688 and 3726 of 8583.
91 Peter Eichstaedt (2014) "Kony 20Never – Inside the Mind of Africa's Most Wanted Man," *Foreign Affairs – Snapshot*, January 14.
92 Myriam Denov (2010) *Child Soldiers: Sierra Leone's Revolutionary United Front*, Cambridge, UK: Cambridge University Press.
93 Rachel Brett and Irma Specht (2004) *Young Soldiers: Why They Choose to Fight*, Boulder, CO: Lynne Rienner Publishers, pp. 9–15, 24, 39–41.
94 UNICEF (2017) "At Least 65,000 Children Released from Armed Forces and Groups over the Last 10 Years – UNICEF News Release," New York and Paris: UNICEF, February 21.
95 Marielle Debos (2016) *Living by the Gun in Chad*, op. cit., Location 1704 of 8583.
96 Ibid., Location 615 of 8583.
97 Roland Paris (2004) *At War's End: Building Peace After Civil Conflict*, op. cit., pp. 191–235.
98 Ibid., pp. 7–8.

Suggested further reading

Fawaz A. Gerges (2016) *ISIS: A History*, Princeton, NJ: Princeton University Press.
Ron E. Hassner (2016) *Religion on the Battlefield*, Ithaca, NY: Cornell University Press.
Siri Hettige and Eva Gerharz (eds.) (2015) *Governance, Conflict and Development in South Asia: Perspectives from India, Nepal and Sri Lanka*, New Delhi, India: Sage Publications.
T. David Mason and Sara McLaughlin Mitchell (eds.) (2016) *What Do We Know About Civil Wars?*, Lanham, MD: Rowman & Littlefield.

Constantine Pleshakov (2017) *The Crimean Nexus: Putin's War and the Clash of Civilizations*, New Haven, CT: Yale University Press.

Philip Roessler (2016) *Ethnic Politics and State Power in Africa: The Logic of the Coup-Civil War*, Cambridge, UK: Cambridge University Press.

Anders Themnér (ed.) (2017) *Warlord Democrats in Africa: Ex-Military Leaders and Electoral Politics*, London, UK and New York, NY: Zed Books.

Gender relations and development   **6**

I Conceptualizing gender

To celebrate changing global attitudes to gender relations, the UN declared the years 1975–1985 to be the Decade for Women. It was a time of optimism, marked by special conferences and projects. The events produced a number of documents outlining three major policy initiatives aimed at addressing systemic obstacles to advancement faced by women around the world. The initiatives dealt with equal rights with respect to pay equity, landownership, and basic human rights. Over 20 years later, the progress has been painfully slow: "even in the most privileged countries, women do not yet enjoy the rights described in the three plans of action of the United Nations Decade for Women."[1] Some may argue that the very act of acknowledging the problems facing women has been a major step in the right direction. As a result, developmental organizations today have policy mandates that include commitments to promote worldwide equality for women.

What interests us most in the context of development is the disproportionate burden of social and economic pressures endured by women in poor and disadvantaged communities. Such pressures often result from socially constructed gender-roles that limit women's ability to break the circle of poverty. It is only by recognizing the developmental issues, which have at their roots deeply seated hierarchies based on gender, that we will be able to transcend some of the most persistent obstacles restricting human advancement.

While some progress has been made, it does not mean that women no longer fear exploitation at work, socio-economic exclusion, forced marriages, or physical violence. Gender-based discrimination is still very much present in contemporary social relations. Feminist scholars define gender as:

> as a set of socially constructed characteristics describing what men and women ought to be. Characteristics such as strength, rationality, independence, protector, and public are associated with masculinity, while characteristics such as weakness, emotionality, relational, protected, and private are associated with femininity.[2]

Feminism contests the acceptance of these dichotomized characteristics as natural, and issues a warning about their impact on politics and various processes of decision making. Feminism also interrogates the work of classical scholars, who are typically silent about women and their contributions to the development of societies.

For example, the previously discussed father of economic liberalism, Adam Smith, admired the efficiency of labor specialization in a factory setting, but never paid much attention to the dynamics of the household. In celebrating enlightened self-interest, he famously wrote: "It is not from the benevolence of the butcher, the brewer, or the baker that we expect our dinner but from regard to their self-interest."[3] Not surprisingly, this sentence gets a pointed critique from a feminist scholar: "Smith neglected to mention that none of these tradesmen actually puts dinner on the table, ignoring cooks, maids, wives and mothers in one fell swoop."[4] Known as a strong opponent of state intervention in the economy, Smith had no problem with the denial of independent property rights to married women or restrictions on access to education.[5] Palpably, Smith's ideas were grounded in the historical context of eighteenth-century Europe. Feminism is known for illuminating such historical omissions. Even those scholars bursting with a revolutionary zeal, Karl Marx and Friedrich Engels, mostly followed the patriarchal attitudes of the time "that limited room for socialist feminist maneuver."[6] In seeking support from a growing male union movement, these fervent critics of capitalism "generated a simplistic theory of socialism that has proven vulnerable to patriarchal and authoritarian appropriation."[7] Feminist scholars offer a welcome bipartisan intervention into the vast field of political economy as they try to reclaim the lost voices of women in the past and the present.

Every community displays a complicated social matrix: individuals with different attitudes and complicated ties to family structures shaped by history and customs. The image of a woman as a mother is interpreted in many cultures by her "natural" role of being a caretaker of children and the elderly. Such a stereotyped persona limits woman's capabilities; it erodes her options to make independent choices, and in the process it restricts woman's agency. Women who are relegated to the sub-servant positions of natural-born caretakers are routinely denied opportunities to learn economically relevant skills. Feminism strives to emancipate women and all individuals from such socially constructed and gender-determined roles.

The feminist cause to empower and emancipate women is not always well received. Criticized as being universalistic and western, feminist ideas frequently clash with religious and cultural values promoted in different parts of the world. In response to such criticism, Martha Nussbaum defends the substance of universal values to the extent that such values allow all human beings to perform the most central human functions with due dignity. Limitations placed on women in this context are considered unjust. On every continent, there are women whose fundamental human rights to be respected and become fully capable individuals are violated simply because they are women. A universal conception of human capability can guide us in pursuing the task:

> legitimate concerns for diversity, pluralism, and personal freedom are not incompatible with the recognition of universal norms; indeed, universal norms are

actually required if we are to protect diversity, pluralism, and freedom, treating each human being as an agent and an end. The best way to hold all these concerns together, is to formulate the universal norms as a set of capabilities for fully human functioning, emphasizing the fact that capabilities protect, and do not close off, spheres of human freedom.[8]

This approach is consistent with the mandate of the UNDP, which stipulates that gender equality:

> Refers to the equal rights, responsibilities and opportunities of women and men and girls and boys. Equality does not mean that women and men will become the same but that women's and men's rights, responsibilities and opportunities will not depend on whether they are born male or female. Gender equality implies that the interests, needs and priorities of both women and men are taken into consideration – recognizing the diversity of different groups of women and men. Gender equality is not a "women's issue" but should concern and fully engage men as well as women. Equality between women and men is seen both as a human rights issue and as a precondition for, and indicator of, sustainable people-centred development.[9]

The goal to achieve gender equality and empower all women and girls is listed fifth in the UN Sustainable Development Goals. There are 17 goals within the 2013 UN Agenda to end poverty by 2030 and to pursue a sustainable future.[10]

II Gender and development: changing approaches

Feminists have long observed that, since theory and practice of international relations had been built by mainly men, the problems of women, including the "feminization of poverty," economic inequality, sexual violence, and political marginalization have been conveniently ignored. There are multiple consequences of looking at the world through the lens of masculinity. First, women have had limited social and political roles and hence their needs are often sidelined. Second, without the ability to take an active role in articulating their problems, the approaches taken to "fix" them are inadequate or misplaced. Finally, the dominant masculine discourse tends to overemphasize "hard issues" like defense, finance, and macroeconomic stabilization, instead of more "feminine" issues like human rights, social welfare, domestic violence, and economic insecurity.

Gradually the international community has begun to focus on women and gender relations within the dominant narratives of development and world affairs.

The change in global attitudes towards women is due to several events:[11]

1   the international feminist movements initiated in the 1970s, including the UN Decade for women from 1975–1985;
2   the end of the Cold War, which ushered in more "soft issues" like the environment, gender-based inequality, and human rights;

3   natural progression of knowledge and understanding;
4   the increasing presence of women in the political arena at an international level;
5   the greater mobilization by women's organizations and civil society groups, aided
    by advancements in communications technology.

Over the decades, there have been five main official approaches to integrating women's
issues into the field of development. They are, in chronological order: 1. the welfare
approach; 2. women in development (WID); 3. The gender and development approach
(GAD); 4. women and development (WAD); 5. mainstreaming gender equality (MGE).
The welfare approach emerged in the time of decolonization and was heavily influenced
by modernization theories. The welfare approach favored the population control. It also
focused on promoting economic growth and consumerism with the help of the international aid organizations, including the World Bank. A number of social welfare programs were introduced in aid receiving countries, aimed at promoting family planning
and combating childhood malnutrition. The welfare approach did not target women
directly. Instead, it used the development approaches for men, as they were considered
the primary providers, assuming that any increase in wealth would be redistributed to
women via their husbands or male providers.

The second approach (WID) was initiated in the late 1970s as a response to the welfare
approach. The welfare approach was criticized for seeing women through the prism of
their main roles within the family. WID, on the other hand, involved focusing on income
generation projects for women specifically. These projects, however, were not considered
very successful because they did not focus on the specific needs of women in their indigenous communities. In fact, Claudia von Braunmuhl observed that "women in the
developing world [were] constructed as a homogeneous mass with few, if any, relevant
differences among them."[12] She also found that the WID projects did not improve the
material wealth of the women and, more importantly, "rarely affected women's subordinated status."[13]

The GAD approach had its roots in the United Kingdom, at the University of Sussex in
the 1980s. While the previous approaches dealt with women as unexamined individuals
with predetermined needs, GAD was more concerned with the social constructs of
gender and gender relations. GAD also "[saw] women as agents of change, and not as
recipients of development assistance."[14] With this approach, development programs took
into consideration the principle of empowerment. By using "gender" as the unit of analysis, "women's issues" became "gender issues" creating a larger area of inquiry. This
approach involved examining the structures of power relations based on gender, while
emphasizing the need of women to organize themselves to increase their political and
economic strength.

The WAD approach, finally, dealt with the issues of North and South divisions. In criticizing the other approaches, women in the South felt that those models did not adequately address the gender relations and the needs of women in the developing world,
especially given the impact of colonialism.[15] Chandra Mohanty has criticized western feminists for making assumptions and constructing superficial knowledge about non-western
women and their social and economic relations.[16]

Since the late 1990s, the MGE approach is a globally accepted strategy for promoting gender equality. The UNDP Gender Equality Strategy 2014–2017 uses the following definition of gender mainstreaming:

> Mainstreaming a gender perspective is the process of assessing the implication for women and men of any planned action, including legislation, policies or programmes, in all areas and at all levels. It is a strategy for making women's as well as men's concerns and experiences an integral dimension of the design, implementation, monitoring and evaluation of policies and programmes in all political, economic and societal spheres so that women and men benefit equally and inequality is not perpetuated. The ultimate goal is to achieve gender equality.[17]

The promotion of gender equality and the empowerment of women is at the heart of the policy mandate of UNDP because is considered intrinsic to the organization development approach. The mandate includes: "advocating for women's and girls' equal rights, combating discriminatory practices and challenging the roles and stereotypes that affect inequalities and exclusion."[18]

The UN Convention on the Elimination of All Forms of Discrimination Against Women was a turning point in recognizing women's issues globally and enshrining their rights within the principles by international law. The convention entered into force on September 3, 1981, and in 2017 it had 99 signatories. One disappointing aspect with respect to implementation concerns a number of exemptions demanded by countries in the name of domestic laws and customs.[19] Another important milestone was the UN 1995 Conference on Women, in Beijing. The resulting Beijing Platform for Action states that "for the girl child to develop her full potential she needs to be nurtured in an enabling environment, where her spiritual, intellectual and material needs for survival, protection and development are met and her equal rights safeguarded."[20] It was also at the Beijing conference that the world recognized the advancement of the goals of "equality, development and peace for all women everywhere in the interest of all humanity."[21] Under section 12, the nations reaffirmed their commitments to

> the empowerment and advancement of women, including the right to freedom of thought, conscience, religion and belief, thus contributing to the moral, ethical, spiritual and intellectual needs of women and men, individually or in community with others and thereby guaranteeing them the possibility of realizing their full potential in society and shaping their lives in accordance with their own aspirations.[22]

After the Beijing Conference, the UN Human Development Report launched its first set of gender related composite indices, the Gender-related Development Index (GDI) and the Gender Empowerment Measure (GEM). The GEM was designed to measure female economic and political participation and their economic power by using the following indicators: women's share of parliamentary seats, proportion of women in top managerial and professional positions, and ratio of female to male estimated earned income. Since these indices were criticized for their limitations in accurately capturing gender disparities and reflecting critical gender issues, in 2010 the UN introduced an alternative pair of gender indices: the Gender Inequality Index (GII) and the new Gender Development Index (nGDI). While the nGDI mainly refines the collection of data regarding the years of schooling, the GII measures the reproductive health of women by gathering data on maternal mortality rates and female labor market participation.[23] Every year, the UNDP releases its annual human development report, updating the available data. The most recent 2016 report is called "Human Development for Everyone," and contains updated data with respect to all the above indicators.[24]

In observing that women in much of the world lack essential support for leading gratifying lives, or "lives that are fully human," Martha Nussbaum proposes the capabilities approach to empowering women, developed in parallel with the previously discussed Amartya Sen development as freedom approach. In her work, Nussbaum simply, but powerfully, reminds us of the thought shared by all feminists "that each person is valuable and worthy of respect as an end" and thus "we must conclude that we should look not just to the total of the average, but to the functioning of each, and every person."[25] Consequently, her theory of "each person as an end" evolves into the "principle of each person's capability." The capabilities approach looks at each individual woman and her life within the context of a list of central human functional capabilities. These capabilities play an important function of individual rights and they should not be interchangeable or excluded. Here is the list of central human functional capabilities:

1   Life – being able to live a normal length life (freedom from dying prematurely).
2   Bodily health – being able to have good health.
3   Bodily integrity – being able to move freely, being able to be secure from bodily harm (including sexual and domestic violence).
4   Senses, imagination, and thought – being able to think freely, reason in an informed way cultivated by adequate education, right to political and artistic free speech, freedom of religion.
5   Emotions – being able to love and show love for others, ensuring that one does not have "overwhelming fear and anxiety" that can hurt the nourishment of this capability.
6   Practical reason – being able to "form a conception of the good and to engage in critical reflection about the planning of one's life."
7   Affiliation – A. being able to engage in social interactions with others in the community. This requires freedom of assembly and political speech. B. being treated as a dignified, equal being through self-respect and non-humiliation. This requires the right to work and be free from all forms of discrimination.
8   Other species – being able to live and engage in a relationship with animals, plants, and nature.
9   Play – "being able to laugh, to play, to enjoy recreational activities."
10 Control over one's environment – A. being able to participate in political choices, and B. material, involving having property rights and employment rights on an equal basis.[26]

This list helps outline political principles based on universal ideas of human capability that can inspire legal and constitutional guarantees. Every country is advised to comply.[27] By drawing upon the insights offered by feminism, Nussbaum problematizes women's failure to attend the required level of functional capabilities by considering it to be a problem of justice. She proposes to abandon the abstract normative theories that characterize utilitarian economics and to seek a philosophy that can make a real difference in international development policy making. We are well advised to pursue a philosophical path that would help understand women's experience of subordination and exclusion.[28]

Nussbaum's feminist contribution to the field of development enriches the capability approach of Amartya Sen by focusing on women. They both contest the limitations of the traditional approach to economic development, which tends to neglect rights, freedoms, and other non-utility concerns. The utilitarian search for satisfaction and happiness has its limits when it comes to making comparisons between individuals with respect to their well-being or deprivation. As Sen observed, peoples' desires and needs can adjust to circumstances, especially to make life bearable in adverse situations. As a result, the utility calculus can be deeply unfair to those who are persistently deprived: for example, the chronically disadvantaged in stratified societies, constantly oppressed minorities in intolerant communities, "routinely overworked sweatshop employees in exploitative economic arrangements, hopelessly subdued housewives in severely sexist societies."[29] The need to survive forces socially and economically marginalized individuals to come to terms with their deprivation, particularly when their grievances cannot be addressed by the due process.

At its core, the capabilities approach is preoccupied with enhancing individual agency. It seeks justice as it stresses the importance of accountability by governments and developmental agencies with respect to programs aimed to address the economic and gender inequality. The capabilities approach also reminds us that "meaningful accountability requires individual rights."[30] During the 1993 UN Conference on Human Rights, participants adopted the Vienna Declaration and Programme for Action, paragraph 18 of which provides a definition of woman's rights that is also used by the UNDP.[31]

> The human rights of women and of the girl child are an inalienable, integral and indivisible part of universal human rights. The full and equal participation of women in political, civil, economic, social and cultural life, at the national, regional and international levels, and the eradication of all forms of discrimination on grounds of sex are priority objectives of the international community.

The capabilities approach has greatly influenced the way developmental agencies now investigate and measure barriers to human development. The approach has also helped to redefine the concept of women's empowerment, as is now used by UNDP:

> Women's empowerment has five components: Women's sense of self-worth; their right to have and to determine choices; their right to have access to opportunities and resources; their right to have the power to control their own lives, both within and outside the home; and their ability to influence the direction of social change to create a more just social and economic order, nationally and internationally.

The concept of empowerment is related to gender equality but distinct from it. The core of empowerment lies in the ability of a woman to control her own destiny. This implies that to be empowered women must not only have equal capabilities (such as education and health) and equal access to resources and opportunities (such as land and employment), they must also have the agency to use those rights, capabilities, resources and opportunities to make strategic choices and decisions (such as are provided through leadership opportunities and participation in political institutions. And to exercise agency, women must live without the fear of coercion and violence.[32]

III Empowering the disadvantaged: towards gender equality

Discrimination impacts women in all societies, but poor women in the developing world are particularly vulnerable. A woman's socio-economic constraints are shaped by her experiences, socialization, and fears. The institutionalization of accepted norms is perhaps the greatest means of identifying with one's gender. Women are reminded of their vulnerabilities from the womb. This was evident in the case of gender choosing in China, where the increasing use of ultrasound and gender detection had led to sex-selective abortions or abandoning of infant girls after birth. The outcome is a severe gender imbalance, as there are now 34 million more men than women in China. It is reported that since the government implemented a one-child family planning policy some 40 years ago, Chinese doctors have performed more than 330 million abortions.[33] In Sri Lanka, women who failed to agree to sterilization after the birth of their child were routinely denied work in the fields or hospitalization should they become pregnant another time.[34] Women disproportionally are victims of domestic violence and face the threat of sexual trafficking. The most recent UN report shows that 49 percent of all detected trafficking victims were adult women and 33 percent were children (12 percent boys and 21 percent girls).[35] There are also significant gender gaps when it comes to the incidence of vulnerable employment, defined as employment that is unpredictable, low paid, and has very little social protection, if at all. In certain countries in Northern Africa, sub-Saharan Africa, and the Arab States, women face a 25 to 35 percent higher risk of being in vulnerable employment than men.[36]

Without a doubt, the family environment is critical for either nurturing strong women or socializing women into acceptance of a patriarchal order. Women find themselves raising children on their own, trapped in low paying, undervalued jobs and vulnerable societal positions. Arland Thornton argues that promoting a number of ideas associated with the concept of a modern family – including family planning, legal marriages among adults only, and the respect for a woman's self-worth – greatly benefit the whole society.[37] While it is important to respect culture and traditions, it is equally as important to ensure that these traditions do not infringe on women's fundamental freedoms and rights. It is difficult to justify the practice of marrying very young girls to much older men and seeing women from arranged marriages left dependent on their husbands' good will. This is certainly the case in some countries, where young brides are forced to marry without any choice or means of escape. Moreover, some girls face punishments such as "acid baths," which are meant to "mutilate and kill" any woman who objects or "disobeys" their husband.[38] In other areas of the world, particularly under customary and religious laws, brothers inherit the wives of their deceased siblings. The women are left without property because they are not allowed to inherit their husbands' belongings.

Land insecurity can devastate women who cultivate their husbands' property only to have the land and harvest taken by the relatives of the husbands. Women routinely perform more agricultural work. As men move away from their traditional roles in farming towards other means of support, the women are left with the farming tasks, household chores, and raising children. Often, older women make up a disproportionate percentage of the population in rural areas, unable to obtain manufacturing or service jobs.[39] Susie Jacobs, conducting research in South Africa, noted that while land security was an important concern to women, issues such as employment, adequate housing, electricity, better health care, and human rights were also at the top of the agenda.[40] Women are also most impacted by changing external conditions. A good example being the structural adjustment programs. Women were adversely affected by "additional burdens imposed on them as a result of decline in real wages, rising unemployment, dramatic increases in prices of household good, and changes in the level and composition of public expenditures."[41] In Cameroon, as a result of the restructuring and privatization of the public sector, women were first to lose their jobs as unskilled laborers. Women working on the palm, rubber, tea, and banana plantations were particularly affected. The consequences also included the loss of social benefits like health care and free housing. As a result, women were forced to find informal employment, many ended up in the agricultural sector, where they worked more hours, for less pay and no benefits. Women were regularly overtaxed and faced extortion by police, travelled long distances, faced possibilities of rape and violent crimes, and had limited access to credit.[42] Overworked and socially marginalized women performing the worst paid jobs, are willing to enter a marriage to survive, even if it is an abusive relationship.

A study conducted by Lisa L. Gezon on marriage, kin, and compensation in Madagascar shows that the capabilities of single women can be severely limited. They remain structurally dependent and vulnerable within the context of their relationships with the men in society.[43] It is observed that women "earn their living through a combination of agricultural day labor, petty buying and selling, and remunerated sexual and domestic relations with men."[44] Gezon argues that marriage can provide the economic security for women, especially where males own or have direct control over agricultural land. If a woman is not married, she faces significant obstacles. First, it is difficult to obtain access to land, and where women do control agricultural land, they often cannot afford to hire male help, which in some cases is essential to maintain the farm. The males in the community are usually the ones with access to herds of cattle, which turn the soil or pull the carts. The men are also stronger, in some cases, more able to perform the heavier agricultural tasks.

In one region of Madagascar, women's activities were restricted because of the system that regulated the flow of water from the canals. The system was controlled by a group of patriarchal families led by dominant men. Single women feared being assaulted if they went into the fields at night. This situation made them vulnerable to manipulations.

The author mentions one specific case where a woman's male neighbor directed the flow of water to by-pass her fields during the nighttime, knowing that the woman would not check on the fields or water supply. As a result, her fields withered, while his harvest grew more prosperous. Some women have turned to domestic and sexual relations with men, giving in to the existing patriarchal social relations that emphasize the economic reliance on males within the community.[45] Such vulnerable women need essential tools facilitating economic independence. More programs like micro-credit would enable the women to break this reliance by leading to the self-sufficiency of individual women in the farming and other economic sectors.

There is no coincidence, the concept of micro-credit was designed with women in mind. Micro-credit has become an important tool of empowering women in the developing world. The founder of the first micro-credit Grameen Bank, Muhammad Yunus, decided to offer small loans only to women. From the start, this idea was very controversial in his native Bangladesh. Both husbands and male community leaders saw the Grameen money-lending as a direct threat to their authority. Yet Yunus believed hunger and poverty are more women's issues than male issues, given women's traditional responsibilities for taking care of children and the elderly. Yunus also knew that poor women were the most vulnerable members of society. He observed:[46]

> A poor woman in our society is totally insecure: she is insecure in her husband's house because he can throw her out any time he wishes.... She cannot read and write, and generally she has never been allowed out of her house to earn money, even if she has wanted to. She is insecure in her in-laws' house, for the same reason as she was in her parents' house: they are just waiting to get her out so they will have one less mouth to feed.

The micro-credit was intended to provide poor women with some degree of financial security. Studies conducted by the Grameen Bank confirmed that women performed better than men in using the money wisely on improving the family's living conditions.[47] Micro-credit also provided poor women with opportunities to start a small business and gain necessary confidence to confront systemic oppression. In the process, it started to change the traditional patriarchal structure of many families.

Women collectively are vulnerable to domestic violence, sexual abuse, political discrimination, and the "feminization of poverty." Individually, they find themselves in unique circumstances that perpetuate gender inequalities. Often, women are without refuge because the family structure permits oppression. According to Nussbaum,

> in many instances, the damage women suffer in the family takes a particular form: the woman is treated not as an end in herself, but as an adjunct or instrument of the needs of others, as a mere reproducer, cook, cleaner, sexual outlet, caretaker, rather than as a source of agency and worth in her own right.[48]

Despite ongoing structural limitations, increasingly we see women taking bold steps towards initiating positive change. Many localized responses to systemic problems, coupled with women's entry into the political arena through protest activities, have shown that women can and do make a difference. The simple act of being visible can induce change. Nussbaum had found during her interviews that merely seeing women organize into groups and demanding attention or services made men look at them differently, as women who are "powerful people who can do things."[49] By collectively dealing with systemic gender inequality, women can lessen the gap between the sexes, aiding in the fulfillment of human capabilities.

With this in mind, we recall three important principles intended to empower women. First, the importance of options, allowing women to exit abusive or economically desolate situations. Policies that increase women's economic options offer a powerful way of promoting their well-being and strengthen women's bargaining position in the family and in the community. Second, the importance of perceived contribution, leading to changing attitudes about domestic work and policies targeting wage discrimination based on gender. Third, the importance of a sense of one's worth, allowing women to reclaim a belief in their own abilities. One way to help operationalize these three principles is to fund and promote women's collectives in order to facilitate their political participation and social empowerment.

In a study by Beatriz Padilla on grassroots socio-political participation and gender identities, it was found that those women who are active in civil society groups find their boundaries as being more fluid, in the sense they move back and forth between the domestic and public spheres. The women who became active in the community also encouraged other women to participate. They also saw themselves as different from the women who relied on men. This was due to the fact that they felt they had "challenged traditional gender roles, they [were] not expecting their husbands or partners to provide, they [were] capable of organizing to change their surroundings and find an answer to their needs."[50]

In a show of defiance, the women of Chipko used their gender roles to initiate change in their community of the Himalayan mountains. Embracing their female stereotypes of being "dutiful and loving sisters, who tie silken thread around their brothers' wrists or as courageous mothers, willing to risk harming themselves to save the trees upon which their children's lives depend,"[51] a group of local women stopped the deforestation in the village. Had they not, they would have had to travel farther away for wood, since their precious ash trees would not have been replanted and the alternatives were not adequate for their household needs.[52]

One of the critical aspects of women's empowerment is the ability to have control over her body. The top three central human functional capabilities on the list identified by Nussbaum concern the abilities of living healthy lives and having choices in matters of reproduction.[53] A path-breaking new study examines an important transition in interpreting and understanding maternal health today. This transition ranges from addressing child and mother mortality risk to addressing the complex causes of maternal risk, such as gender inequality, racism, and legacy of colonialism. The research demonstrates that major variations in attitudes towards childbirth exist between women in the Global North and the Global South. While women of the Global South prefer full access to new medical technologies as a way of diminishing risk and the stigma of poverty, the women of the Global North increasingly desire natural births at home with little medical interventions.[54]

The World Health Organization (WHO) Department of Reproductive Health and Research recognizes how important it is that people "be empowered to exercise control over their sexual and reproductive lives, and have access to related health services." This is an essential step in dealing with health-related problems like HIV/AIDS, unwanted pregnancies, female genital mutilation, or sexual violence.[55] One researcher who studied programs aimed at helping prostitutes to be protected against HIV/AIDS in urban Senegal found out that sadly those implementing such programs did not live up to their duties. For example, the clinic staff would demand money from prostitutes for condoms that were supposed to be free of charge, or the clinics would suddenly close, leaving the women without any supply.[56]

Prostitutes are socially marginalized although most women enter prostitution as a last resort, often escaping abuse and violence at home. Women can be abandoned by their husbands or their children, and since they do not have adequate skills to find work they end up on the streets. In other cases, women are sold into the trade. The United Nations reveals the growing international illicit business of human trafficking.[57] Women are often lured under false pretenses to foreign nations, where they find themselves isolated and treated as the property of traffickers. They are usually illegal migrants and have their passports taken from them; such women believe that the police would be unwilling to help and are subjected to violent and exploitive conditions, physically and emotionally abused, raped or violated without the use of condoms, hence being at high risk of contracting HIV/AIDS. Consequently, the United Nations Population Fund (UNFPA, formerly the United Nations Fund for Population Activities) has consistently advocated that violence against women and girls is a human rights violation and a public health priority.[58]

According to the 1995 Beijing Conference's Platform for Action, "violence against women" is defined as

> any act of gender-based violence that results in, or is likely to result in, physical, sexual or psychological harm or suffering to women, including the threats of such acts, coercion or arbitrary deprivation of liberty, whether occurring in public or private life.[59]

Section 113 notes, specifically, violence occurring in the family and the community. However, dowry-related violence, sexual abuse, female genital mutilations, and marital rape continue to take place in families. Family life in many traditional communities is deemed to be a private matter, and the laws protecting women are weak in many countries. For example, according to the 2005 report by Amnesty International, in some areas of Nigeria more than two-thirds of women have experienced some form of violence within the family.[60] Martial rape was not considered a crime under their penal code, and even in areas that were within the law, the justice system did not offer justice or protection for women seeking help.[61]

In the case of Guatemala, women in distress are often abandoned in their time of need by the government, which fails "to provide judicial redress or adequate medical attention."[62] From 2001 to 2006 in Guatemala over 2,200 women were brutally killed. Among the killed were students and poor homeless women, but most of the murdered were sex trade workers. Stigmatized as they are, these women are afraid to come forward and accuse the perpetrators. There were only two cases that ended with convictions in 2006. Amnesty International recorded faulty investigations by the police, missing information and poor evidence and data collection.[63]

Gender based violence and discrimination are symptoms of a fragmented and dysfunctional society. There is no place for gender-based inequality in the democratic state where individual rights are respected. Therefore promoting gender equality is the guiding principle behind the most recent UNDP strategy. It outlines three main areas of work, shown below.[64]

Box 6.1 UNDP strategy main areas of work

1. Sustainable development pathways:

The sustainable development pathways area of work provides an opportunity to address inequalities and reshape policies to empower women and girls in all their diversity, so that they can become catalytic agents of change and equal partners with men in the quest to promote growth that is inclusive, just, equitable and sustainable. With women's engagement, success in eradicating poverty, promoting sustainable consumption and production patterns and sustainable management of natural resources can be achieved.

2. Inclusive and effective democratic governance:

Ensuring women's and men's equal participation in governance processes, and their equal benefits from services, are preconditions for the achievement of inclusive and effective democratic governance. The democratic governance area of UNDP work provides an opportunity to advance women's legal rights and empowerment, strengthen their access to justice, ensure gender responsive and equitable service delivery, and promote their equal participation in decision making.

3. Resilience building:

Gender equality and women's empowerment are integral to building individual, institutional and societal resilience. Systemic inequalities overall, and especially those between women and men in the economic, social and political spheres, exacerbate the impact of economic, disaster and climate-related and political shocks and impede sustainable development and durable peace. Women need to be engaged at all stages of formal and informal peace processes and their priorities must inform the agenda for conflict prevention, early recovery from crises, durable peace, resilience and sustainable development.

The UNDP gender equality strategy provides practical policy recommendations and lists a number of measures to be put in place by national partners.[65] To facilitate the implementation of the organization's commitments, the UNDP has partnered with other UN and civil society groups around the world. Over the years, however, such attempts to promote gender equality and empower women through developmental assistance programs have yielded mixed results. Measurable progress has been made in areas supported by a consistent level of funding, such as health and education. Globally, between 1995 and 2015, maternal mortality rates fell by nearly half and the gender gap in primary education virtually disappeared. However, in the areas lacking substantial developmental funding, such as economic and political participation and gender-based violence, the status of women remained unchanged, or even suffered a setback. For example, in Latin America and the Caribbean, the ratio of women to men living in poverty increased from 1997 to 2012, despite declining net poverty rates for the region. Scholars observe that developmental agencies are not sufficiently prioritizing the advancement of women and girls.[66]

IV Combating the barriers: women as leaders and doers

The UNDP gender equality strategy acknowledges the existence of structural barriers to gender equality that need to be eliminated.[67] The 2016 Human Development Report is an eye opener by identifying the mutually reinforcing gender barriers that deny women the empowerment necessary to realize the full potential of their lives. For example, women are legally discriminated against with respect to opportunities in more than 150 countries, in 18 countries women must obtain their husband's approval to get a job, in 32 countries procedures for women to obtain a passport differ from those for men, and in 100 countries women are prevented from pursuing some careers solely because of their gender. The report notes that some of those barriers "are deeply embedded in social and political identities and relationships – such as blatant violence, discriminatory laws, exclusionary social norms, imbalances in political participation and unequal distribution of opportunities."[68]

Women continue to occupy a relatively small place on the political and economic stage of many countries. There are certainly exceptions. For example, India, Pakistan, Bangladesh, Sri Lanka, Trinidad and Tobago, Peru, Guinea Bissau, Thailand, Senegal, and Mali have all had female prime ministers, a feat still to elude the governments of the United States, and to some extent, Canada.[69] Some nations in the developing world have higher levels of female political representation than the developed countries, as measured by the share of seats in parliament held by women, including Costa Rica at 33.3 percent, Nicaragua 39.1 percent, Cuba 48.9 percent, Ecuador 41.6 percent, South Africa at 40.7 percent, Bolivia 51.8 percent, Rwanda 57.5, and Uganda at 35 percent.[70] The numbers for Canada and the United States are 28.2 percent and 19.4 percent, respectively. It is difficult to generalize, but is a central theme with respect to women's bargaining power within different political, social, and cultural environments: "it is uneven and not conducive to women's participation."[71]

Still, the rates of women's participation in parliament when compared with the HDI statistics indicate that the group of countries with Very High Human Development is also characterized by a high level of female political representation. Among the top 15 countries on the 2015 HDI list, female parliamentary participation ranges from over 43 percent in Sweden to over 19 percent in the United States and Ireland.[72] There is also another interesting difference among different regions of the world. For example, the Nordic nations have the highest percentage of seats held by women, while most Middle Eastern countries have quite low female parliamentary participation, ranging from 0 percent in Qatar, and 1.5 percent in Kuwait, to 19.9 percent in Saudi Arabia. In sub-Saharan Africa, the picture is mixed, with some of the poorest countries having high women participation, such as Mozambique with 39.6 percent, while others, such as Mali with 9.5 percent, are lower.[73] Researchers who have studied these matters for some time observe cross country variations, often reflecting the importance of cultural factors in electoral politics and that more egalitarian nations have more women in power.[74]

Variations also occur as the nature of governance changes. Post-communist Poland and Hungary are good examples; the reformist transitions in these countries, and in the entire region, were far from being gender neutral. The reforms disproportionally affected women, who became overburdened with working multiple jobs and taking care of family members when social programs disappeared. The number of women politicians decreased. Previously both countries had instituted a quota system to mandate female participation. After the elimination of the quota system, the number of parliamentary seats held by women in Poland immediately dropped from over 20 percent to 13 percent. In post-communist Hungary, women parliamentary participation equaled 7 to 11 percent, a stark difference from over 20 percent under the communist regime.[75] Part of this drop was attributable to the scope of economic changes during the transitional phases of democratic reforms. The volatile economic situation particularly placed women in precarious positions, marked by the fear of unemployment and persistence of low paid jobs, given the changing labor market and economic uncertainty. Although international NGOs tried to encourage women to actively participate in politics, they found that women could not afford to volunteer. In fact, western NGOs were sometimes criticized for imposing their strategies on local women's groups without understanding the domestic politics, and hence contributing to the marginalization of such groups.[76]

One could expect that once the economy improved, women's political participation would increase. Indeed, the situation has improved in Poland. In 2015, women occupied 22.1 percent of parliamentary seats and Poland ranked thirty-sixth on the HDI list, yet the same year in Hungary women's parliamentary participation was only 10.1 percent, although Hungary placed forty-fourth on the same HDI list. One interesting difference between these countries concerns the GII. Here Poland scores better than Hungary: for 2014, Poland is placed twenty-eighth, while Hungary is placed forty-second.[77] It is possible that greater gender equality in Poland is reflected by greater political participation by women.

In more traditional communities around the world, cultural norms can play a powerful role in socializing women into accepting gender stereotypes of their specified roles within a society "from religiously based exclusion in the Middle East to patriarchal communalism in the South Pacific."[78] Girls are often socialized and encouraged to participate in raising children and tending the family farm. Boys, on the other hand, are encouraged to be the leaders in society, both within the household and the public sphere. Moreover, women are often taught to be seen and not heard, reinforcing their lack of confidence within the public realm and realizing that their opinions are not as significant as those of their male counterparts.

> In many countries, traditions continue to emphasize, and often dictate, women's primary role as mothers and housewives. A traditional, strong, patriarchal value system favors sexually segregated roles, and so-called "traditional cultural values" militate against the advancement, progress and participation of women in any political process.[79]

Despite the advances made over the years, studies have shown that women in some nations still face discrimination with respect to culturally informed family laws. For example, women can be constrained in their abilities to travel, work, and participate independently in the public realm.[80] According to Abu-Zayd, the traditional norm in nations like Egypt, Jordan, and Lebanon, particularly among indigenous tribes, dictates that women are not to mix with men and, consequently, an Egyptian MP detailed that "the head of a certain tribe told her that he would mobilize his entire tribe to prevent a woman from running in the elections."[81]

There is a possible linkage between women's economic progress and the diminished role of culture in a society. Women who break the traditional barriers and become part of the labor force can alter the societal dynamics already altered by technological advancements, widely accessible social media, and educational opportunities for women. These changes lead to a weakening of traditional values and challenging of existing customs. In the words of one scholar:

> Moving out of the house and into the workforce appears to have a consciousness-raising effect on women. Greater development increases the number of women who are likely to have formal positions and experience, for example in labor unions or professional organizations. Culture is related to development, and as development increases, women's standing in society relative to men becomes more equal. On the other hand, two countries could be quite similar in terms of development, but women may have come substantially further in terms of equality in one country than in the other. While culture consistently has been believed to be important, it has been difficult to test directly for an effect.[82]

Women continue to face significant obstacles when it comes to their participation in the economic decision-making processes. Many women are dependent on their husbands and fathers, who oftentimes do not support women's desires to be active politically and independent socio-economically. Multiple studies examining the extent of poverty in different regions of the world have demonstrated that women routinely represent a disproportionate share of the poorest people in the communities of both developed and developing nations. This phenomenon, called "feminization of poverty," was in fact first coined in the United States following a study that demonstrated the high level of poverty among the fast-growing number of female-headed households. By the mid-1980s, it was believed that almost half of all the poor in the United States lived in families headed by women.[83]

Additional analyses researching this phenomenon confirmed that women are often at the economic bottom of many societies. They face the danger of being unable to change their circumstances. This is due to the systemic barriers women face, such as social impediments to acquiring job related skills, or legal constraints disallowing women to acquire capital and making them ineligible to own property.

There is evidence that women are most affected during times of financial crises and economic downturns. Poor economic conditions can be exacerbated by structural adjustment policies in countries where women "confront social barriers that crowd them into some industries and occupations, foreclose entry into others, and generally push them onto the margins of economic life."[84] Women tend to be particularly vulnerable economically in those parts of the world where family structure is patriarchal. In such communities, they cannot pursue the employment of their choice or distribute the fruits of their labor as they please. While they are expected to perform considerable unpaid domestic labor, they are also expected to produce crafts or other items for sale at the market. Such items are often appropriated and sold by their husbands or fathers, who are the sole financial decision makers in the households.

The study lists the following additional reasons why women remain economically vulnerable: 1. Customary biases and intra-household inequalities lead to lower consumption by, and fewer benefits for, women and girls among lower-income groups; 2. Women's geographic and occupational mobility is constrained by family and childrearing responsibilities; 3. The legal and customary frameworks often do not treat women as autonomous citizens but rather as dependents or minors – with the result that in many countries, women cannot own or inherit property, seek a job, remain employed, or take out a loan without the permission of husband or father; 4. Labor-market discrimination and occupational segregation result in women being concentrated in the low-wage secondary employment sectors, in the informal sector, and in the contingent of "flexible labor."[85] The study concluded:

> although the claim that the majority of the world's poor are women cannot be substantiated, the disadvantaged position of women is incontestable. We may conclude that globally, women are especially severe victims of poverty in at least three ways. First, gender inequalities and the underachievement of women's entitlements and capabilities in many countries put women at a distinct disadvantage vis-à-vis men and in the face of a range of impoverishing conditions. They are also more vulnerable to highly exploitative conditions. Second, they work longer hours than men do at both productive and reproductive activities, and still earn less than men. Third, their capacity to lift themselves out of poverty is circumscribed by cultural, legal, and labor-market constraints on their social and occupational mobility.[86]

Throughout history, women of all backgrounds have been disadvantaged with respect to access to education. Being educated is one of the three key dimensions of the HDI, together with the ability of living a healthy life and having a decent standard of living. The HDI's education component is measured by mean of years of schooling for adults over 25 years old, and by total expected years of schooling for children under 18 years of age. The statistical data collected and released in the context of the HDI indicates the proportion of female and male population with at least some secondary education. Indicators gathered for the most recent period (2005–2014) confirm the correlation between a high level of human development and women's extended years of schooling.

In Canada, Austria, Finland, and Estonia, 100 percent of women over the age of 25 have some secondary education. This is closely followed by the United Kingdom, Norway, Switzerland, the United States, and other developed nations. All the mentioned countries also show over 60 percent participation of women in the labor force, except for Estonia and Finland, where numbers were close to 56 percent. In Saudi Arabia, 60.5 percent of women had some postsecondary education, but only 20 percent were in the labor force. This is quite different from Argentina, where the numbers were 56.3 percent and 47.5 percent, respectively. In some developing countries with relatively high HDI, still less than 60 percent of women have postsecondary education, such as Panama, Venezuela, Mexico, Brazil, and China, but women's participation in the labor force is very high, ranging from 49 percent in Panama to 59.4 percent in Brazil and 63.9 percent in China. One can speculate that many women working in these countries occupy low skilled and hence low paid positions.[87]

The picture is predictably bleaker among countries with medium and low HDI. If we use the range between 40 percent and 20 percent of women with some postsecondary education, the following countries appear: Indonesia, Paraguay, Bangladesh, India, and Kenya. Yet, these numbers are a poor predictor of female labor force participation, which ranges from a very high 62.2 percent in Kenya, to 27 percent in India. And then there are low HDI countries, where less than 10 percent of women have some postsecondary education, such as Cambodia, Yemen, Mauretania, Rwanda, Senegal, Papua New Guinea, Afghanistan, and Tanzania.[88] Again, in terms of female labor force participation, the differences can be staggering: with the highest for Tanzania (88.1 percent), Rwanda (86.4 percent), and Cambodia (78.8 percent), to a low in Yemen (25.4 percent), and Afghanistan (15.8 percent).[89] Some of these differences can be explained by the cultural factors and the prevalence of traditional family structure that discourages women from seeking employment outside the home. Overall, these statistics tell us very little, however, about the type of jobs that women engage in. We can only hypothesize that in countries with medium and low HDI, where access to postsecondary education for women is a problem, a large percentage of working women perform difficult low paid jobs.

Based on a study examining the obstacles to human development in South Asia, two scholars found that the opportunity costs for women to seek education were very high and, hence, women frequently stopped going to school.[90] Even in areas where education was free, families were often unable or unwilling to pay for textbooks, uniforms, and other outside costs. Moreover, from a cultural perspective, parents were unlikely to invest in daughters' education because they expected them to leave the household after marriage and be taken care of by their husbands. The distance girls were required to travel was also a dissuading factor. Where schools were far from the home it was expensive to send children away to attend them, with the priority given to sons in the traditional societies.

Normally, there is a combination of obstacles hindering a woman's ability to become independent. Access to education alone is not sufficient for women to become a strong community leader. In the case of South Africa, women's empowerment came in the form of creating daycare centers that allowed mothers to be active in politics and go to work. More interestingly, new office facilities had to be built with female washrooms included.[91] The new South African Constitution provided women with a sense of security and confidence by introducing new affirmative action policies aimed at hiring more women in the public sector. Specific gender-focused departments were also created, linked to the Office on the Status of Women. This office was tasked with paying attention to the so-called "soft issues," and ended up creating legislation for increased immunization programs, housing subsidies, and renovating schools.[92]

In the case of the Republic of Korea, women made up only 3 percent in the Korean National Assembly in the 1990s. Despite the nation's consistent movement towards democratization during this period, there existed a degree of a masculine nature of institutionalized politics.[93] In 2014, South Korea was considered a very high HDI country, ranked twenty-third on list by share of seats in parliament, remaining relatively low at 16.3 percent, while 50.1 percent of women participated in the labor force.[94] South Korea is not alone. The gender roles within a traditionally oriented society dictate the types of placements within the government that women attain. However, the example of Korea also shows that democratization increases opportunities for more equal participation of women but that it takes time to change societal attitudes.

Women, without a doubt, are an essential part of any political system, yet they remain underrepresented worldwide. The UN Resolution on Women, Peace and Security (Resolution 1325) calls on "Member States to ensure increased representation of women at all decision-making levels in national, regional and international institutions and mechanisms for the prevention, management, and resolution of conflict." Furthermore, Article 60 of the declaration makes the following statement:

Box 6.2 Article 60 statement

Women's empowerment reshapes their image by focusing on their roles and contributions to peace and security. From this perspective, women cease to be confined to an image of themselves as victims and social assistance beneficiaries. They become citizens who enjoy the right to:

- Participation, representation and decision making
- Equitable access to resources and production inputs
- Autonomy
- Control of their own body and protection from violence.[95]

The above principles guided the life of environmental and women's rights activist, Wangari Maathai (1940–2011). She is an inspiring example of a woman who managed to overcome multiple obstacles to become the first African woman to win the Nobel Peace Prize, in 2004. This former parliamentarian from Kenya won international recognition for her contributions to sustainable development, democracy, and peace.

Wangari Maathai embodies the essence of an activist woman in a developing nation. She was initially delayed from starting school when living on a farm where her father worked, but she was determined to pursue education. From early on she believed that education was a way out of poverty, and graduated with top marks from both primary and secondary school in Kenya. These achievements allowed her to obtain a rare opportunity to study in the United States, under the program created by John F. Kennedy in 1960. Using her education and strong will, she took up the role of advocate and women's rights activist in, what some would call, unconventional ways. Lobbying governments throughout Africa to invest more in education would remain one of her life passions. Upon her return to Kenya she was the founder of the Green Belt Movement, an organization that was dedicated to responding to the needs of rural women, by replanting forests and repopulating abandoned lands. The primary needs of the women, she notes, were a "lack of firewood, clean drinking water, balanced diets, shelter and income."[96] By planting trees, she notes that women were able to "gain some degree of power over their lives, especially their social and economic position and relevance in the family."[97] Maathai also took an active role in Kenyan politics, defending the rule of law and organizing political rallies. She ran for a parliamentary seat in 2002, winning it with 98 percent of the votes, and later served as the assistant minister for environment.[98] Maathai's courage and determination to empower disadvantaged individuals highlight her human-centered approach to development. In her own words:

> This work with local women to stimulate self-reliance reminded me how essential it is that political leaders, no matter what is happening at international or national levels, recognize the importance of improving conditions in people's daily lives.[99]

Practitioners and scholars have long maintained that "the most significant way to build lasting solutions to global problems and achieve economic development" is to enlist local women activists and make them equal partners in all endeavors.[100] Despite these calls, developmental agencies have been slow in prioritizing the goal of eliminating the barriers that prevent the advancement of women and girls.

> The gender gap in development assistance persists despite a substantial body of evidence confirming that investment in women yields high returns on poverty eradication and economic growth. Research demonstrates, for example, that reducing barriers to women's economic participation decreases poverty and increases GDP. Promoting gender equality also enhances food security: a study by the Food and Agriculture Organization shows that equalizing women's access to productive resources increases agricultural output and could reduce the number of hungry people in the world by 150 million. In addition, improving women's health has demonstrable economic effects: access to family planning, for instance, helps fuel economic growth, and increased female educational attainment not only raises household income but also lowers national health expenses and rates of infant and child mortality.[101]

In recognizing the pivotal role played by women in the economic development of societies, Canada's minister for international development announced a ground breaking shift in Canada's aid delivery. This new strategy is called the Feminist International Assistance Policy, and places women at the front of Canada's development agenda. Its principles aim at improving the effectiveness of aid delivery while assisting women in combating the key barriers and forms of exclusions that prevent them from becoming agents of positive change in the development of their communities. In summary, Canada's new Feminist International Assistance Policy prioritizes the following six action areas:[102]

1   Gender equality and the empowerment of women and girls
    This area of work is aimed at reducing sexual and gender-based violence. The assistance will be offered to organizations who actively empower women, provide services to survivors of violence, and work to advance the rights of women. Additional resources will be directed towards building recipient governments' capacity to provide specialized health and social services for women and girls.

2   Promotion of human dignity
    Under this action area, the focus is on timely and targeted humanitarian assistance to protect the most vulnerable, while recognizing particular needs of women and girls. The emphasis is on access to fundamental social services such as health care, education, adequate nutrition. Additional resources will be allocated to prevent gender-based violence during humanitarian crises, involving support for reproductive health needs and for counseling.

3   Fostering economic growth that works for everyone
    This action area is consistent with the feminist principle, promoting inclusive economic progress. The assistance will be focused on creating and strengthening relevant economic opportunities for every member of a society. Special attention will be devoted to ensuring that women and girls can become economically independent.

4   Promotion of environmental protection and encouragement of climate action
    Women, especially those women who are sole caretakers of the children and the elderly, are often most impacted by the depletion of natural resources and environmental problems. This action area encourages the participation of women as leaders in processes of designing and implementing policies dealing with environmental protection and climate change.

5   Working towards inclusive governance
    This action area recognizes the importance of good governance. In this context, it aims to support reforms strengthening the rule of law, inclusive institutions, and accountable government that respects basic human rights while it helps advancing social and economic opportunities for all citizens. Special attention will be paid to ending any form of discrimination against women while encouraging their greater participation in the political processes.

6   Strengthening global peace and security
    This action area means supporting "greater participation of women in peacebuilding and post-conflict reconstruction efforts, help to increase women's representation in the security sector and enforce a zero-tolerance policy for sexual violence and abuse by peacekeepers."[103]

It is important to note that all six action areas present interconnected sets of challenges, necessitating often simultaneous responses and concerted work on all of them in order to help improve the well-being of the most vulnerable individuals.

This comprehensive strategy has very practical suggestions for advancing a human-centered approach to economic development. Its emphasis on transformative change with respect to aid delivery challenges the agencies that deliver aid where it is most needed. The policy encourages gender-sensitive solutions that are innovative and effective. It also includes new monitoring services to avoid waste and duplication of services. The policy was well received worldwide, suggesting this could be a new model for the delivery of foreign aid.

Notes

1 Judith P. Zinsser (2002) "From Mexico to Copenhagen to Nairobi: The United Nations Decade for Women, 1975–1985," *Journal of World History*, Vol. 13, No. 1, p. 143.

2 J. Ann Tickner and Laura Sjoberg (2016) "Feminism," in Tim Dunne, Milja Kurki, and Steve Smith (eds.), *International Relations Theories – Discipline and Diversity* (4th edition), Oxford, UK: Oxford University Press, p. 180.

3 Cited on page 59 in Nancy Folbre (2009) *Greed, Lust, and Gender*, Oxford, UK: Oxford University Press.

4 Nancy Folbre (2009) *Greed, Lust, and Gender*, op. cit., p. 59.

5 Ibid., p. 60.

6 Ibid., p. 186.

7 Ibid., p. 223.

8 Martha C. Nussbaum (2000) *Women and Human Development – The Capabilities Approach*, New York, NY: Cambridge University Press, p. 106.

9 UNDP (2014) *The Future We Want: Rights and Empowerment – UNDP Gender Equality Strategy 2014–2017*, New York, NY: United Nations Development Programme, p. 27.

10 UNDP (2016) *Human Development Report 2016 – Human Development for Everyone*, New York, NY: UNDP, p. 46.

11 Diana Thorburn (2000) "Feminism Meets International Relations," *SAIS Review*, Vol. 20, No. 2, pp. 3–4.

12 Claudia von Braunmuhl (2002) "Mainstreaming Gender – A Critical Revision," in Marianne Braig and Sonja Wolte (eds.), *Common Ground or Mutual Exclusion? Women's Movements and International Relations*, London, UK: Zed Books, p. 57.

13 Ibid., p. 58.

14 Linda Cardinal, Annette Costigan, and Tracy Heffernan (1994) "Working Towards a Feminist Vision of Development," in Huguette Dagenais and Denis Piche (eds.), *Women Feminism and Development*, Montreal, Canada: McGill-Queen's University Press, p. 414.

15 Janet Henshall Momsen (2004) *Gender and Development*, London, UK: Routledge, p. 14.

16 Chandra T. Mohanty (1988) "Under Western Eyes: Feminist Scholarship and Colonial Discourse," *Feminist Review*, Vol. 30, No. 3, pp. 61–88.

17 UN (2001) *Supporting Gender Mainstreaming*, New York, NY: The Office of the Special UN Adviser on Gender Issues and Advancement of Women, March.

18 UNDP (2014) *The Future We Want: Rights and Empowerment – UNDP Gender Equality Strategy 2014–2017*, op. cit., p. 3.

19 UN (2017) *The Convention on the Elimination of All Forms of Discrimination Against Women*, New York, NY: The United Nations Treaty Collection. Online, available at: https://treaties.un.org/pages/Home.aspx?clang=_en accessed May 15, 2017.

20 UN (1995) *Fourth World Conference on Women Platform for Action*, Section 39. Online, available at: www.un.org/womenwatch/daw/beijing/platform/plat1.htm#concern.
21 Beijing Declaration and Platform for Action (1995) *Fourth World Conference on Women*, September 15, 1995, A/CONF.177/20 (1995) and A/CONF.177/20/Add.1 (1995), point 3. Online, available at: www1.umn.edu/humanrts/instree/bejingdec.htm.
22 Ibid., Section 12.
23 UNDP (2015) *Gender Equality in Human Development – Measurement Revisited*, Issue Paper Prepared for the Expert Group Meeting, New York, NY: UNDP-Human Development Report Office.
24 UNDP (2016) *Human Development Report 2016 – Human Development for Everyone*, New York NY: UNDP.
25 Martha C. Nussbaum (2000) *Women and Human Development*, op. cit., p. 56.
26 Ibid., pp. 79–80.
27 Ibid., p. 298.
28 Ibid., pp. 299–301.
29 Amartya Sen (1999) *Development as Freedom*, New York, NY: Random House, pp. 62–63.
30 William Easterly (2010) "Democratic Accountability in Development: The Double Standard," *Social Research*, Vol. 77, No. 4, p. 1076.
31 UNDP (2014) *The Future We Want: Rights and Empowerment – UNDP Gender Equality Strategy 2014–2017*, op. cit., p. 27.
32 UNDP (2014) *The Future We Want: Rights and Empowerment – UNDP Gender Equality Strategy 2014–2017*, op. cit., p. 27.
33 Simon Rabinovitch (2013) "Data Reveal Scale of China Abortions," *Financial Times*, March 15.
34 Betsy Hartmann (1995) *Reproductive Rights and Wrongs: The Politics of Population Control* (2nd edition), Boston, MA: South End Press, p. 42.
35 UNODC (The United Nations Office on Drugs and Crime) (2014) *The Global Report on Trafficking in Persons*, New York, NY: UN, p. 5.
36 UN-Habitat (2016) *World Cities Report 2016 – Urbanization and Development: Emerging Futures*, Nairobi, Kenya: UN-Habitat p. 16.
37 Arland Thornton (2001) "The Development Paradigm, Reading History Sideways, and Family Change," *Demography*, Vol. 38 No. 4, p. 454.
38 Amnesty International (2005) *Nigeria: Unheard Voices: Violence Against Women in the Family*, May 31.
39 Janet Henshall Momsen (2004) *Gender and Development*, London, UK: Routledge, p. 141.
40 Susie Jacobs (2004) "Livelihoods, Security and Needs: Gender Relations and Land Reform in South Africa," *Journal of International Women's Studies*, Vol. 6, No. 1, p. 12.
41 Bharati Sadasivam (1997) "The Impact of Structural Adjustment on Women: A Governance and Human Rights Agenda," *Human Rights Quarterly*, Vol. 19, No. 3, pp. 648–649.
42 Lotsmart Fonjong (2004) "Challenges and Coping Strategies of Women Food Crops Entrepreneurs in Fako Division, Cameroon," *Journal of International Women's Studies*, Vol. 5, No. 5, pp. 3–12.
43 Lisa L. Gezon (2002) "Marriage, Kin and Compensation: A Socio-Political Ecology of Gender in Ankarana, Madagascar," *Anthropological Quarterly*, Vol. 75, No. 4, pp. 675–706.
44 Ibid., p. 676.
45 Ibid., pp. 682–684.
46 Muhammad Yunus with Alan Jolis (2001) *Banker to the Poor – The Autobiography*, Dhaka, Bangladesh: The University Press Limited, p. 88.
47 Ibid., pp. 88–90.
48 Martha C. Nussbaum (2000) *Women and Human Development*, op. cit., p. 243.
49 Ibid., p. 287.
50 Beatriz Padilla (2004) "Grassroots Participation and Feminist Gender Identities: A Case Study of Women from the Popular Sector in Metropolitan Lima, Peru," *Journal of International Women's Studies*, Vol. 6, No. 1, pp. 109–110.

51 Tamma Kaplan (2001) "Uncommon Women and the Common Good: Women and Environmental Protest," in Sheila Rowbotham and Stephanie Linkogle (eds.), *Women Resist Globalization: Mobilizing for Livelihood and Rights*, London, UK: Zed Books Ltd., p. 35.

52 Ibid., p. 32.

53 Martha C. Nussbaum (2000) *Women and Human Development*, op. cit., p. 78.

54 Candace Johnson (2014) *Maternal Transition: A North–South Politics of Pregnancy and Childbirth*, London, UK: Routledge.

55 WHO (2004) Department of Reproductive Health and Research of the WHO "Sexual Health – a New Focus for WHO," *Progress*, No. 67. p. 2.

56 Michelle Lewis Renaud (1997) *Women at the Crossroads: A Prostitute Community's Response to AIDS in Urban Senegal*, New York, NY: Routledge, pp. 79–80.

57 UN.GIFT (2010) *Human Trafficking and Business: Good Practices to Prevent and Combat Human Trafficking*, Report Prepared by United Nations Global Initiative to Fight Human Trafficking (UN.GIFT).

58 UNFPA (2008) *UNFPA Strategy and Framework for Action to Addressing Gender Based Violence 2008–2011*, New York, NY: The United Nations Population Fund (UNFPA).

59 UN (1995) *Fourth World Conference on Women Platform for Action*, Section 113. Online, available at: www.un.org/womenwatch/daw/beijing/platform/violence.htm.

60 Amnesty International (2005) *Nigeria: Unheard Voices: Violence Against Women in the Family*, May, 30 London, UK. Index number: AFR 44/004/2005. Online, available at: www.amnesty.org/en/documents/afr44/004/2005/en/

61 Ibid., p. 3.

62 Amnesty International (2005) *Guatemala: No Protection, No Justice: Killings of Women in Guatemala*, Report, June.

63 Ibid.

64 UNDP (2014) *The Future We Want: Rights and Empowerment – UNDP Gender Equality Strategy 2014–2017*, op. cit., p. 9.

65 Ibid., pp. 10–13.

66 Rachel Vogelstein (2016) "Development's Gender Gap – The Case for More Funding for Women and Girls," *Foreign Affairs – Snapshot*, July 26.

67 UNDP (2014) *The Future We Want: Rights and Empowerment – UNDP Gender Equality Strategy 2014–2017*, op. cit., p. 5.

68 UNDP (2016) *Human Development Report 2016 – Human Development for Everyone*, p. 6.

69 Kim Campbell briefly held office as Canadian prime minister in 1993 after the resignation of Brian Mulroney, but she lost the subsequent election.

70 UNDP (2015) *Human Development Indicators for 2015*, New York, NY: UNDP, table 5.

71 Nadezdha Shvedova (2002) "Obstacles to Women's Participation in Parliament," in *Women in Parliament*, Stockholm: International IDEA.

72 UNDP (2015) *Human Development Indicators for 2015*, op. cit., table 5.

73 Ibid.

74 Pippa Norris and Ronald Inglehart (2001) "Cultural Obstacles to Equal Representation," *Journal of Democracy*, Vol. 12, No. 3. p. 134.

75 Patrice C. McMahon (2002) "International Actors and Women's NGOs in Poland and Hungary," in Sarah E. Mendelson and John K. Glenn (eds.), *The Power and Limit of NGOs: A Critical Look at Building Democracy in Eastern Europe and Eurasia*, New York, NY: Columbia University Press, p. 32.

76 Ibid., pp. 50–51.

77 UNDP (2015) *Human Development Indicators for 2015*, op. cit., table 5.

78 Andrew Reynolds (1999) "Women in the Legislatures and Executives of the World: Knocking at the Highest Glass Ceiling," *World Politics*, Vol. 51. No. 4, p. 551.

79 Nadezdha Shvedova (2002) "Obstacles to Women's Participation in Parliament," op. cit.

80 Gehan Abu-Zayd, "In Search of Political Power – Women in Parliament in Egypt, Jordan and Lebanon," in *Women in Parliament*, Stockholm, Sweden: International IDEA, p. 4.

81 Ibid., p. 5.

82 Nadezdha Shvedova (2002) "Obstacles to Women's Participation in Parliament," op. cit.
83 Valentine M. Moghadam (2005) *The "Feminization of Poverty" and Women's Human Rights*, Paris, France: The Gender Equality and Development Section of UNESCO, p. 6.
84 Ibid., pp. 21–22.
85 Ibid., pp. 22–23.
86 Ibid., pp. 31–32.
87 UNDP (2015) *Human Development Indicators for 2015*, op. cit., table 5.
88 Ibid. These countries are placed in order ranging from 9.9 percent in Cambodia to 5.6 percent in Tanzania.
89 Ibid., table 5.
90 Mahbub ul Haq and Khadija Haq (1998) *Human Development in South Asia*, Oxford, UK: Oxford University Press, p. 87.
91 Mavivi Myakayaka-Manzini (2002) "Women Empowered – Women in Parliament in South Africa," in Nadezdha Shvedova, *Obstacles to Women's Participation in Parliament*, op. cit.
92 Ibid.
93 Seungsook Moon (2002) "Women and Democratization in the Republic of Korea," *Good Society*, Vol. 11, No. 3, p. 37.
94 UNDP (2015) *Human Development Indicators for 2015*, op. cit., table 5.
95 UN (2000) *United Nations Security Council Resolution 1325 on Women, Peace and Security*. Online, available at: www.un.org/womenwatch/osagi/cdrom/documents.
96 Wangari Maathai, Nobel Lecture, December 10, 2004, Oslo, Norway. Online, available at: http://nobelprize.org/peace/laureates/2004/maathai-lecture-text.htm.
97 Ibid.
98 BBC News (2004) "Wangari Maathai Rose to Prominence Fighting for those Most Easily Marginalized in Africa – Poor Women," Friday, October 8. Online, available at: http://news.bbc.co.uk/1/hi/world/africa/3726084.stm.
99 Wangari Maathai (2009) *The Challenge for Africa*, New York, NY: Pantheon Books, p. 127.
100 Danny Glenwright (2017) "Ottawa's Focus on Aid Delivery for Women is Smart – and Will Save Money," *Globe and Mail*, June 11.
101 Rachel Vogelstein (2016) "Development's Gender Gap – The Case for More Funding for Women and Girls," op. cit.
102 Government of Canada (2017) *Canada's Feminist International Assistance Policy*. Online, available at: http://international.gc.ca, accessed June 9, 2017.
103 Ibid.

Suggested further reading

Dara Kay Cohen (2016) *Rape During Civil War*, Ithaca, NY: Cornell University Press.
Helene Cooper (2017) *Madame President: The Extraordinary Journey of Ellen Johnson Sirleaf*, New York, NY: Simon & Schuster.
Jana Everett and Sue Ellen M. Charlton (2013) *Women Navigating Globalization: Feminist Approaches to Development*, Lanham, MD: Rowman & Littlefield.
Namulundah Florence (2014) *Wangari Maathai: Visionary, Environmental Leader, Political Activist*, Brooklyn, NY: Lantern Books.
Ann Marie Leshkowich (2014) *Essential Trade: Vietnamese Women in a Changing Marketplace*, Honolulu, HI: University of Hawaii Press.
Janet Momsen (2013) *Gender and Development* (2nd edition), London, UK: Routledge.
Aili Mari Tripp (2015) *Women and Power in Postconflict Africa*, Cambridge, UK and New York, NY: Cambridge University Press.

Environment, sustainability, and development   7

1 Conceptualizing sustainable development

For the most part of the world's history, humans have pursued technological advancements and used the available land with little attention paid to the conservation of natural environments. Scarcity is one of the defining aspects of economic analysis, but this term was not born out of concern for nature. The dilemma of scarcity is about limited resources for a society with unlimited wants. Classical theories of economics were never preoccupied with the issue of Earth's limited natural resources. Only over the last few decades, have policy makers started to acknowledge environmental harms caused by rapid industrialization, resource extraction, and overuse of land. The prerogative to consume natural resources wisely has become more pronounced as the global population continues to grow.

There are inescapable linkages among economic activities, state policies, natural resources, the ecosystem, and how societies function. The concept *sustainable development* was coined in recognition of how connected the issues at stake are. Formal debates about the environment first emerged at the 1972 United Nations Conference on the Human Environment in Stockholm, Sweden. The central themes of the conference were:

- the interdependence of human beings and the natural environment;
- the links between socio-economic development and environmental protection;
- the need for a global vision and common approach to environmental problems.

Building upon these themes, the World Commission on Environment and Development released a report in 1987 defining sustainable development as: "development that meets the needs of the present without compromising the ability of future generations to meet their own needs."[1]

It took another five years until a comprehensive set of multilateral documents dealing with the issue of sustainable development were adopted by more than 178 governments. The acceptance took place at the 1992 UN Conference on Environment and Development, held in Rio de Janeiro, Brazil. At the same time, the UN created a special Commission on Sustainable Development, tasked with monitoring and reporting on the implementation of the landmark Rio agreement. In 2013, this commission was replaced by the UN high-level political forum (HLPF) on sustainable development, which is currently responsible for adopting negotiated declarations and for helping to define the developmental agenda of the UN.

The concept of sustainable development has been politically successful in raising awareness about the fragility of our planet and its natural resources, but it has also generated criticism. Scholars and policy makers complain about the broad scope of the term, which endorses a generalized approach to sustainability. Specifically, they observe that the conceptual framework of sustainable development fails to meet rigorous meta theoretical criteria, which include analytical definitions of the objectives and the constructs employed to fulfill them. As a result, the concept has limited ability to engender research and policy that can lead to sustainable development.[2] The Millennium Development Goals are a case in point, falling short of meeting the stated aspirations by the target date of 2015. In a move to strengthen the global consensus to eradicate poverty, the UN General Assembly adopted in September 2015 a new comprehensive package of environmental, social, and economic objectives, known as Sustainable Development Goals. There are now 17 goals, which are expected to galvanize international efforts to end poverty, protect the planet, and ensure the prosperity of all over the next 15 years. These goals are part of the UN's new sustainable development agenda.[3]

The saliency of the environmental issues is increasingly recognized by institutional economists, who argue in this context for relevant reforms of economic organizations. The need for reforms arises from three interrelated factors: 1. the growing interdependence of the international economy and the physical environment; 2. the explosive impact of demography and technology; 3. the tight physical constraints on governments and developmental agencies.[4] Global market integration provides many opportunities, but also causes externalities prompted by rapid economic and technological progress. Individual states, and even regional clusters of states, cannot effectively address global warming, acid rain, pollution of air and oceans, public health crises, large-scale migration of people, and shortage of clean water. Such problems have an environmental dimension transcending the political boundaries of states, and this is why they require collective action.

Over the years, several international conventions have been signed by concerned countries to promote cooperation on environmental issues. One of the most notable is the 2015 Paris Agreement, negotiated under the UN Framework Convention on Climate Change. The deal committed its signatories to reduce greenhouse gas emissions by targeted dates.[5] However, the anticipated withdrawal of the United States from the agreement under the new US administration in 2017 threatened the deal and revealed its limitations. While environmental issues are gaining importance, there is only a fragile consensus about what kind of multilateral policy measures should be put in place. Furthermore, there is no enforcement mechanism to ensure that such measures are implemented. The 2017 World Development Report focuses on promoting good governance, assessed in terms of governments' capacity to deliver: peace, justice, and strong institutions. The report begins with a normative statement:

every society cares about freeing its members from the constant threat of violence (*security*), about promoting prosperity (*growth*), and about how such prosperity is shared (*equity*). It also assumes that societies aspire to achieving these goals in environmentally sustainable ways.[6]

In a progressive move, members of the WTO established the Committee on Trade and Environment: "to identify the relationship between trade measures and environmental measures in order to promote sustainable development."[7] To date, however, there is no global agreement linking trade and the environment. Sustainable development is a widely used concept informing the agendas of the UN, the IMF, the World Bank, and the WTO, but none of these organizations contain a legally binding treaty on the issue.

II The sustainability of economic advancement as the population grows

The idea of sustainable development began to take shape in the early 1970s as signs of the environmental damage were becoming more visible across the planet. It was also a reaction to the cornucopian perspective on growth in the latter part of the twentieth century. The term cornucopian is derived from the ancient Greek "horn of plenty." Cornucopians reject the notion that Earth has finite resources that need to be protected. They expound the belief that technological innovation will obviate any potential environmental catastrophe and remedy any ecological imbalance that poses harm to humanity.[8] With respect to the future, cornucopians tend to be libertarians maintaining that the global economy will continue to grow as long as free markets are allowed to operate with minimum restrictions. Nonetheless, following the oil crises of the 1970s and greater awareness of environmental degradation around the globe, cornucopian dismissal of environmental barriers to growth began to be challenged. Scholars and economists alike started to wonder whether the expansion of the global economy based on the principle of unlimited growth was indeed sustainable. In the developing world, where rainforests were slowly disappearing and fertile lands were lost to desert, there was greater impetus for a novel approach that would balance economic interests with environmental needs.

When the report "Our Common Future" defined sustainable development it also provided a foundation for including environmental matters in the debate about developmental policies. The report is often called the Brundtland Report after the chairperson in charge of preparing the document, then prime minister of Norway, Ms. Gro Harlem Brundtland. She wrote in the foreword:

> Many critical survival issues are related to uneven development, poverty, and population growth. They all place unprecedented pressures on the planet's lands, waters, forests, and other natural resources, not least in the developing countries. The downward spiral of poverty and environmental degradation is a waste of opportunities and of resources. In particular, it is a waste of human resources.[9]

The report is one of the seminal documents of the twentieth century. It represents global recognition of the environmental problems facing the planet and the connection between poverty, inequality, and environmental degradation.

Greater need for sustainable development came when the forces of economic globalization appeared to clash with fears over the shortage of natural resources and the continuing dependence on ecologically noxious fossil fuels for energy. However, the philosophy behind sustainable development is not veiled in pessimism. Environmental pessimists are radical in advocating major slowing down of the world economy. In direct opposition to cornucopians, they believe that there are limits to growth and the current rate of depleting the resources is leading us to an inevitable catastrophe. On the other hand, the premise of sustainable development supports the notion that an economic growth is possible but it has to be carefully reexamined. As the Brundtland Report concludes:[10]

> The concept of sustainable development implies limits – not absolute limits, but limitations that the present state of technology or social organization and the capacity of the biosphere to absorb the effects of human activities impose on the resources of the environment –, but both technology and social organisation can be organised and improved so that they will open the way to a new era of economic growth.

The Brundtland Report remains as relevant as ever today for pointing out the main areas where the work has to be done in order to achieve sustainable development. There must be greater environmental conservation and technologies should be innovated to acquire greater energy efficiency and be used in ways friendly to the environment. The report pays special attention to poor countries of the Global South. It is concerned with securing a worldwide fairness for future generations. The report suggests that social equality, economic growth and environmental sustainability are not mutually exclusive, but rather should be balanced in a way that each society can achieve economic progress while respecting the environment and allowing all people to meet their fundamental needs. Some of the more controversial findings of the report recognize that achieving sustainable economic growth requires technological and social change.

The Brundtland Report recommended urgent action on eight key issues to ensure sustainability. These eight issues are:

1 Population and human resources

The Brundtland Report called for urgent action to address population growth, referring to the fact that only in 1985 some 85 million people were added to the then existing global population of 4.8 billion. Fast forward to the current UN report. As of August 2016, the UN estimated that the global population reached 7.4 billion and could pass the eight billion mark by 2025. The UN also points out that a worrisome gap between the number of people on the planet and the available resources "is all the more compelling because so much of the population growth is concentrated in low-income countries, ecologically disadvantaged regions, and poor households."[11] There is no equity when it comes to the resource consumption per capita since the developed countries use a disproportionally larger amount of the Earth's resources.

2 Food security

The Brundtland Report called for urgent action to combat the widespread hunger that continued to exist in many parts of the world, despite the dramatic increase in the global production of cereal, meat, and milk since 1950. The report observed that this increase in food production has been due to new methods of farming, including the use of new seed varieties, and increased irrigation, but the cost of new inventions placed them beyond the reach of most small farmers in developing countries. Farm subsidies in the North resulted in over-production that relies on the new methods of farming. These methods have had detrimental effects on the environment as the overuse of chemical pesticides and fertilizers has led to widespread pollution of water and biological magnification of these chemicals in food chains, which in turn contributes to widespread soil degradation. The report concluded that food security requires good conservation programs, which can be undermined by inappropriate agricultural, economic, and trade policies contributing to the overuse of land and other natural resources. In 2017, agriculture remains the most distorted area of international trade.[12] Millions of people still go to bed hungry. Food security is listed second on the UN's 2015 list of Sustainable Development Goals.[13]

3 Species and ecosystems

The Brundtland Report stresses the vital importance of preserving the world's plants, animals, and micro-organisms. According to the report, only 4 percent of the Earth's land area is dedicated to the conservation of species and eco-systems. It also noted that the most bio-diverse ecosystems are the wet tropical forests. The forests of Latin America were estimated to contain over one million species of plants, animals, birds, and insects. The destruction of these forests would have serious implications for the world since a substantial proportion of the production of medicines depends on species found in the tropical forests. Over the last few decades, the extinction of many species has accelerated. Many protected areas are becoming commercialized due to population pressures and economic considerations. In 2016, the UN reported that in tropical countries the annual net forest loss equaled seven million hectares, or the size of Ireland. Protection of ecosystems and forests is listed fifteenth on the 2015 Sustainable Development Goals list.[14]

4 Energy

The Brundtland Report stressed the fact that our primary sources of energy are non-renewable: natural gas, oil, coal, peat, and conventional nuclear energy. The dependence on fossil fuels has had four major consequences: 1. the serious probability of climate change resulting from the emissions of carbon dioxide (Greenhouse Effect); 2. urban and industrial air pollution from the combustion of fossil fuels; 3. acidification of the environment from the same causes; 4. health risks to the workers of nuclear plants involved in the production and disposal of wastes, and a possibility of a catastrophic accident leading to the destruction of communities around the power plants. As a result, the report called on governments to act in concert and to invest in renewable energy sources. The report estimated that, by the year 2025, global energy consumption would increase by 40 percent over 1980 figures. The contemporary situation fits the worst scenario predicted by the report. Global demand for oil in 2014 was about 92.6 million barrels a day (mb/d), as the International Energy Agency (IEA) reported. In 2015, growth in global demand was revised up by 1.4 mb/d to a daily average of 94 million barrels. The IEA approximates that fossil fuels make up more than 75 percent of this increase, with coal and oil remaining central to the primary fuel mix. The same agency also estimates that US$26 trillion (in 2008 dollars) of investment is needed through to 2030 to address the growing energy demand, especially in view of better standards of living in the emerging economies.[15] Ensuring access to affordable, reliable, and sustainable energy for all is seventh on the UN's 2015 Sustainable Development Goals list.[16]

5 Industrialization

The Brundtland Report criticized reckless industrialization, which brought economic growth but had a devastating environmental impact. The report predicted a slow decrease in manufacturing production among the developed nations and a corresponding increase in the developing world. Composition of the merchandise trade of developing countries showed the steady growth of exports of manufactured goods relative to primary goods from 13.3 percent of total non-oil exports in 1960 to 54.7 percent in 1982.[17] The study by UNCTAD from 2014 shows an interesting continuation of this trend. Most importantly, developing countries remain, as they were at the time of the Brundtland Report, the main suppliers of primary products and natural resources for international markets. Developing countries' exports represent about two-thirds of international trade in primary products, and about three-quarters of that in natural resources.[18] The ability of countries in the Global South to ensure the responsible extraction of natural resources, the disposal of hazardous wastes and industrial pollution is limited. Although the developed countries have relevant technology, the prohibitive costs and technological inequalities between countries prevent the developing world from taking advantage of technological solutions. The Brundtland Report has already called for urgent action to help the poor nations in this respect. Sustainable industrialization is listed as ninth on the UN's 2015 list of Sustainable Development Goals.[19]

6 The urban challenge

The Brundtland Report projected that by the turn of the century about 50 percent of the world's population would be living in urban centers. Indeed, according to a recent UN study, 54 percent of people on Earth in 2014 lived in urban areas, and this number is expected to rise to 66 percent by 2050. The incoming growth of mega-cities with 10 million inhabitants or more will be mainly concentrated in Asia and Africa.[20] The rapid spread of cities has resulted in inadequate urban infrastructure, weak social services, inadequate housing, and environmentally related health concerns. The Brundtland Report noted that fast-growing cities in the developing world would experience shortages of basic services – clean water, sanitation, schools, and transport due to the lack of resources. The report called for national strategies on the issue of sustainable urban development. It also advocated strengthening local communities to stimulate a geographically balanced economic progress.[21] Making cities inclusive, safe, resilient, and sustainable is listed eleventh on the UN's 2015 list of Sustainable Development Goals.[22]

7 Managing the commons

The Brundtland Report called for concerted international actions with respect to the *global commons*. It also recognized that the realities of growing economic and ecological interdependence have challenged the principle of state sovereignty with respect to the oceans, Antarctica, and outer space. They are part of the global commons, for which all nations have joint responsibility. Sustainable development in the global commons "can be secured only through international cooperation and agreed regimes for surveillance, development, and management in the common interest."[23] The report noted that marine environmental problems, such as over-fishing and marine pollution, are increasing rapidly. The 2016 UNDP report offers some devastating data about the current condition of the Earth's oceans and seas. For example: every year, 300 million tons of plastic are manufactured, but only 15 percent is recycled, leaving 46,000 floating pieces of plastic per square mile of ocean. The goal to conserve and sustainably use the oceans, seas, and marine resources ranks seventeenth on the 2015 UN list of Sustainable Development Goals.[24]

8 Peace, security, development, and the environment

In one of its concluding chapters, the Brundtland Report identified environmental stress as a source of conflict. The scarcity of life-sustaining resources, such as fresh water, over-exploitation of the land, hazardous pollution, overall depletion of natural resources, and climatic changes lead to famine and deepening of poverty, which in turn contribute to social unrest and conflict. The report also warned that excessive military expenditures could divert funds from the urgent environmental problems facing developing nations. Both the developing and the developed nations were asked to re-consider their spending priorities in view of the report's findings. Promoting peaceful societies is sixteenth on the 2015 UN list of Sustainable Development Goals.[25]

The findings of the Brundtland Report sound surprisingly relevant today, although 30 years have passed since its publication. The report placed environmental concerns at the forefront of debates about the future direction of developmental policies, but also received some noteworthy criticism. Critics asked how to assess sustainable development? They pointed out that the Brundtland Report's definition of the term was too simplistic and belied the complexities regarding the issue. Another major criticism stemmed from the report's insistence on the importance of technology, which privileged material well-being and the economic status of developed economies. The report also failed to provide concrete targets or guidelines, allowing policy-makers to determine what sustainability actually meant in a particular context.[26]

Notwithstanding the difficulties of reaching consensus among different societies on strategies aimed at achieving sustainable development, the present generation must weigh the opportunity costs of short-term economic growth and the long-term prospects for maintaining a healthy planet. The developmental agendas coined by the major international organizations hope to eradicate poverty everywhere, but the overall world's population is continuously expanding, with poor developing countries registering the highest increases. The UN estimates that by 2100, the population will quadruple in one of the poorest regions of the world, sub-Saharan Africa.[27] The issue of population growth is complicated because of its ethical dimension. Most of the world's religions openly prohibit measures to control natural reproduction processes.

Thomas Robert Malthus (1766–1834), was an English demographer and political economist, best known for his pessimistic but highly influential theory about population growth. In his paper, written during the Industrial Revolution in England, "An Essay on the Principle of Population," he predicted that unrestrained population growth would eventually overwhelm the food supply and thus lead humanity into catastrophe.[28] His views challenged the traditional understanding of high fertility in positive terms as an economic plus since it increased the number of workers available to the economy.

Malthus never gave a timeframe for when his catastrophe would occur, but it is safe to say that it has not happened yet. One explanation can be found in the demographic transition theory, which in a nutshell suggests that a society's transition from high mortality and high fertility to low mortality and low fertility rates as economic development progresses.[29] Such transition was observed in the twentieth century, when the most technologically advanced countries of Europe saw their population stabilize, with low birth and death rates, because of a range of factors, including rising living standards, availability of contraception, education of women, and advances in medicine. In poorer societies, where infant mortality rates are great and the availability of advanced medicine is low, birth rates are high because families feel uncertain about the lasting survival of their infants. The theory suggests that economically advanced modern societies tend to feel little necessity to bear many children and so fertility rates will fall. The present asymmetrical growth of the world's population gives some credence to this argument.

Differing attitudes about the accelerating growth of the global population have divided the development community. The corollary to Malthus' dire predictions was delineated in a report that continues to arouse debate on the topic of overpopulation to this day. The study called *Limits of Growth*, prepared by a global think-tank called the Club of Rome in 1972, used a computer simulation to calculate the alarming consequences of an expanding human population, given a finite pool of resources available on the planet. Some 30 years later, a group of scientists revisited the findings of the original report, to conclude that humanity was indeed going deeper into unsustainable territory. The study pointed to rising sea levels, depletion of fisheries, deforestation, soil loss, and polluted water supplies.[30]

There has been much criticism of the gloomy forecasts of *Limits to Growth*, but others think that we may have already witnessed societal collapse due to overstretched natural resources. In his book *Collapse: How Societies Choose to Fail or Succeed*, Jared Diamond postulates that Rwanda's high population density, and shortages of food brought on by soil erosion, droughts, and local deforestation contributed to the eruption of the 100-day genocide of 1994. Diamond writes how agricultural production and the Rwandan population both grew unequally following Rwanda's independence from Belgium. Compounding the problem of the inefficient and traditional agricultural methods still prevalent in Rwanda was that the high population density would squeeze out potential farmland. Furthermore, farmland in itself became highly coveted in Rwanda as wealthier farmers bought more land from the poor and land disputes became more commonplace.[31]

Another scholar who shares the pessimism about population growth suggests there should be strict controls on reproductive choice. A microbiologist by training, Garrett Hardin became famous for his essay, "The Tragedy of the Commons," which evoked the case of common land where everyone had the right to graze their cattle. The essay argues that even when the common land becomes overgrazed, people continue to feed their animals and further destroy the fields. According Hardin, individuals see no point in preserving the commonly available resource if others continue to use it. People may be aware of the risk that comes with destroying the land, but the mix of selfishness, greed, and unregulated exploitation eventually makes the land unusable for all.[32] Hardin's argument supports the promotion of private property ownership, and strict governmental regulations over fertile land.

Peter Bauer, a professor of political economy at the London School of Economics, admonishes against what he considers to be erroneous reasoning on the subject of population growth. The problem, as Bauer sees it, lies not in population-pressure, but rather in incompetent economic policies. With sound policies that mirror western economies and contain western attitudes on fertility, the attainment of higher incomes will eventually lead to a decline in fertility rates in the developing world.[33]

Most of the issues identified in the Brundtland Report remain top developmental priorities today, as identified by the UN's 2015 Sustainable Development Goals. There is one notable exemption. The Brundtland Report placed concerns over global population growth at the top of its agenda for action. In contrast, the 2015 UN Resolution that adopted the Sustainable Development Goals hardly even mentions this issue. This may be surprising, given the fact the world in 2017 has over seven billion people and is projected to reach eight billion by 2025. The ethical considerations and recognition that reproductive rights are individual rights make it difficult to address population-related concerns through universal goals. It is also true that technological advances make it possible to feed every person on the planet. The biggest challenge of today is not necessarily the population growth but how we utilize the available natural resources in a sustainable way.

III Non-renewable resources, energy security, and the power of oil

Population growth is only one of the issues that highlight the necessity of preserving the environment. Another is the reliance of modern economies on oil and gas. Our dependence on those non-renewable energy sources further complicates the relationship between socio-economic development and sustainable use of natural resources. In his seminal book on the subject, Daniel Yergin identifies three important themes that underlie the global history of oil. First, oil has played a pivotal role in the development of the modern capitalist economy. Second, oil is closely linked with national developmental strategies and global politics and power. Third, we have become a hydro-carbon civilization so dependent on oil that we often tend to overlook the environmental issues that come with this dependence.[34] From its early days, petroleum allowed the growth of the car industry and accelerated the design of the first planes. With cars came demands for networks of roads and highways, forever changing the landscapes of communities. Today both oil and natural gas have boundless commercial applications – from transportation and a power source for most industries, to being essential components in fertilizers, plastics, and chemicals – our civilization would collapse without oil. Anxieties over ensuring an adequate supply have resulted in international agreements, but have also provoked conflicts. Oil companies remain the biggest multinational corporate entities in the global economy. However, this dominant position of oil in the global economy is challenged by the environmental movement. The story of oil is indeed intertwined within the dynamics of international relations, and this is why the sustainability of our economic development has to be taken into account in the context of energy security.

Demand for oil keeps rising, as industrialization and economic growth has occurred internationally. Table 7.1 offers a brief historical look into our continuing dependence on oil. The numbers come from the most recent report prepared by the IEA and the

Table 7.1 Historical comparison of oil consumption in selective regions of the world (in thousand tons)

Regions	1971	1990	2013
Africa	29,273	67,948	141,693
Central and South America	72,042	117,932	214,870
Asia (excl. China)	52,225	146,079	399,109
China (incl. Hong Kong)	36,849	86,165	433,147
World	1,954,198	2,529,664	3,584,794

OECD.[35] The table focuses on the developing regions of the world, which are particularly interested in having an adequate supply of oil to advance their economic development.

The same report also shows an incredible increase in the production of electricity from environmentally destructive fossil fuels over the same period. In 1971, the world produced 3,905,386 gigawatt hours (GWh) of electricity from fossil fuels. Where 1 GWh equals one billion watt hours or 3.41 billion British thermal units (Btu). In 2013, that number has risen to 15,715,235 GWh. The biggest chunk of this growth occurred in the emerging economies. China, for example, only produced 114,636 GWh of electricity from fossil fuels in 1971, but in 2013 its production expanded to 4,235,474 GWh.[36] Also, despite the invention of a more environmentally friendly gasoline and better engines for cars, ships, and planes, the world's consumption of oil in transport continues to rise from 871,050 thousand tons in 1971 to 2,287,853 thousand tons in 2013.[37]

Between 1990 and 2008, global oil consumption grew almost 30 percent, from 67 million to 86 million barrels a day.[38] According to the most recent data included in "International Energy Outlook 2016," prepared by the US Energy Information Administration (EIA), there will be a significant increase in worldwide energy demand over the period from 2012 to 2040. Total consumption of energy is expected to expand from 549 quadrillion Btu in 2012 to 629 quadrillion Btu in 2020, and to 815 quadrillion Btu in 2040, which is a 48 percent increase from 2012 to 2040.[39] Such data explains why energy security has long dominated the agendas in economic policy-making.

Concern for energy security was in fact at the heart of the first talks leading to the European integration. In the words of Jacques de Jong:

> The EU project started with an energy source, with the creation of the European Coal and Steel Community (ECSC) in 1952. This ECSC Treaty, signed in Paris in 1951 by France, Germany, Italy and the three Benelux-countries, was a desire to unite these countries by controlling steel and coal, which were fundamental to the war industries. It therefore had a strong basis in post WW-II thinking on peace building, using a "peace-through-energy" approach.[40]

Despite the enormous importance of petroleum for energy security, there is no international legally binding treaty encouraging countries to coordinate their energy policies. There is no agreement to ensure stability and sustainability when it comes to the exploration for and distribution of oil. On the contrary, the history of oil has been characterized by conflicts, fierce competition, and uncertainly of prices.[41]

Since the dawn of man, people have cut down trees and gathered wood so they can provide themselves with warmth and the requisite heat for cooking. Coal, a fossil fuel, fueled the Industrial Revolution in Europe and made way for important technological innovations that helped advance the sciences. Coal remains a cheap source of energy in many developing countries, including China and India. The customary resistance of developing countries to make a commitment on curbing fossil fuel emissions is explained by pointing out that the developed world was allowed to industrialize with no limits attached.

Though coal and wood are still used to varying degrees around the world, the most important source of energy that helps drive modern economies and societies is petroleum

– a type of oil that is known colloquially as black gold and crude oil. It is most fitting that petroleum is considered to be black gold when unrelenting demand for it has sparked a massive expansion of exploratory technologies, allowing for deep water oil exploration and the extraction of oil from kerogen shale, otherwise known as tarsand fracking (or hydraulic fracturing).

For the better part of the twentieth century, the global trade in oil was dominated by the so-called Seven Sisters, or seven major oil companies. The Seven Sisters formed in the wake of the breakup by the US Government of Standard Oil. Standard Oil of New Jersey was established in 1870 by John D. Rockefeller, who was a chairman and major shareholder. Early on, Rockefeller recognized the importance of petroleum, which was soon to change the developmental trajectories of countries around the world. It is well documented that Rockefeller was ruthless in expanding Standard Oil, often through the use of questionable tactics that included outright demands for concessions from the railroads, manipulation of freight rates, industrial espionage, and even threats against competitors. These tactics worked, because by 1879, the Standard Oil Company of New Jersey controlled 90 percent of US refining capacity.[42] In 1906, the Roosevelt Administration brought suit against Standard Oil under federal anti-trust law (the Sherman Anti-trust Act of 1890) for sustaining a monopoly and restraining interstate commerce. The case ended in 1911, when the US Supreme Court dissolved Standard Oil into several separate corporate entities. The largest of the original Seven Sisters was Standard Oil of New Jersey, which later became Exxon, and never lost its lead, followed by Standard Oil of New York, which became Mobil and Standard Oil of California, which eventually became Chevron.[43]

The way Standard Oil was divided did not prevent its former "parts" from working in harmony, and the new companies carried on their old commercial relationships. In March of 1938, oil was discovered by Standard Oil of California in the Saudi sands. Socal and Texaco formed the joint venture called Aramco (Arabian American Oil Company) to develop these resources.[44] By the 1940s, the US based oil companies became powerful players in the Middle East, a group that included also Dutch and British firms. The American discovery in Saudi Arabia was a momentous event since it initiated American companies into the geopolitical game of oil and Middle Eastern politics. The fact that American companies were even allowed to dig for oil in the Middle East was a result of hard negotiations and signing of the so-called Red Line Agreement in July 1928.[45] Until this agreement was signed, European, mainly British, oil companies had a virtual monopoly in places such as Iran, Iraq, and several smaller Gulf States. They grudgingly allowed US companies somewhat limited access inside *the red line*, which included Saudi Arabia. The American discovery there was a beginning of a long-standing US–Saudi relationship.[46]

From a developmental point of view, it is a fascinating story. When in 1933 Standard Oil of California reluctantly agreed to offer the Saudis a small up-front payment for the concession to look for oil in their unpredictable desert, the country consisted mostly of nomadic communities. By the late 1970s, Saudi Arabia was on the path to become one of the richest countries in the world, although its wealth was totally dependent on oil. In principle, Aramco was nationalized by Saudi Arabia in 1976, but an agreement was reached with the foreign partners to ensure the continuation of the commercial relationship. The Saudis still relied on foreign processing, transportation, and marketing networks.[47]

Until the 1970s, the Seven Sisters dominated their home and international markets "through vertical integration, that is, controlling supply, transportation, refining, and marketing as well as exploration and refining technologies."[48] They were well organized, and hence able to negotiate successfully with oil producers in developing countries. The outcome was stable supply markets and continuing low prices. To counter the influence of the Seven Sisters over the oil prices and production, four Persian Gulf nations (Iran, Iraq, Kuwait, and Saudi Arabia) and Venezuela decided to create the OPEC in September 1960, with aim to obtain higher prices for crude oil.[49] By 1973, eight other nations (Qatar, Indonesia, Libya, the United Arab Emirates, Algeria, Nigeria, Ecuador, and Gabon) had joined OPEC. Ecuador suspended its membership in 1992, but returned as a full member in 2007. Indonesia suspended its membership in January 2009.[50]

The abundant global supply of oil rendered OPEC largely irrelevant during its first decade of existence. However, as the United States started to experience dwindling domestic reserves, its reliance on foreign oil became an issue. In 1973, the United States experienced a serious energy crisis, when the price of oil went from around US$2.00 to over US$11.00 per barrel. The spike in oil prices followed the decision by OPEC to place an embargo on oil exports to the United States in the wake of the Arab–Israeli Yom Kippur War. The combined petroleum revenues of the oil exporters rose from US$23 billion in 1972 to US$140 billion by 1977.[51] The disruption of oil supplies and the threat of unilateral actions by OPEC heightened concerns about energy security worldwide. The era dominated by the Seven Sisters was over.

By the mid-1970s, OPEC became an influential player on the geopolitical stage, known for fixing global oil prices and dictating sales rules. While the mainstream western discourse habitually refers to OPEC as a cartel, its members describe their organization as "a modest force for market stabilization." The concept of cartel frames OPEC within a narrative of negativity and contempt. Surprisingly, OPEC's anti-competitive practices were never challenged under GATT/WTO competition rules, and most likely they never will.[52] In 1979, the US District Court ruled that OPEC activities are "governmental acts" of state, rather than "commercial acts," and for that reason they are protected by the Sovereign Immunities Act of 1976. This legal decision effectively removed OPEC from the reach of US anti-competitive laws. There were several attempts in the US Congress to overturn this decision, which all failed.[53]

In 1998, the OECD launched an anti-cartel program with the adoption of the council recommendation concerning effective action against "hard core" cartels but there was never any action taken against OPEC. The recommendation formulated the following definition of a cartel:

> a "hard core cartel" is an anticompetitive agreement, anticompetitive concerted practice, or anticompetitive arrangement by competitors to fix prices, make rigged bids (collusive tenders), establish output restrictions or quotas, or share or divide markets by allocating customers, suppliers, territories, or lines of commerce.[54]

Right after the 1973 oil crisis, Europe responded to the invitation from the United States to establish the 1974 Agreement on an International Energy Program, leading to the creation of the IEA. The main point of the agreement was an automatically triggered mechanism that would ensure Atlantic oil allocation in the case of a defined supply shortfall. Afterwards, nothing much happened in the European Commission until the release of a 1985 report, "Towards an Internal Energy Market," and the 1987 Brundtland Report connected the issue of energy with growing concerns over environmental sustainability.[55]

Over the decades, countries had employed various national strategies intended to ensure stable supplies of oil and gas. Most of these strategies resulted in bilateral arrangements, which have done little to stabilize global energy markets. Even the most substantial international treaty on energy, chapter 6 of the North American Free Trade Agreement (NAFTA), the so-called Energy chapter, remains a two-country deal between Canada and the United States, despite being part of a regional trade agreement that also includes Mexico. Mexico opted out of chapter 6, citing its constitution.[56]

Mexico's constitution declares all underground resources to be a property of the Mexican state, which has provided an excuse for the government to restrict competition, nationalize the property of foreign operators in 1938, promote mismanagement, and hamper innovation in the sector. Despite the large oil supplies, Mexico's developmental strategy did not pay off. The strategy too heavily relied on oil revenues. Once oil prices went down in the early 1980s, the national oil company of Mexico, Pemex, was not capable of rescuing the country from incoming financial problems. Muddled in political corruption and economic mismanagement, by 1982 the country had run an enormous foreign debt of US$84 billion. This was the first sign of a debt crisis that affected many developing countries, although in the case of Mexico, its oil reserves should have been enough to ensure a steady economic growth and prosperity. The overreliance on the oil industry, which first helped to develop Mexico's economy, became the main factor in undermining it.[57] Under the terms of the emergency loans that rescued Mexico from its crushing debt, the country changed its developmental strategy towards liberalization. In 1994, it also signed the NAFTA with the United States and Canada.

NAFTA was the first trade agreement that explicitly linked trade to sustainable development in its preamble. Under the deal, all three contracting parties also agreed to negotiated environmental side agreements, which led to the creation of environmental institutions in all countries. It also established the Commission for Environmental Cooperation, a tri-national body based in Canada, a mechanism for investigating NAFTA's adverse impact on the environment and allegations of non-enforcement of national environmental law. Critics, however, were fast to point out that NAFTA's environmental agencies and provisions were institutionally weak and poorly funded by respective governments. For example, a number of provisions in NAFTA's dispute settlement allow firms to challenge environmental regulations. In short, the environmental aspect of NAFTA should be strengthened to be effective.[58]

Environmental provisions are increasingly placed within trade agreements. Such deals, however, tend to favor deregulation, and privilege investment and trade. Arguably NAFTA's most important part was the energy deal between the United States and Canada. Under NAFTA's proportional clause (Articles 315 and 605), Canada is required to maintain oil and natural gas flows to the United States at a level that averages the three previous years, "which, in effect, makes Canadian exports of oil and other energy resources to the US compulsory."[59] Other scholars posit that NAFTA's proportionality clause simply acts as a guarantee of the ongoing energy cooperation between Canada and the United States.

> The tar sands of Alberta and natural gas deposits in the Mackenzie Delta are prom-
> ising sources of future Canadian production. At a minimum, Canadian oil and
> natural gas deposits should play a role as part of North American "insurance policy"
> (in addition to the Strategic Petroleum Reserve) against acute shortages.[60]

The most ambitious attempt to establish a legal framework for international energy cooperation is the Energy Charter Treaty. It emerged as a European initiative in the early 1990s, but has only 48 contracting parties as of 2016 that actually signed and ratified it. Furthermore, except for Kazakhstan, none of the signatories is a major oil-producing country. Countries such as China, Nigeria, Canada, the United States, Saudi Arabia, Iran, Kuwait, and Venezuela, remain merely observers, while the Russian Federation, Australia, and Norway, never ratified the treaty. One can argue that the treaty is meaningless; in addition, it does not contain a treaty-specific dispute settlement arrangement that could serve as an enforcement mechanism and deter non-compliance.

Within the last two decades, however, OPEC's influence is waning. Though a large share of the world's oil supply is still controlled by OPEC's member-states, a greater amount of oil is increasingly found and produced in countries such as the United States, Russia, Brazil, and Canada. Equally important to world prices are new drilling technologies, championed by Canada and the United States. Some even call it the US energy revolution, impacting both gas and oil production in the United States. One industry expert notes that the United States may become the world's largest gas exporter, given its production has risen by 25 percent since 2010. Meanwhile US oil production has grown by 60 percent since 2008.[61] As a result, some see the influence of OPEC as continuing to shrink.

Another problematic aspect of OPEC stems from a surprisingly poor developmental record of some of its members. Critics argue that overreliance on lucrative sources of revenue, such as oil, in fact discourages countries from diversifying their economies and leads to predatory exploitation of those vital natural resources. For example, Nigeria is a vast country that contains some of the world's greatest reservoirs of oil. Yet it is also a case of squandered opportunities, having some of the biggest environmental problems on the planet. The paradox of being a leading oil exporting country and failing to translate this capacity into developmental success led scholars to coin the so-called "resource-curse" hypothesis, discussed earlier in the context of armed conflicts. A stream of secure revenues from natural oil and gas endowments too often prevent governments from seeking innovation, diversification, and accountability. Such countries experience the "resource curse," since an abundance of crucial resources impedes sustainable development.[62]

Close to a third of the world's oil production is located in the Middle East. It is a land of conflict, beset by acrimonious relationships among neighbors and characterized by authoritarian rule. The Saudi royal family rules Saudi Arabia, the largest producer of oil in the world over the last several decades. Only recently has the country started to diversify its economy. However, while Saudi Arabia has accrued enormous profits from its oil supply, the price of maintaining the oil-dependent state has been growing, and there are concerns of instability in the kingdom.[63] An important empirical study has determined that most resource-rich countries are not wealthy. Even Saudi Arabia, with its enormous production of oil at US$60 per barrel would have a problem to lift all its citizens above the poverty level comparable to the United States. The study argues that by not taking into account the depletion of non-renewable assets, resource-rich countries are misrepresenting their revenue base: "it is like augmenting the family income by selling the family silver. It cannot last and is really a form of asset disposal – not a source of income."[64]

Concerns over the world's supply of oil have renewed calls to examine alternative forms of energy. And even if we keep perfecting new technologies of oil extraction, there are questions about the environmental capacity to absorb the damage done by them. Hydraulic fracturing (fracking), which is the injection of high-pressure streams of sand, water, and mixed chemicals, is known for creating air pollution from drilling sites known as "well pads," contamination of water, and the devastation of vast amounts of soil. Economically speaking, the fracking method of extracting gas and oil is very expensive as it consumes huge amounts of water and requires a vast amount of land. Economists do not like it for costs and externalities that environmental activists see as dangerously undermining sustainable development.[65] One of the worst environmental disasters to date had to do with deep water drilling for oil in the Gulf of Mexico by BP (British Petroleum). On April 20, 2010, the offshore drilling rig called Deepwater Horizon exploded, causing the largest oil spill in history. In the immediate aftermath, BP was accused of violating a number of US environmental regulations, triggering court proceedings that would last for years.[66] This tragedy resonated with memories of the March 1989 *Exxon Valdez* oil spill, which happened when the tanker owned by Exxon ran into a reef off the coast of Alaska. This event undermined the legitimacy of the powerful Exxon, an oil company with vast influence exercised globally.[67] The environmental impact of both disasters was unprecedented. These developments, coupled with the falling prices of crude oil, have led to the questioning of the current business model as practiced by big oil companies. From the coast of Australia to Canada's oil sands, across the industry some US$400 billion in expected investment was cancelled or delayed:

> "nightfall is coming" for big oil companies, threatened on one side by the rise of renewable energy and climate policies that will curb the growth of fossil fuel demand, and on the other by the smaller, nimbler companies that lead the shale oil and gas industry.[68]

One technology that can provide energy in lieu of fossil fuels and shale fracking is nuclear energy. Nuclear energy, however, is extremely controversial. Not only may the costs to pursue a nuclear program for energy purposes be prohibitive for a number of developing countries, the risks associated with such a scheme can be enormous. Three Mile Island, Chernobyl, and Fukushima conjure up images of the possible hazards that can arise when accidents happen at a nuclear power plant. When one of the nuclear reactors exploded at the power plant of Chernobyl, the problem soon became international as radiation crossed national boundaries in Eastern Europe. The catastrophe took place on April 26, 1986 at the Chernobyl nuclear power plant in Ukraine (then part of the Soviet Union). The environmental impact of this accident was devastating. The explosion produced radioactive debris that drifted for miles over parts of Europe and Scandinavia, permanently contaminating land in the proximity of the disaster, leading to the evacuation and resettlement of roughly 200,000 people. Some 30 years after the disaster, Greenpeace reported that the large area around the disaster remained a wasteland and thousands of people continued to suffer from life-threatening illnesses. After all, the accident released 100 times more radiation than the two nuclear bombs dropped in 1945 on Hiroshima and Nagasaki.[69] As a result, several countries have changed their policy objectives and decided to stop any further expansion of nuclear power plants. Regardless of the means for energy, the perennial issue when it comes to energy production and consumption is the effects it may have on the environment.

IV Deforestation, pollution, and the politics of climate change

Other serious ecological issues particularly impacting developing countries relate to deforestation and soil erosion leading to desertification. The loss of forests and the expansion of deserts around the world not only means the loss of natural resources, but also threatens agricultural production, and even the way of life for many communities. Moreover, because trees and most plants absorb carbon dioxide from the atmosphere by a process called photosynthesis and then release oxygen back into the atmosphere, deforestation is regarded to be a contributing factor to air pollution. The 2016 Human Development Report brings up some startling data:

Box 7.1 2016 Human Development Report data

Every year, 24 billion tons of fertile soils are lost to erosion, and 12 million hectares of land are lost to drought and desertification, affecting the lives and livelihoods of 1.5 billion people. Desertification could displace up to 135 million people by 2045. Biodiversity is below safe levels across more than half the world's lands....

In 2012 an estimated 8.4 million people died from air, water or land pollution. At least 6.5 million people a year are believed to be dying from air pollution, with many more injured. The cost of air pollution in welfare losses has been estimated at $5 trillion, 60 percent of which is in developing regions. About 2.7 billion people still depend on wood or waste fires that cause indoor air pollution, affecting women and children the most. Indoor air pollution leads to around 3.5 million deaths a year. Forests and trees provide vital resources to 1.3 billion people, and in developing countries, forest income is second only to farm income among rural communities. Between 60 million and 200 million indigenous peoples rely on forests for survival. Acting as the lungs of the world, forests also slow climate change, and acting as carbon sinks, they increase resilience."[70]

Though deforestation can occur naturally through forest fires and unpredictable weather patterns, the vast extent of deforestation in the modern age is attributable to human activities involving logging for the expansion of living spaces, industrialization, and agriculture. With the world population rising, the consequent demand for land, wood products, and food has had a negative impact on the Earth's forests. In the case of Brazil, sections of the rainforests in the Amazon basin have given way to provide room for cattle ranching. Forests are cleared to provide wood to accommodate the needs of construction, the expansion of cities, and the creation of new networks of roads.

The cutting down of mangroves, trees typically found in tropical coastal areas around the world, is seen by some environmentalists as a contributing factor to the severity of the tsunami caused by the 2004 Indian Ocean earthquake. Vastly reduced in some regions to provide space for hotels, shrimp aquaculture, and housing, mangroves provide a natural barrier between coastal areas and the possible harshness of wind and waves. Had the mangrove forests remained intact, some environmentalists believe, the devastating impact of the tsunami that claimed over 200,000 lives may have been lessened. The removal of forests may not only mean increased vulnerability to hurricanes and tsunamis, but also to deluge. Deforestation in the Amazon region, Africa, and Southeast Asia increasingly causes worrisome changes in rainfall patterns, with floods in some areas while droughts in others.[71]

The proliferation of diseases such as malaria is also believed to be an effect of deforestation. Malaria is a particularly troublesome problem for the African continent. Transmitted by mosquitoes carrying the deadly virus, malaria infects over 300,000 people and kills about one million each year. Studies have shown that deforestation aids the proliferation of the disease because the deforested area can be congenial to mosquitoes if the proper conditions are met. For example, in the African region under examination, following massive deforestation efforts conducted to make room for a mining development, malaria transmission increased by three-quarters. There were also other diseases that have enjoyed proliferation after deforestation such as sleeping sickness and sand fly disease, which can be devastating for poor communities with no adequate medications to combat their effects.[72]

Together with farming methods that further damage the land, deforestation has led to warming temperatures and fewer incidences of rainfall. The result of which is a landscape afflicted by dryness. Furthermore, since trees consume carbon dioxide and provide oxygen, deforestation has been partly blamed for the rise of carbon dioxide in the Earth's atmosphere. Environmentalists believe that climate change has been exacerbated by the amount of deforestation that has occurred in modern times. The issue central to the argument on climate change is global warming – a phenomenon whereby average atmospheric and oceanic temperatures rise over a period a time. To account for the apparent climate change, several theories have been expounded, the most popular of which being that of the greenhouse effect. The enhanced (anthropogenic) greenhouse effect is amplified by human activities leading to greenhouse gases potentially causing global warming. The said greenhouse gases include carbon dioxide, nitrous oxide, chlorofluorocarbons, and methane. Of all the emissions of these gases due to human activity, carbon dioxide overwhelmingly accounts for the largest share because the combustion of fossil fuels is a source of its release into the atmosphere.[73] Together, the effect these greenhouse gases have on the world is that the sun's radiation becomes trapped in the lower atmosphere and the surface of the Earth is warmed like a greenhouse. Global temperatures rise and climates are subsequently affected.

Related to deforestation is the phenomenon of desertification in which deserts expand over land that was once useful for agriculture and wildlife. These issues have become a problem for pastoralist societies of East Africa. Such societies have traditionally used the land to manage livestock and pastures. Customarily, pastoralist groups would migrate elsewhere in the region to obviate overgrazing and avoid famine and drought. Unfortunately for them, however, the modern age has imposed restrictions on their mobility. They have seen the availability of land diminish in recent times as a result of urbanization and the privatization of communal lands, and political boundaries have been delineated. Due to their practice of utilizing public land for their livestock, politicians and international development agencies sought the integration of such groups in modern society, with the hope of changing their way of life. The argument for integration stipulated that allowing the practices of pastoralist societies would amount to something similar to the tragedy of the commons. If each resident sought to increase their livestock to attain more wealth, and there are no laws to restrict their actions, the green would be destroyed since each resident seeks to maximize his gains at everyone's cost. While the tragedy of commons was avoided historically by pastoralist communities due to their nomadic nature, the permanent settlement of these groups on land might make it a reality. The Maasai of the Kajiado district of Kenya typified this unfortunate fact, as overgrazing and traditional farming methods have led to food and water shortages and further environmental degradation in the region.[74]

The causes of desertification are not restricted to only overgrazing and overuse of land. The salination of soil by over-irrigation and deforestation has contributed to desertification. Irrigation in agriculture has been extremely helpful in providing farmland with extra water and thus leading to increases in food supply. However, too much irrigation can have harmful effects on the land. This may happen if the irrigation system is poorly designed and managed and the irrigation water is thus not drained properly, if at all. When the excess water accumulates, it will gradually come close to the ground's surface whereby it causes salination in the soil due to minerals such as salt contained within the water. The buildup of salts in the soil thus leads to soil degradation and can lead to a loss of productivity, and may even render arable land infertile. Deforestation can worsen these conditions as the removal of watersheds can lead to the expansion of salt flats. So is the case in Senegal and Brazil, where river estuaries have been damaged by the clearance of forests. Lack of proper drainage is not the only source of over-irrigation. Over-irrigation may also occur if there are too many wells in place. When the water table, or the underground surface of groundwater, falls in an extreme manner as a result of an excess number of wells, wells will go dry and the land may be rendered unusable. In summary, a combination of poor policies and weak institutional oversight lead to cropland expansion, overgrazing, and infrastructure extension, which in turn contribute to desertification.[75]

The consequences of desertification are more far-reaching than the loss of fertile lands. Deserts are not just areas of little or no vegetation. They are areas bereft of water. Water is important, not only for daily human consumption and hygiene, but also for agriculture and industry. Humans are presently using over half of the accessible water, and consumption is rising. Moreover, groundwater in places such as sub-Saharan Arica, India, Bangladesh, Pakistan, and China are overexploited for irrigative purposes. China's economic progress, especially, has caused devastating environmental pollution of its rivers, soil, and air, with ongoing impact on the neighboring countries.[76] The demand for water has already tripled in the mid-twentieth century and will only grow in the future.

Despite the saliency of the environmental issues there is still an ongoing debate over the evidence concerning climate change. Though opponents to the theory form only a small minority in the academic community, there are several prominent scientists who question the science behind global warming. Richard S. Lindzen of MIT is one such greenhouse skeptic. In one article, based on a study conducted by a team of scientists from MIT and NASA, he demonstrated how changes in cloud formation can offset the warming effect of greenhouse gases. In another article, he agreed that there has been a change in temperature in recent years but questions the accuracy and uncertainties of forecasts that see global warming and the resultant climate changes in this century. Because carbon dioxide accounts for about 0.04 percent of the Earth's atmosphere, any augmentation of its concentration in the atmosphere caused by human activities is believed by some scholars to be marginal, as is the consequent rise in temperature.[77]

Whether average temperatures in the world rise persistently within the century or not, the costs of sustained global warming are high according to the scientists who support the theory. These include the loss of forests, the reshaping of agricultural land, and the rise in sea level due to the melting of mountain glaciers and ice caps. An average rise in sea level across the planet will have several consequences. One is inundation and may threaten low-lying coastal areas, particularly countries that are composed solely of islands. Another undesirable outcome of higher sea levels is an increase in flooding of coastal areas. This in itself can pose a new array of challenges to developing countries, other than the immediate damage incurred. Erosion would be hastened among cliffs and beaches and thus do harm to the tourism industries of affected countries. Moreover, freshwater ecosystems could be contaminated with increased salt seeping into groundwater systems and rivers.

Environmental degradation is a serious developmental problem, whether exacerbated by global warming or not. In fact, it is seen by some to be a potential cause of societal collapse. In the study of international relations, the role that the environment per se plays has been historically small. Lately, however, the concept of environmental security has been introduced in recognition that threats to the environment can undermine national security, and even humankind. Some say that the lack of environmental security is already undermining peace and order in parts of the developing world. In an influential book called *The Coming Anarchy*, Robert D. Kaplan wrote of how population pressure, environmental degradation, and scarcity of resources can lead to societal collapse. His forecasts are filled with the following understanding of the environmental concerns:

> It is time to understand "the environment" for what it is: the national security issue of the early twenty-first century. The political and strategic impact of surging populations, spreading disease, deforestation and soil erosion, air pollution, and, possibly,

rising sea levels in critical over-crowded regions like the Nile Delta and Bangladesh – developments that will prompt mass migrations and, in turn, incite group conflicts – will be the core foreign-policy challenge from which most others will ultimately emanate.[78]

Kaplan is not alone in predicting that future conflicts between peoples and societies will be fought over resources. In his book *Environment, Scarcity, and Violence*, Thomas F. Homer-Dixon develops a detailed model of the sources of environmental scarcity and their relation to the growing armed conflicts.[79] He talks about water shortages in China, population growth in sub-Saharan Africa, and land distribution in Mexico, for example, to show that scarcities arise from the degradation and depletion of renewable resources, the increased demand for these resources, and their unequal distribution. He shows that these scarcities can lead to deepened poverty, large-scale migrations, sharpened social cleavages, and weakened institutions. The prognosis is grim as the author describes how these effects eventually lead to violence. He argues that conflicts in Chiapas, Mexico and ongoing turmoil in many African and Asian countries, for example, are already partly a consequence of environmentally related problems. Presently, the Puntland area of Somalia is in the middle of an internal armed conflict, and one of the accelerants fueling the hostilities is the devastating drought that has engulfed the region for the past two years. Close to 6.2 million people, or almost half of Somalia's total population, have been experiencing acute water and food shortages.[80]

The 2016 World Development Report reveals the complexity of the environmental matters. Whether we consider a seemingly wide range of issues, ranging from the concerns over global inequality, energy security, shrinking areas of fertile lands and forests, possible climate change, and conflicts over limited natural resources, environmental sustainability remains firmly in focus. The report links economic growth with the environment and with the ways societies live and relate to the world around them in the context of three interdependent paths that should complement each other: human development, environment, and economic advancement. The realization that these three should be addressed in concert and that they should never be placed in opposition to each other is perhaps the most important lesson of the report. In this context, it is useful to summarize the new development goals, officially adopted by the UN General Assembly in September 2015.[81]

Box 7.2 2015 Sustainable Development Goals

Goal 1

End poverty in all its forms everywhere

Goal 2

End hunger, achieve food security and improved nutrition and promote sustainable agriculture

Goal 3

Ensure healthy lives and promote well-being for all ages

Goal 4

Ensure inclusive and equitable quality education and promote lifelong learning opportunities for all

Goal 5

Achieve gender equality and empower all women and girls

Goal 6

Ensure availability and sustainable management of water and sanitation for all

Goal 7

Ensure access to affordable, reliable, sustainable and modern energy for all

Goal 8

Promote sustained, inclusive and sustainable economic growth, full and productive employment and decent work for all

Goal 9

Build resilient infrastructure, promote inclusive and sustainable industrialization and foster innovation

Goal 10

Reduce inequality within and among countries

Goal 11

Make cities and human settlements inclusive, safe, resilient and sustainable

Goal 12

Ensure sustainable consumption and production patterns

Goal 13

Take urgent action to combat climate change and its impacts

Goal 14

Conserve and sustainably use the oceans, seas and marine resources for sustainable development

Goal 15

Protect, restore and promote sustainable use of terrestrial ecosystems, sustainably manage forests, combat desertification, and halt and reverse land degradation and halt biodiversity loss

Goal 16

Promote peaceful and inclusive societies for sustainable development, provide access to justice for all and build effective, accountable and inclusive institutions at all levels

Goal 17

Strengthen the means of implementation and revitalize the Global Partnership for Sustainable Development

Smart environmental policies make sense. With this understanding, America's Obama Administration actively supported the new round of talks on an agreement replacing the expired Kyoto Protocol on climate change. The final stage of acrimonious negotiations took place in December 2015 in Paris, where 195 countries managed to adopt the first-ever universal and legally binding global climate action plan. The difficulties of negotiating the treaty stemmed from the diverse interests of countries. For example, oil producers, struggling economies with large forested areas, land-locked countries relying on natural resources, poor countries facing issues of desertification, islands with eroding shore lines, and mature developed states all face distinct environmental challenges. In contrast to the Kyoto Protocol, the new agreement was expected to require binding commitments from all its signatories. Some developing countries joined India in resisting this idea. India feared that a binding agreement would restrict their economic growth. China was specifically concerned about the intrusive monitoring of their domestic policies. In the end, the deal reflected the ambitions of the EU and the United States working towards a strong deal for all.

Officially called the Paris Agreement, it commits its signatories to limit the rising global average temperatures by undertaking rapid reductions in global emissions of harmful gasses. The agreement also commits governments to come together every five years to set the ambitious targets for such reductions and report their progress to each other when it comes to implementation of the promised targets. The Paris Agreement is an imperfect document, but both the United States and the EU promised to provide funding to help poor developing countries deal with the commitments made. In the end, the deal reached in Paris showed that there was a strong coalition of countries committed to maximizing global efforts to combat climate change, even at the price of increasing aid for poorer countries.[82] This coalition, however may be undermined by the changed direction of the US environmental policies under the new Trump Administration. The Paris Agreement entered into force on November 4, 2016, after it was ratified by a required number of countries.[83]

In recognition of the environmental issues, the government of China is adopting stricter emission regulations and promises to clean the polluted rivers.[84] In 2014, a new agreement was signed by the United States and China to reduce emissions in both countries. Under the deal, China promised to prevent its carbon emissions from growing after 2030, and the United States promised to reduce its emissions by 26 percent (from 2005 levels) by 2025. This agreement eventually paved the way for the 2015 multilateral Paris Treaty on climate change. India continues to view the treaty as an unfair plan that can negatively impact its economy. India has long maintained that the industrialized countries should take deeper cuts in their emissions. With a quarter of its population still lacking proper access to electricity, India insists on fulfilling the domestic demand first, even at the price of fossil fuels emissions.[85]

The Paris Treaty, however, is facing an uncertain future given the US shift in its environmental policy approach. This shift involves revoking many environmental regulations with the hope to stimulate the economy and bringing back jobs in coalmines. This hardly makes sense even from the economic point of view. Coal, a terrible pollutant, is expensive to extract and not very efficient. Even China recognized the destructiveness of its toxic emissions by deciding to invest in new clean technologies. Starting in 2015, its new strategy involves the installation of thousands of wind turbines and solar panels. The Chinese government is on record as cancelling the construction of more than 100 coal-fired power plants in 2017. Such initiatives, supported with President Xi's commitment to the 2015 Paris Treaty, signal a pro-environment policy change in Beijing.[86]

The commitments made within the framework of the Paris Agreement advance the cause of environmental sustainability, but the international treaties only go so far. Most important policies concerning the environment are made domestically in concert with other countries because of the growing international consensus on the issue. A study conducted by the Breakthrough Institute shows that countries have been instituting policies to curb emissions in a way that is consistent with the changing societal attitudes towards environmental issues. Such regulatory measures were instituted even prior to the 1997 Kyoto Protocol, suggesting that countries would introduce them in any event. Consequently, even if the new US administration abandons the Paris Treaty, environmental matters will continue to inform economic policy making. On the condition that the United States continues to invest in the nation's nuclear power plants, maintains tax incentives for the deployment of wind and solar energy, and the shale revolution continues, it is estimated that the US carbon emissions outcomes may even outperform the US commitments made in Paris.[87] No matter what the politics of the day, decision makers would be smart to reward those who use our limited resources wisely with a minimum environmental footprint.

Notes

1 UN World Commission on Environment and Development (1987) *Our Common Future*, August 4. The full text of the Brundtland Report can be downloaded as UN General Assembly Document A/42/427. The UN website is online, available at: www.un.org.

2 Merle Jacob (1994) "Toward a Methodological Critique of Sustainable Development," *Journal of Developing Areas*, Vol. 28, No. 1, January, pp. 237–252.

3 UN (2015) *Resolution Adopted by the General Assembly on 25 September 2015* (A/RES/70/1), New York, NY: The United Nations.

4 Robert Picciotto (1997) "From Participation to Governance," in Christopher Clague (ed.), *Institutions and Economic Development*, Baltimore, MD: Johns Hopkins University Press, p. 363.

5 UN/FCCC (Framework Convention on Climate Change) (2015) *Adoption of the Paris Agreement* (FCCC/CP/2015/L.9/Rev.1), Paris, 12 December.

6 World Bank (2017) *World Development Report – Governance and the Law*, Washington, DC: The World Bank Group, p. 4.

7 WTO (1994) *WTO Legal Texts – 1994 Marrakesh Ministerial Decision on Trade and Environment*. Online, available at: www.wto.org.

8 Robert A. Jackson (1995) "Population Growth: A Comparison of Evolutionary Views," *International Journal of Social Economics*, Vol. 37, No. 1, p. 3.

9 UN World Commission on Environment and Development (1987) *Our Common Future*, op. cit.

10 Ibid.

11 Ibid., chapter 4.

12 Fiona Smith (2009) *Agriculture and the WTO: Towards a New Theory of International Agricultural Trade Regulation*, Cheltenham, UK: Edward Elgar.

13 UNDP (2016) *Human Development Report 2016 – Human Development for Everyone*, New York, NY: UNDP, p. 46.

14 Ibid., p. 46.

15 Llewelyn Hughes and Phillip Y. Lipscy (2013) "Politics of Energy," *Annual Review of Political Science*, Vol. 16, May, pp. 449–469.

16 UNDP (2016) *Human Development Report 2016*, op. cit., p. 46.

17 UN World Commission on Environment and Development (1987) *Our Common Future*, op. cit., chapter 8, table 8.2.

18 UNCTAD (2015) *Key Statistics and Trends in International Trade 2014*, Geneva, Switzerland: UNCTAD, p. 5.

19 UNDP (2016) *Human Development Report 2016*, op. cit., p. 46.

20 UN (2014) *World Urbanization Prospects – The 2014 Revision*, New York, NY: UN DESA Population Division.

21 UN World Commission on Environment and Development (1987) *Our Common Future*, op. cit., chapter 9.

22 UNDP (2016) *Human Development Report 2016*, op. cit., p. 46.

23 Ibid., chapter 10, paragraphs 1 and 2.

24 Ibid., p. 46.

25 Ibid., p. 46.

26 Becky J. Brown, Mark E. Hanson, Diana M. Liverman, Robert W. Merideth Jr (1987) "Global Sustainability: Toward Definition," *Environmental Management*, Vol. 1, No. 6, November, pp. 713–719.

27 UNFPA (2016) *State of World Population 2016*, Report Published by the United Nations Population Fund (UNFPA). Online, available at: www.unfpa.org/swop (accessed March 18, 2017).

28 Thomas R. Malthus (1798) *An Essay on the Principle of Population*, Published in 1958, New York, NY: E.P. Dutton and Co.

29 John C. Caldwell (2006) *Demographic Transition Theory*, The Netherlands: Springer Publishing.

30 Donella Meadows, Jorgen Randers, and Dennis Meadows (2004) *The Limits to Growth: The 30-Year Update*, White River Junction, VT: Chelsea Green Publishing Company.

31 Jared Diamond (2004) *Collapse: How Societies Choose to Fail or Succeed*, New York, NY: Viking Press, pp. 320–323.

32 Garrett Hardin (1968) "The Tragedy of the Commons," *Science*, Vol. 162, No. 3859, pp. 1243–1248.

33 Peter Bauer (2000) *From Subsistence to Exchange*, Princeton, NJ: Princeton University Press.

34 Daniel Yergin (2008) *The Prize: The Epic Quest for Oil, Money and Power*, New York, NY: Free Press, prologue.

35 OECD/IEA (2015) *Energy Statistics for Non-OECD Countries – 2015 Edition*, IEA: Paris, France, pp. 567–569 (tables: 61–63).

36 Ibid., pp. 524–525 (tables: 18–19).

37 Ibid., p. 590 (table: 84).

38 Daniel Yergin (2008) *The Prize*, op. cit., prologue.

39 EIA (Energy Information Administration) (2016) *The International Energy Outlook*, Washington, DC: The US Department of Energy.

40 Jacques de Jong (2008) "The 2007 Energy Package: The Start of a New Era?," in Martha M. Roggenkamp and Ulf Hammer (eds.), *European Energy Law – Report V*, Antwerp, Belgium: Intersentia Publishers, p. 95.

41 Daniel Yergin (2008) *The Prize*, op. cit.

42 Ibid., p. 26.

43 Ibid., pp. 91–94.

44 Daniel Yergin (2008) *The Prize*, op. cit., p. 392.

45 Ibid., p. 189.

46 Rachel Bronson (2006) *Thicker Than Oil: America's Uneasy Partnership with Saudi Arabia*, New York, NY: Oxford University Press.

47 Daniel Yergin (2008) *The Prize*, op. cit., pp. 632–633.

48 Joan E. Spero and Jeffrey A. Hart (2003) *The Politics of International Economic Relations*, Belmond, CA: Thomson & Wadsworth, p. 301.

49 M. Roggenkamp, C. Redgwell, I. del Guayo, and A. Rønne (eds.) (2007) *Energy Law in Europe – National, EU, and International Regulation* (2nd edition), Oxford, UK: Oxford University Press, p. 7.

50 OPEC (2017). Online, available at: www.opec.org.

51 Daniel Yergin (2008) *The Prize*, op. cit., p. 616.

52 It can be argued that OPEC violates GATT Article XI, which prohibits nations from maintaining quotas or any other quantitative restriction on exports.

53 Dan Weil (2007) "If OPEC Is a Cartel, Why Isn't It Illegal?" *Newsmax American*, November 25.

54 OECD (1998) *Recommendation of the Council Concerning Effective Action against Hard Core Cartels*, C(98)35/FINAL, p. 3.

55 Jacques de Jong (2008) "The 2007 Energy Package: The Start of a New Era?," op. cit., pp. 96–98.

56 Maxwell A. Cameron, Brian W. Tomlin (2000) *The Making of NAFTA – How the Deal was Done*, Ithaca, NY: Cornell University Press.

57 Daniel Yergin (2008) *The Prize*, op. cit., pp. 254, 712–714.

58 Stephen P. Mumme (1999) "NAFTA and Environment," *Foreign Policy in Focus*, October 1.

59 Tony Clarke (2008) *Tar Sands Showdown: Canada and the New Politics of Oil in an Age of Climate Change*, Toronto, Canada: Lorimer, p. 138.

60 Gary C. Hufbauer and Jeffrey J. Schott (2005) *NAFTA Revisited – Achievements and Challenges*, Washington, DC: Institute for International Economics, p. 59.

61 Edward L. Morse (2014) "Welcome to the Revolution – Why Shale is the Next Shale," *Foreign Affairs*, Vol. 93, No. 3, pp. 3–7.

62 Macartan Humphreys, Jeffrey D. Sachs, and Joseph E. Stiglitz (eds.) (2007) *Escaping the Resource Curse*, New York, NY: Columbia University Press.

63 Ian Bremner (2004) "The Saudi Paradox," *World Policy Journal*, Vol. 21, No. 3, p. 25.

64 Geoffrey Heal (2007) "Are Oil Producers Rich?," in Macartan Humphreys, Jeffrey D. Sachs, and Joseph E. Stiglitz (eds.), *Escaping the Resource Curse*, op. cit., p. 170.

65 Fred Krupp (2014) "Don't Just Drill, Baby – Drill Carefully – How to Make Fracking Safer for the Environment," *Foreign Affairs*, Vol. 93, No. 3, pp. 15–20.

66 FT (2010) "BP: Eagles and Vultures, *Financial Times*, July 1.

67 Steve Coll (2012) *Private Empire – Exxon Mobil and American Power*, New York, NY: Penguin Group.

68 Ed Crooks and Chris Adams (2016) "Oil Majors' Business Model Under Increasing Pressure," *Financial Times*, February 16.

69 Greenpeace (2016). Online, available at: www.greenpeace.org/international/en/campaigns/nuclear/nomorechernobyls/, accessed March 2017.

70 UNDP (2016) *Human Development Report 2016*, op. cit., p. 38.

71 Deborah Lawrence and Karen Vandecar (2014) "Effects of Tropical Deforestation on Climate and Agriculture," *Nature Climate Change*, Vol. 5, pp. 27–36.

72 Jonathan A. Patz, Thaddeus K. Graczyk, Nina Geller, and Amy Y. Vittor (2000) "Effects of Environmental Change on Emerging Parasitic Diseases," *International Journal of Parasitology*. Vol. 30, No. 12: pp. 1395–1405.

73 William D. Nordhaus (1991) "To Slow or Not to Slow: The Economics of the Greenhouse Effect," *Economic Journal*, Vol. 101, No. 407, pp. 920–921.

74 Elliot Fratkin (2001) "East African Pastoralism in Transition: Maasai, Boran, and Rendille Cases," *African Studies Review*, Vol. 44, No. 3, pp. 2–6.

75 Helmut Geist (2005) *The Causes and Progression of Desertification*, Surrey, UK: Ashgate Publishing.

76 Judith Shapiro (2016) *China's Environmental Challenges* (2nd edition), Cambridge, UK: Polity Press.

77 Christopher Essex and Ross McKitrick (2002) *Taken by Storm – The Troubled Science, Policy and Politics of Global Warming*, Toronto, Canada: Key Porter Books Limited, pp. 41–43.

78 Robert D. Kaplan (2001) *The Coming Anarchy – Shattering the Dreams of the Post-Cold War*, New York, NY: Vintage Books, pp. 19–20.

79 Thomas F. Homer-Dixon (2001) *Environment, Scarcity, and Violence*, Princeton, NJ: Princeton University Press.

80 Vanda Felbab-Brown (2017) "Puntland Problems – It's Not Just Al Shabab That Threatens the Region's Stability," *Foreign Affairs – Snapshot*, June 15.

81 UN (2015) *Resolution Adopted by the General Assembly on 25 September 2015* (A/RES/70/1), op. cit.

82 Nick Mabey (2015) "The Geopolitics of the Paris Talks," *Foreign Affairs Snapshot*, December 13.

83 UN/FCCC (2015) *Adoption of the Paris Agreement* (FCCC/CP/2015/L.9/Rev.1), op. cit.

84 Kenneth Rapoza (2016) "China's Tougher Environmental Policies Not Only Good for The Locals," *Forbes*, December 26.

85 Scott Moore (2014) "Delhi Dilemma: India Is Now the Biggest Barrier to a Global Climate Treaty," *Foreign Affairs – Snapshot*, November 20.

86 Fred Krupp (2017) "Trump and the Environment: What His Plans Would Do," *Foreign Affairs*, Vol. 96, No. 4, July/August.

87 Ted Nordhaus and Jessica Lovering (2016) *Does Climate Policy Matter? – Evaluating the Efficacy of Emissions Caps and Targets Around the World*, Oakland, CA: The Breakthrough Institute, November 28.

Suggested further reading

Paul Collier (2010) *The Plundered Planet: Why We Must – and How We Can – Manage Nature for Global Prosperity*, Oxford, UK: Oxford University Press.

Felix Dodds, Ambassador David Donoghue, Jimena Leiva Roesch (2017) *Negotiating the Sustainable Development Goals: A Transformational Agenda for an Insecure World*, London, UK: Routledge.

Sarah Harper (2016) *How Population Change Will Transform Our World*, Oxford, UK: Oxford University Press.

Geoffrey Heal (2016) *Endangered Economies: How the Neglect of Nature Threatens Our Prosperity*, New York, NY: Columbia University Press.

Stephen J. Macekura (2015) *Of Limits and Growth: The Rise of Global Sustainable Development in the Twentieth Century*, New York, NY: Cambridge University Press.

J.R. McNeill and Peter Engelke (2016) *The Great Acceleration: An Environmental History of the Anthropocene since 1945*, Cambridge, MA: The Belknap Press of Harvard University Press.

Joseph Romm (2015) *Climate Change: What Everyone Needs to Know*, New York, NY: Oxford University Press.

Rural development and food security

8

I Conceptualizing food security

Food is perhaps the most fundamental source of comfort and survival in the world. And yet, while there is no dispute regarding the vital importance of adequate nutrition in the lives of societies, millions of people around the world suffer from chronic hunger. As one author notes:

> No human right has been so frequently and spectacularly violated in recent times as the right to food, despite the fact that it is one of the most consistently enshrined rights in international human rights law, as constantly reaffirmed by governments.[1]

These conflicting realities, the basic need for food to stay alive, and the violation of this right, lead to problems associated with deprivation, famine, and poverty. In this chapter, we seek to understand the persistence of such predicaments in the Global South. We also examine new initiatives to coordinate agricultural and social protection policies and programs that aim at helping poor famers in the developing world break the cycle of poverty.[2] It is important to note that most of the 800 million hungry people in the world today live in rural areas of sub-Saharan Africa and their livelihoods depend largely on agriculture.[3]

The main theme of the chapter concerns food security, which just like development itself is a contested concept. Early attempts of the UN to define food security reflected somehow limited understanding of the problem. At the time, the discourse was primarily focused on the adequate production of food at national and international levels. Following years of scholarly debates and interventions from practitioners, we finally observed a paradigm shift in approaching the issue. Just at the turn of the new millennium, food security became understood as a matter of both limited availability of food and restricted access to food. Amartya Sen is widely credited with changing our way of thinking about famines and deprivation by focusing on the importance of access and entitlement to food.

During the World Food Summit of 2002, summit representatives of the most vulnerable countries asked for immediate actions aimed at eliminating global hunger. They articulated the need to provide practical objectives on how this fundamental human right can be realized. As a result, a set of guidelines intended to transform the right to food from an abstract aspiration to an operational tool for action was adopted by the UN's Food and Agriculture Organization (FAO) in 2004. With this move, an overwhelming majority of countries reaffirmed their commitment to ensuring food security for all people. Properly known as the *Voluntary Guidelines to Support the Progressive Realization of the Right to Adequate Food in the Context of National Food Security*, the text defines our concept under examination:

> Food security exists when all people, at all times, have physical and economic access to sufficient, safe and nutritious food to meet their dietary needs and food preferences for an active and healthy life. The four pillars of food security are availability, stability of supply, access and utilization.[4]

The document clarifies the pillars supporting the concept of food security, to ensure that the guidelines serve

> to guarantee the availability of food in quantity and quality sufficient to satisfy the dietary needs of individuals; physical and economic accessibility for everyone, including vulnerable groups, to adequate food, free from unsafe substances and acceptable within a given culture; or the means of its procurement.[5]

What is especially interesting about these definitions is that they compel us to look at food security in a broad context, involving both global trends and local developments. The global point of view underlines the factors influencing international agricultural trade, while the local focus assesses the changes impacting country-specific agricultural sectors.

Traditionally, the countries in the Global South were predominantly agricultural. The waves of colonization, together with the impact of industrialization, have altered these societies' ways of organizing themselves. In addition, new technologies have influenced the farming methods around the world, especially given the invention of fertilizers and, more recently, genetically modified seeds. The access to new technologies and genetically modified seeds presents a particular dilemma for poor farmers in developing countries. Such seeds constitute a patented invention and cannot be simply re-used, despite the long established tradition of *farmers' rights* that habitually gave farmers the right to keep and re-use the products of their work. Instead, a license has to be purchased to legally use the industry generated genetically modified seeds. Therefore, many farmers in the developing world are not able to access these seeds because of their high costs. Paradoxically, the fact that genetically advanced seeds are known for their high efficiency and resistance makes them the most desirable in poor countries with only limited access to water and good soil. The issue of food security is thus increasingly connected with technological inventions and trade policies concerning IPR. There is evidence that agricultural policies that combine interventions in the form of access to credit and access to technology produce better long-term results in terms of lessening rural poverty.[6]

II Global hunger, rural poverty, and problematic agricultural trade

According to Article 25, paragraph 1, of the 1948 Universal Declaration of Human Rights:

> Everyone has the right to a standard of living adequate for the health and well-being of himself and of his family, including food, clothing, housing and medical care and necessary social services, and the right to security in the event of unemployment, sickness, disability, widowhood, old age or other lack of livelihood in circumstances beyond his control.

Ensuring adequate supplies of food for all citizens of any country, together with safety, is considered to be the primary responsibility of every government. Unfortunately, the condition of chronic hunger prevails in many parts of the world. According to the FAO's 2015 report "The State of Food Insecurity in the World":[7]

> About 795 million people are undernourished globally, down 167 million over the last decade, and 216 million less than in 1990–92. The decline is more pronounced in developing regions, despite significant population growth. In recent years, progress has been hindered by slower and less inclusive economic growth as well as political instability in some developing regions, such as Central Africa and western Asia.

The FAO was created in 1945 as a specialized United Nations agency.[8] In 1951, its headquarters were permanently established in Rome, Italy. The FAO leads international efforts to defeat world hunger, acting as a neutral forum where all countries meet as equals to negotiate agreements and debate policy.

The following are the FAO's strategic objectives:

- Help eliminate hunger, food insecurity and malnutrition
- Make agriculture, forestry and fisheries more productive and sustainable
- Reduce rural poverty
- Enable inclusive and efficient agricultural and food systems
- Increase the resilience of livelihoods to threats and crises.

Despite the progress made, millions of people around the world remain undernourished. In parallel with concerns over steady population growth, it has been suggested that the world can actually run out of food. Perhaps the most famous of the pessimistic prognosticators are followers of Thomas R. Malthus, who believe that the current situation is untenable and it will eventually lead to a catastrophe. A consistent lack of food has harmful long-term effects:[9]

Undernourishment and deficiencies in essential vitamins and minerals cost more than 5 million children their lives every year, cost households in the developing world more than 220 million years of productive life from family members whose lives are cut short or impaired by disabilities related to malnutrition, and cost developing countries billions of dollars in lost productivity and consumption.

The developmental costs are devastating. From the economic point of view, there can be no further progress or prosperity if people go hungry. From the social point of view, lack of adequate nourishment leads to divisions and growing inequalities within societies. From a political point of view, food shortages or more pronounced struggles over access to food, can lead to social unrests or civil wars.

When dire predictions about future food shortages are voiced, some scholars claim that we have the resources and technology to eradicate hunger. The problem stems from the imbalanced usage and distribution of the relevant tools and resources. One the most vocal supporters of this argument is Per Pinstrup-Andersen, the former director of the International Food Policy Research Institute, who observes:

> Global food supplies are sufficient to meet the calorie requirements of all people if food were distributed according to needs. Per capita food supplies are projected to increase further over the next 20 years. Thus, the world food problem now and in the foreseeable future is not one of global shortage. Instead, the world is faced with three main food-related challenges: widespread hunger and malnutrition, mismanagement of natural resources in food production, and obesity.[10]

Still, despite these claims that the world is capable of feeding everybody, food insecurity remains a serious global problem. The year 2015 marked the end of the monitoring period for the Millennium Development Goals' targets. This positive trend, however, is not progressing rapidly enough for millions of people. In terms of meeting the Millennium Development Goals' hunger reduction targets, the FAO report shows that a total of 72 developing countries out of 129, or more than half the countries monitored, have reached them.[11]

Most of the countries that managed to reach the Millennium Development Goals enjoyed stable political conditions and economic growth, often accompanied by social protection policies targeted at vulnerable population groups. Unfortunately, progress has been uneven. Some regions, such as Latin America, the east and southeastern regions of Asia, the Caucasus and Central Asia, and the northern and western regions of Africa, significantly reduced the proportion of their undernourished populations. For example, in Latin America there were about 34 million undernourished people in 2015 in comparison to 66 million during the 1990–1992 period. In Eastern Asia, the corresponding numbers are approximately 145 million in 2015 and 295 million in the early 1990s. While in Southeastern Asia there were about 61 million undernourished people in 2015, in contrast to about 138 million in the early 1990s. Progress was also recorded in Southern Asia, Oceania, the Caribbean, and Southern and Eastern Africa. But in two regions, southern Asia and sub-Saharan Africa, it is difficult to talk about progress. The FAO report estimated that there were as many as 281 million undernourished people in Southern Asia and about 220 million in sub-Saharan Africa during the 2014–2016 period. While the number of undernourished people marginally decreased in Southern Asia, the numbers actually rose for sub-Saharan Africa, from 176 million in the early 1990s.[12] In many countries that failed to reach the Millennium Development Goals' targets, natural disasters, economic crises, political instability, and armed conflicts increased vulnerability and food insecurity for large parts of populations.

In somewhat strange contradiction, empirical studies confirm that across developing countries, the prevalence of food insecurity is generally higher in rural areas than in urban areas.[13] How can one explain this situation? There are several factors that contribute to the disadvantaged position of rural populations living in places characterized by poverty and subsistence:[14]

> Most of the populations living in these areas are poor and lack sufficient housing, infrastructure and services that can mitigate the impact of a disaster. They may also live in flood-prone or geologically unstable areas, or farm marginal lands. Demographic changes, environmental degradation, changes of river, dam and land management and other factors increase vulnerability. Susceptibility to natural hazards aggravates the adverse effects of these natural events, particularly in the least developed and conflict-ridden states.

Farmers depend on many unpredictable variables, like weather and access to water, making their profession one of the most difficult and labor intense. Demographic changes see a growing number of young people unwilling to farm the land and moving to the cities in search for better jobs. The dire situation that many farms endure is also partially due to the unbalanced competition they face from the big agribusiness firms able to use economies of scale and the newest technology to maximize the efficiency of food production. Small farmers routinely place at the lowest level on the socio-economic ladder, and suffer the gravest challenges when it comes to economic advancement. In addition, "because many peasants are illiterate or semiliterate, need others to transport their crops, and lack easy access to credit, they are dependent on government bureaucrats, merchants, and money lenders who frequently exploit them."[15] Further, this group generally has no ability to effectively lobby the government for new developmental initiatives aimed at improving their lives. Farmers in the developing world frequently are short of formal education, have difficultly accessing urban centers, do not know the regulations and, consequently, have problem with convincing the politicians to hear their pleas.

> Despite their vast numbers, peasants usually play a rather muted role in Third World politics. Because most LDCs don't have competitive national elections, those numbers do not convert readily into political influence. The peasantry's political leverage is also limited by their poverty, lack of education, dependency on outsiders, and physical isolation from each other and from centers of national power.[16]

Small farmers in the developing world tend to be widely dispersed throughout regions where customarily flawed infrastructure is further crippling their ability to organize into groups. In contrast, farmers in the developed world are well organized, connected by roads and technology and routinely protected by state subsidies. In fact, farmers in the EU and the United States form some of the most powerful lobbying groups, capable of effectively influencing the agricultural policies of their countries.

The most widely cited example of agricultural subsidy is perhaps the EU Common Agricultural Policy (CAP), which at some point consumed well over 50 percent of the EU's entire budget. The policy was launched in the early 1960s to fulfill the objectives defined in one of the founding documents of the EU, the 1957 Treaty of Rome. The CAP is a price support program that had long benefited European farmers by artificially inflating prices they received for their agricultural products. Over the years, its growing costs and mounting criticism resulted in several waves of reforms aiming at scaling down the harmful impact of CAP policies. For example, the 2003 reform was marked by decoupling most direct payments from production to give clearer market signals to farmers. These reforms have led to some changes in practice. While the CAP budget remained at around 50 billion throughout the 1990s, it decreased as a percentage of the EU budget from 70 percent in 1985 to around 40 percent in 2009.[17] Additional reforms were negotiated in 2013 and implemented in 2015. However, as of 2016 the CAP still accounted for approximately 39 percent of the EU's budget, with costs of over €58 billion (approximately US$61 billion at current exchange rates). The largest chunk of this money went to farmers, who represent roughly 3 percent of the EU's population, in the form of direct financial support.[18]

Protectionist agricultural policies, such as those of the EU and other developed nations, most notably the United States, are accused of stifling competition, hampering productivity gains, favoring big agribusiness firms, and distorting international agricultural markets. The developmental opportunities of many poor countries become compromised by high import tariffs and high safety standards imposed on their agricultural products. Subsidies also encourage an oversupply of certain agricultural products, which are then sold, or donated in the form of aid, to countries in the Global South. As a result, the agricultural protectionism of the North contributes to the deepening of rural poverty in the developing world by driving poor farmers out of business as they are unable to compete on the world's markets with heavily subsidized products. According to the 2003 UN Human Development Report, in 2000 every cow in the EU received about US$913 in subsidies, compared with an average of US$8 annual aid per person sent from the EU to sub-Saharan Africa. Still, the country known for the highest agricultural subsidies was Japan, with a price tag of US$2,700 per subsidized cow. Japan sent even less aid to Africa, corresponding to about US$1.47 per person. For the record, in 2000, the average per capita income in sub-Saharan Africa was US$490.[19]

The issue of agricultural subsidies was expected to be resolved during the Uruguay Round of trade negotiations. This historic round ended with the creation of a formal trade organization in 1995, replacing the provisional GATT. When the talks were launched in 1986, one of the main objectives was to address problematic agricultural trade. After difficult negotiations, which at times threatened to derail the completion of the round, the final WTO Agreement on Agriculture was reached, but it fell short of expectations. The deal turned out to be a modest move towards the liberalization of agricultural trade. Its main significance stemmed from the multilateral attempts to bring fair competition to this vastly distorted sector of the world economy. In summary, the WTO Agreement on Agriculture accomplished three things. First, it converted all existing quotas (non-tariff barriers) and unbound tariffs (volatile because unpredictable) on agricultural imports into bound (fixed and predictable) tariffs. Second, it prohibited new export subsidies, and in principle obliged WTO members to work on cutting existing ones. And third, it began to tackle domestic subsidies and other support measures, which protect domestic farmers against foreign competition. The agreement was conceived as part of a progressive process of negotiations aimed at securing substantial reductions in support and protection in the agricultural sector.[20] Following the WTO's establishment, developing countries pushed for fair access for their agricultural products to fulfill the WTO mandate of facilitating free trade for all nations. They argued that the WTO has maintained unfair rules in favor of developed economies because it permits them to effectively protect their domestic agriculture markets at the expense of the poorer nations. This view claims that the entire WTO system should be reformed to reflect the connection between development and trade.

Developing countries expected that at least some of the imbalances of the agricultural trade rules would be remedied during the WTO Doha round of negotiations.[21] Specifically, three sets of issues were important to developing countries: meaningful reduction in domestic support measures, significant improvement of market access, and phasing out of all forms of export subsidies. From the outset of the talks WTO members were divided, as two contentious views emerged, one arguing that "food trade liberalization would put food security in jeopardy;" while the other argued that "food trade liberalization would bring more opportunities to food security."[22] While both sides of the argument hold firm positions, the reality is that food security is a complex subject, involving more than just the introduction of more equitable trade rules: "Apart from productivity in agriculture, income level, and the food market, factors such as ecological balance, resource preservation, population growth, and the political and legal system may also affect food security."[23] Still, the existing multilateral trade rules as they relate to agriculture are far from fair. The limited WTO Agreement on Agriculture permits big trading nations to keep the existing domestic subsidies, but it prohibits the introduction of new subsidies in the developing world. Furthermore, the developed economies made good use of the WTO Agreement on Sanitary and Phytosanitary Measures, which allows countries to impose trade restrictions on agricultural imports if imported products are considered to be below the expected standards. On several occasions, major trading nations closed their doors to agricultural products from the developing world, citing food safety concerns, and used the provisions of the WTO Agreement to defend their actions.

The Doha Declaration launching the round of negotiations at the fourth WTO ministerial meeting in Qatar in 2001, acknowledged the firm link between trade and development. However, the progress of the round has been slow, and as of 2017, negotiations have not been finalized. Developing countries insist that agricultural subsidies in the rich countries make it impossible to trade on fair terms, and fair trade is something that they see as essential to achieve food security. The exporting of food would permit poor farmers to gain incomes and facilitate greater agricultural production. From the point of view of developing countries, this is the path to independence with respect to food, and possibly other products. For its part, the EU justifies agricultural subsidies by explaining that since food is a basic human need, it cannot be treated as any other commodity, and given its "cultural significance" it must be protected.

Frustrated with the sluggish progress of the round, an Oxfam report gave it a failing mark in 2009. The report demonstrates how the Doha Round gradually turned to be more about market access than development, favoring the position of the industrialized countries. Although some progress on agriculture was made at the WTO ministerial conference in Hong Kong in 2005, it came with a price for developing countries of losing policy space in non-agricultural market access. And, despite the fact that the Hong Kong meeting ended with a political commitment to phase out agricultural export subsidies, both the EU and the United States reintroduced them in 2009 in response to domestic pressures.

The first serious signs of trouble emerged during the 2003 WTO ministerial meeting in Cancun, when a group of cotton-producing African countries (Benin, Burkina Faso, Chad, and Mali) teamed together to demand immediate action on ending cotton subsidies maintained by the developed economies, most notably the United States. Despite the vital importance of cotton trade to some of the poorest countries in Africa, this proposal was never seriously addressed during the Cancun ministerial meeting.[24] Unable to reach consensus on a range of issues, especially on agriculture, this fifth WTO ministerial meeting ended without a formal declaration, and was considered a serious setback in the attempts to bridge differences between the developed and developing countries. The position of the EU and the United States during the meeting demonstrated the enduring power of domestic agricultural interest groups in the industrialized world. In an attempt to rescue the failing Doha Round, limited progress was made at the 2005 WTO meeting in Hong Kong. The concluding declaration specified an end date of 2013 for eliminating agricultural export subsidies, with an additional commitment that developed countries' export subsidies on cotton should be eliminated by 2006.[25] Yet the declaration did not call for the elimination of the trade-distorting domestic subsides for cotton, which for example, allows US producers to destroy competition from the poor countries by selling cotton on international markets at below production cost prices. It is estimated that this form of subsidy accounted for 80 to 90 percent of all US subsidies for cotton (approximately US$3.8 billion) in 2004. Furthermore, the declaration made a point of asking WTO members to develop proper disciplines on food aid, which can be objectively classified as a subsidy for domestic farmers, but did not require specific implementation results under its mandate.[26]

There was very little movement in the Doha Round until the 2013 ministerial meeting in Bali. The new director general of the WTO, Roberto Azevêdo, was determined to push negotiations forward. This goal was partially achieved by signing the first multilateral agreement since the establishment of the WTO. Although, it was a welcome achievement, the agreement on trade facilitation was not enough to make the meeting successful. The recurring controversy over food stock purchases to be exempted from WTO disciplines classifying them as subsidies pitted India against the United States. The final ministerial decision does not go as far as directly allowing government managed food stockholding for food security reasons, but instead it states that WTO members should not challenge such programs under the WTO DSU for the time being.[27]

Reforming agricultural domestic support policies remained a crucial issue for developing countries at the 2015 Nairobi WTO ministerial meeting. Nevertheless, the opposing views on the matter were not reconciled during the difficult negotiations. On the positive side, the so-called Nairobi package included a decision on export competition in agriculture, with a commitment by countries who still use agricultural export subsidies to eliminate them, subject to different time schedules. This legally binding decision commits developed countries to eliminate such subsidies by 2018. With respect to cotton, the Nairobi decision called for allowing LDCs free duty and free-quota access. At the Nairobi meeting the WTO members also agreed to continue the practice legitimized at Bali of not challenging developing countries in maintaining food stockpile programs, normally considered a form of subsidy. This is hardly a preferable outcome from the point of developing countries. It also increasingly questions the effectiveness of the WTO and the stability of multilateral trade rules. The negotiators in Nairobi were given a set of options as outlined in the report on public stockholding for food security purposes. The report examines the issue of food security by looking at specific programs introduced by a number of developing countries while it seeks a permanent multilateral approach to the issue of food security.[28] However, instead of reforming multilateral trade rules to allow fair competition in international agricultural markets to help poor countries advance their rural economies, the WTO succumbed to the pressures of the agricultural interests in the developed world.

III IPRs, GMOs, and TRIPS vs. Farmers' rights

One of the important subjects in the debate about food security is the use of new technologies and the right of farmers to freely use novel inventions to improve the quality and supply of agricultural products. The discovery of pesticides and chemical fertilizers significantly improved the efficiency of farming. However, the most revolutionary event that altered the way food is grown was the discovery of commercially produced genetically modified organisms (GMO). Presently, a growing variety of crop plants, farm animals, and soil bacteria are subject to genetic engineering.

There are two main issues linked to the introduction of such technologies in agriculture: safety and access. First, there is a question of food safety. The prevalence of chemical fertilizers and pesticides casts doubts about the environmental sustainability of land and the quality of food grown with excessive help of chemical enhancements. In this context, GMO also provoke debate about the safety of genetically engineered food, given the unprecedented scope of modification. In reality, farmers have been engaged in the genetic altering of plants and animals for generations using traditional breeding techniques. Artificial selection for specific, desired traits has resulted in a variety of organisms, ranging from sweetcorn to different breeds of dogs. But this kind of genetic engineering was restricted in scope as it was only applied to naturally occurring variations. In recent decades, however, technological advances in the field of genetic engineering have allowed for more elaborate manipulation of genetic materials and precise control over the genetic changes introduced into an organism. Scientists can now incorporate new genes from one species into a completely unrelated species through genetic engineering, optimizing agricultural performance or facilitating resistance to diseases. Crop plants, farm animals, and soil bacteria are some of the more prominent examples of organisms that have been subject to genetic engineering. Agricultural plants are one of the most frequently cited examples of GMO. Some benefits of genetically enhanced seeds are increased crop yields, reduced costs of food production, reduced need for pesticides, enriched nutrient composition, resistance to disease, and greater food security. A number of animals have also been genetically engineered to increase yield and decrease susceptibility to disease. For example, salmon have been engineered to grow larger and mature faster, and cattle have been enhanced to exhibit resistance to mad cow disease.[29]

> At the 1992 UN Earth Summit in Rio de Janeiro, Brazil, 150 countries signed *the Convention on Biological Diversity*, which entered into force on December 29, 1993. This pact has three main objectives: 1) the conservation of biological diversity; 2) the sustainable use of the components of biological diversity; 3) the fair and equitable sharing of the benefits arising out of the utilization of genetic resources. In January 2000, the Cartagena Protocol on Biosafety was adopted in response to new technological advancements.

The critics of GMO are not convinced about their benefits. Since genetic engineering is an example of artificial human intervention in the natural processes, those who oppose it warn that the long-term consequences of genetic engineering are unclear. Some indicate that precisely because of their unknown effects, humanity should stay on the side of caution and either completely prohibit the use of GMO or limit their advance and application. The precautionary principle in fact has become a norm in international environmental law, accepted by a number of countries as a basis for policy making. With respect to GMO the precautionary approach was incorporated within the Cartagena Protocol on Biosafety, which is an international treaty regulating the safe transfer, handling, use, and cross-border movement of GMO. Article 1 specifies the objective of the Protocol:

> In accordance with the precautionary approach contained in Principle 15 of the Rio Declaration on Environment and Development, the objective of this Protocol is to contribute to ensuring an adequate level of protection in the field of the safe transfer, handling and use of living modified organisms resulting from modern biotechnology that may have adverse effects on the conservation and sustainable use of biological diversity, taking also into account risks to human health, and specifically focusing on transboundary movements.

Countries can refer to the Cartagena Protocol when restricting the use of GMO in situations of scientific uncertainty with regard to potentially harmful ecological and health effects. It is important to note that the use of the precautionary principle shifts the burden of proof to the proponent (exporter) who now has to demonstrate that the GMO in question is safe. The precautionary principle is reflected in the EU regulatory frameworks.[30]

However, US lawmakers have followed a somewhat different approach, favoring cost–benefit analysis and quantitative risk assessments to guide health, safety, and environmental policies. Critics allege that the persistent skepticism towards the precautionary principle in the United States is motivated by the government's desire to accommodate industrial and producer interests.[31]

The second issue concerns access to new inventions. Both chemical formulas used in fertilizers and GMO are patented, meaning that a title-holder, who is legally considered an owner of the patented invention, has a right to set a price for access (license) to a product that is derived from this invention. Historically, the importance of such inventions, together with the costly patenting process, has privileged the industrialized countries. The majority of new technological inventions that are now legally protected by patents are developed by large western corporations, who often work closely with renowned universities and are eager to invest in top research and development facilities. New technologies mean greater profit and a greater share of the market. On the negative side, the accelerating race in search for novel inventions can lead to the monopolization of new technologies by firms from developed countries. Such monopoly means that the costs of using the patents can be set very high, making new technologies prohibitively expensive for underprivileged consumers in the developing world.

Poor countries, however, have strong and growing interests in accessing new technological discoveries that improve their developmental prospects and agricultural outputs. The already impoverished farmers in the Global South fear the monopolization of the world food market by large multinational companies (agribusiness) that control the distribution of genetically modified seeds. The technological divide between the Global South and the industrialized North with respect to research and development in many different fields is a source of concern. This situation puts the developing world at a serious disadvantage because many inventions, such as life-saving medications, chemical fertilizers, and GMO, are now protected by IPR. In fact, IPR are viewed as one of the drivers behind economic progress. With the increased significance of the knowledge-based economy, the creation, use, and dissemination of intellectual property has become a key determinant of new opportunities for advancement.

The subject of IPR resonated globally with the introduction of the TRIPS, which became an integral part of the WTO. From the beginning, the agreement was controversial and it soon came to symbolize the friction between the technologically advanced countries and the rest of the world. Developing countries resented the fact that TRIPS required all members of the WTO to adopt and fully comply with four international conventions on the protection of intellectual property. Two of these conventions go back to the nineteenth century, when a group of industrialized nations decided to legally protect intangible creations and inventions. Until the establishment of the WTO in 1995, it was optional for any country to join them. The most famous are the Paris Convention, relating to patents, and the Berne Convention, which deals with copyrights. Both conventions have been modified several times and are administered by the World Intellectual Property Organization (WIPO).

Under the TRIPS Agreement, every WTO member has to become a signatory to all four existing conventions[32] and has to ensure protection relating to the following major categories of IPR: patents, trademarks, copyrights (and neighboring rights), industrial designs, geographical indicators, integrated circuits, and trade secrets.[33] Supporters of TRIPS point out that every country has a relative freedom to design its IPR laws as long as these laws guarantee the minimum level of IPR protection specified by the agreement. For example, the minimum term of patent protection under TRIPS is 20 years from the date of application. However, these minimum standards are consistent with the level of protection afforded by the IPR laws in the developed world, while such laws were often nonexistent in many developing countries.

Article 7 of the TRIPS Agreement stipulates that the protection of IPR should contribute to "technological innovation" as well as to the "transfer and dissemination of technology" – both of which are matters of major concern to developing countries. Reference to the concepts of "mutual advantage," "social and economic welfare," and "balance of rights and obligations" indicates that the recognition and enforcement of IPR are subject to higher social values. However, the first six years of implementing TRIPS were characterized by attempts by industrialized nations on ensuring an adequate level of IPR protection for their firms engaged in foreign direct investment activities. Very little attention was paid to the promises to transfer and disseminate technological knowledge to meet the developmental objectives of less developed nations. Only in 2001, during the Doha ministerial meeting, did the WTO members adopt the declaration on the TRIPS Agreement and Public Health that responded to some of the most urgent demands from the developing world by addressing the problem of high-priced and patent protected medications.[34] The commitment to ensure the affordability of an agreed list of medications involved a complex scenario, not limited to: negotiating legal exemptions from patent rights, the establishment of systems to deliver and monitor distribution, and alternative mechanisms to finance purchases of these medications by developing countries. The declaration, however, did not provide the resolution on the matter of technology transfer. There is no agreement among the WTO members on other subjects related to the high prices set by the owners of intellectual property, most notably GMO.

Critics complain that excessive focus on protecting the rights of titleholders residing in the industrialized WTO member states has turned attention away from the protection of products of direct relevance to many developing countries. For example, Vanuatu, a small least developed country, was unable to complete the WTO's accession process for several years. Vanuatu was resisting the excessive liberalization demands and the request for the immediate establishment of an IPR regime. And while the costs of creating a new elaborate system of IPR laws under the TRIPS guidance were obvious and difficult to meet, the benefits were hard to envision. Vanuatu is the biological home of the root crop "kava." Kava is a traditional beverage in Vanuatu and has been consumed by the people of Vanuatu long before the arrival of Europeans. However, large pharmaceutical companies in the United States and the EU now produce, and have patented, kava pills.[35] Apart from exporting the unprocessed kava root for pennies, Vanuatu gains very little of the international sales of the increasingly popular natural sedative. The TRIPS Agreement does not protect the indigenous traditional rights that Vanuatu claims to have to kava. The agreement does not provide international standards for the protection of such knowledge, or native formulas, or traditional indigenous artistic designs.

Despite that fact that indigenous knowledge has been the basis for much of the development of modern agriculture and medicine, the present-day aboriginal and native communities have to pay for patented products that are either derived from such knowledge or based on the plants grown in these communities. One suggestion in this regard is that TRIPS should accord legal recognition to traditional knowledge and communal rights holders. Despite the criticism, TRIPS does not stipulate criteria and does not contain obligations to protect indigenous plant varieties, allowing for the exploitation of biological resources in developing countries by the powerful pharmaceutical corporations. TRIPS stipulates that substances, organisms, and plants existing in nature are a discovery and not an invention and thus cannot be patented. They are simply considered commodities for sale. Yet Article 27.3 of the TRIPS Agreement guarantees that formulas, seeds, and micro-organisms created with these natural ingredients can be fully protected by patents and thus only accessible after a payment of a licensing fee. The provisions of the Convention on Biological Diversity allow countries to exercise sovereign rights over their biological resources, but for the convention to be legally binding it would have to be included within the TRIPS framework. Much to the dismay of developing countries, the framers of the TRIPS Agreement refused to take into account the principles of the Convention on Biological Diversity.

The fact that it is administratively challenging to comply with all provisions of the agreement further adds to the complaints of developing countries. For example, the enforcement provisions of TRIPS require that every WTO member provide civil as well as criminal remedies for the infringement of IPR. Judicial systems, regulatory and administrative frameworks, and enforcement procedures must be developed or modernized in many countries to comply with this aspect of TRIPS. The technical complexity of many IPR rules means that judges, customs officials, and others involved in enforcing them may require additional technical training. The enforcement provisions also obligate members to provide the means by which right-holders can obtain the cooperation of customs authorities to prevent imports of counterfeited goods. In summary, it is difficult for developing countries to strike a balance between the WTO's requirement to protect IPR in the name of promoting innovation, and the overwhelming costs and administrative requirements of implementing the agreement.

The Convention on Biological Diversity remains outside the legal umbrella of the WTO, and countries who decide to adopt policies informed by its provisions do so on a voluntary basis. Nevertheless, the convention ignited debate about the need to develop a system for the global protection of indigenous communities and their natural environment. The convention established the obligation to "respect, preserve and maintain knowledge, innovations and practices of indigenous and local communities" (Article 8). Under the terms of the convention, access to genetic resources is granted by the "country of origin" on the basis of mutually agreed terms. Article 2 states that the country of origin of genetic resources "means the country which possesses those genetic resources in in-situ conditions." And in-situ conditions "means conditions where genetic resources exist within ecosystems and natural habitats, and, in the case of domesticated or cultivated species, in the surroundings where they have developed their distinctive properties."[36]

The convention also recognizes an important concept of *farmers' rights*. Farmers' rights have been developed to acknowledge the contribution of farmers as guardians of existing seeds and species as well as creators of new plant varieties and pioneers of domesticating wild species. For example, tomatoes could not be popularized around the world and grown commercially, if it were not for the replanting, conservation, and disease resilience modification made by farmers in different countries over the centuries. International negotiations on formally recognizing farmers' rights took place in the FAO with a deal reached in 1989. The outcome was a resolution asking that farmers and their communities participate on equal terms in the benefits derived from the use of plant genetic resources. The resolution defines farmers' rights as

> arising from the past, present and future contributions of farmers in conserving, improving, and making available plant genetic resources, particularly those in the centres of origin/diversity. These rights are vested in the International Community, as trustee for present and future generations of farmers, for the purpose of ensuring full benefits to farmers, and supporting the continuation of their contributions (FAO Resolution 5/89).[37]

The above definition highlights a crucial distinction between plant genetic resources and other resources, or raw materials, treated as tradable goods. This distinction unfortunately has problematic consequences for how farmers can assert their rights. By not specifying what genetic resources are covered and who can claim ownership, the definition further adds to the confusion over the meaning of farmers' rights.[38] There is nothing in the text of the FAO resolution that objects to the practice of patenting GMO. Since the innovation component or alteration of plant genetic resources creates a new GMO, such a novel *product* can then be patent protected and commercially restricted under IPR laws. Thus, if a farmer wants to re-use or enhance genetically engineered seeds that were originally patented and sold to him with a license, he can be in violation of patent laws. He must always buy a license to be able to use such seeds. Farmers see this approach not only as unfair, because it allows patent owners to control the price of often essential food, but that it also contradicts the Convention on Biological Diversity. Farmers in developing countries find this issue particularly problematic since most of the genetic diversity found in modern agricultural crops was cultivated and perfected by traditional rural communities.

There is also another convention, the International Union for the Protection of New Varieties of Plants (UPOV), which awards certificates with reciprocal breeders' rights for its 74 member countries.[39] This intergovernmental organization (IGO) is based on the International Convention for the Protection of New Varieties of Plants signed in Paris in 1961. Unfortunately, plant varieties created in the past by indigenous farmers often cannot meet the UPOV criteria of distinctness to be given a certificate. However, just like the Convention on Biological Diversity, the UPOV's effectiveness is limited because it also operates largely on a voluntary basis. The only organization that by the nature of its legally binding status can make a difference in compelling countries to comply with both conventions is the WTO. But to make it happen, the WTO members would have to agree to bring these two conventions under the legal umbrella of the organization. It is difficult to envision this at the moment. Currently, the WTO's TRIPS agreement does not incorporate the principles of the Biodiversity convention and it has a tenuous relationship with the UPOV.

Looking from the position of developing countries, the global landscape of contemporary agricultural production and trade is vastly asymmetrical. International agreements do very little to introduce fairness in agricultural trade at a time when food security continues to be a problem. Then there are accusations about *bio-piracy*, where unscrupulous firms engage in the search for previously unknown biological resources and traditional knowledge (ancient medical formulas) to appropriate them for commercial purposes in patents without compensating or even informing the providing community. The argument about *bio-piracy* makes a claim that rich industrialized countries continue to exploit developing countries by seeking ownership rights over the biodiversity of the Global South. The connection is made to the colonial era, when the European powers took control of crop resources in their colonies to enrich themselves and control world trade in cotton, sugar, tea, rubber, and spices.

Another concern over traditional knowledge relates to the obligation under Article 27.1 of TRIPS to grant patents "in all fields of technology," privileging those inventors and businesses who not only have developed new inventions but who also have legal expertise to write complicated patent applications. In contrast, many local communities in developing countries are not aware and do not have the capacity to protect their rights under a patent system. Misappropriation of indigenous knowledge with its formulas based on species found in the developing world is commonly explained by the argument that they exist in the public domain with no legal claims attached. The need to protect traditional knowledge in intellectual property courts inspired non-governmental organizations to partner with local indigenous communities in order to contest the attempts of perceived bio-piracy in courts. The joint efforts of such groups resulted in establishing the Traditional Knowledge Resource Classification (TKRC), an innovative structured classification system for the purpose of systematic arrangement, dissemination, and retrieval was evolved for about 5,000 subgroups against the few subgroups available in the International Patent Classification (IPC), related to medicinal plants. A great example of how this works in practice is the Traditional Knowledge Digital Library (TKDL) of India, created to prevent misappropriation of a country's traditional medicinal knowledge.

The TKDL of India records a number of successful, and some failed, attempts of fighting bio-piracy in the courts. One successful case concerns turmeric, best known as a traditional spice used in many Indian dishes. Turmeric is also valued for its special properties that make it a popular ingredient in medical formulas and cosmetic applications. In 1995, a US patent was granted to scholars from the University of Mississippi in the United States that allowed them to use turmeric in a medication to heal wounds. Immediately, the government of India filed a complaint with the USPTO, requesting the re-examination of the case. India maintained: "that turmeric has been used for thousands of years for healing wounds and rashes and therefore its medicinal use was not a novel invention."[40] In support of the argument, India provided documentary evidence of traditional knowledge, including ancient Sanskrit text and an article on the subject published in 1953 in the

Journal of the Indian Medical Association. Although the American scholars contested the Indian claim, the patent was revoked in 1997. This was an important victory for the advocates of traditional knowledge. For the first time, a patented medication based on the traditional formula used for generations in a developing world was effectively challenged in the US courts.

The TKDL of India also reports an unsuccessful case when, despite an organized legal campaign, a US patent based on a mythologized plant used in traditional South American formulas was upheld. This case concerns ayahuasca, a vine like drink derived from the bark of *Banisteriopsis caapi*, which grows throughout the Amazon basin of South America (Peru, Colombia, Ecuador, and Brazil). The indigenous tribes in the region have been known to use ayahuasca as a ceremonial drink, but also for its medical properties. In 1986, the US Patent Office granted a patent to a scientist who claimed IPR over a new and distinct variety of the plant, based on the specimens he collected in the Amazon region. The patent allowed him to produce ayahuasca-like products commercially. The patent generated outcry among traditional communities. The Coordinating Body of Indigenous Organizations of the Amazon Basin, which represented more than 400 indigenous Amazon tribes in the Amazon region, challenged the patent with the help of two organizations, the Center for International Environmental Law and the Coalition for Amazonian Peoples and their Environment. They argued that the person requesting the patent was not the original breeder since ayahuasca had been widely used in traditional medicine and cultivated for this purpose for generations.[41] After years of legal arguments the United States Patent and Trademark Office (USPTO) revoked the patent in 1999, with an understanding that the claimed plant variety named Da Vine was neither distinctive nor novel. The scientist, however, was able to effectively appeal the decision and the patent rights were restored to him in 2001.

Between 2009 and 2011 the TKDL team submitted 571 third party observations against patent applications submitted by companies of the United States, Great Britain, Spain, Italy, and China. These legal challenges resulted in 53 of such patent applications to be either set aside or withdrawn and cancelled. The TKDL of India has become a model for other countries.[42] Governments in the developing world increasingly collaborate with civil-society groups on initiatives aimed at protecting traditional knowledge and local plants because it takes time and it is very expensive to fight well-organized legal teams representing major corporate entities.

It took Mexico almost ten years of legal proceedings to obtain a ruling that canceled in April 2008 the patent on *Enola Bean*, a bean commonly grown by Latin American farmers, by the USPTO.[43] The International Center for Tropical Agriculture (known by its Spanish acronym, CIAT) championed the legal challenge to the patent. Despite the length of the process, the CIAT and Mexico never gave up the fight, not only because they were concerned about the immediate economic impact of the Enola Bean patent, more broadly they were concerned "that the patent would establish a precedent threatening public access to plant germplasm – the genetic material that comprises the inherited qualities of an organism – held in trust by CIAT and research centers worldwide."[44] There are 11 such gene-banks maintained worldwide. They contain crop materials such as seeds, stems, and tubers held in trust with the UN FAO. Collectively, these gene-banks house a total of about 600,000 plant varieties, with the aim to conserve agriculture biodiversity and maintain global food security.

The opening agenda of the WTO Doha development round included the undertaking to review the TRIPS agreement in the context of the perceived imbalances it created. The special ministerial declaration stated the importance of access to public health and stated that the issue should be addressed with relation to the high costs of patented medications. A special declaration extended some flexibilities to least-developed countries when it comes to implementing TRIPS in this context.[45] The declaration also promised to work on establishing a system of protection for geographical indications other than wine and spirits in acknowledgment that developing countries have region-specific products, like basmati rice. The TRIPS agreement, at the request of France, protects the name of the product like the champagne associated with one designated region in France. As a result, any other bubbling wine in the world must be called by another name, such as a sparkling-wine. The TRIPS agreement does not afford this kind of geographical indication[46] protection to other products.

The Doha Declaration also responded to the calls of developing countries by initiating the work on examining the relationship between TRIPS and the Convention on Biological Diversity, the protection of indigenous knowledge, folklore, and other issues raised by concerned WTO members.[47] Unfortunately, the dismal progress made at the subsequent WTO ministerial meetings has proven to be particularly slow with respect to IPR and TRIPS. These matters are barely mentioned in the 2005 Hong Kong Declaration.[48] The Bali Declaration is even more disappointing, issuing a special decision on extending a "temporary" moratorium, which was first put in place in 1995, on initiating trade disputes with respect to TRIPS regarding non-compliance with the agreement even if there is no direct violation of its terms.[49] The 2015 Nairobi Declaration repeats the same procedure.[50] In summary, the WTO as a multilateral forum has failed to address the demands of less developed members of the organization on issues of vital importance to them. After over a decade of slowly moving negotiations, trade rules on agriculture continue to be distorted in favor of the industrialized nations, who still maintain their domestic support measures and export subsidies and restrict agricultural imports from the developing world. The global standards guiding the domestic regimes of IPR remain under the influence of the TRIPS threshold of standards, although the legal power of the agreement may be somewhat weaker given the state of the WTO. This means that we may observe a rise in bilateral disputes concerning agricultural policies and the protection of IPR to the detriment of the developing world.

IV Access to food and domestic agricultural policies

Although problems of hunger and malnutrition were first associated with the inadequate production of food, this view was soon challenged by Amartya Sen. He was particularly influential in shifting the discourse about food insecurity, from viewing it simply as a failure of the state to produce a sufficient amount of food, to viewing it as a failure of governance. He observed that famine happens when the government fails to respond to the different needs of communities under its jurisdiction when it comes to land and food access and their ability to sustain themselves. According to Sen: "the focus has to be on the economic power and substantive freedom of individuals and families to buy enough food, and not just on the quantity of food in the country in question."[51] Even if the government in charge expects an adequate food supply in the future, the main challenge will be to ensure fair distribution and affordability.

Sen's extensive research into the 1984–1985 famine in Ethiopia's Wollo region convinced him that the abysmal government policies and lack of democratic oversight directly contributed to the crisis. Over one million people died but arguably it was a tragedy that could have been prevented since overall in the country there were enough food supplies to avert disaster. The controversial governmental response to the drought and famine was the policy to move about 600,000 people from their communities in the affected region to the southern part of the country. People were transported by the military and many died during this forced resettlement. Yet as the drought situation in the Wollo region deteriorated, the wealthy landowners elsewhere were benefiting from keeping the price of food artificially high. Poor farmers in the Wollo region were denied their entitlement over an adequate amount of food. They lost their lands and thus could not use their labor in a productive way and in addition were faced with unfair exchange conditions.[52]

Sen's analysis of famines arises from the paradox: we live in a time when global food production outpaces global population growth and yet millions suffer from hunger. During the Bengal famine of 1943, the Ethiopian famine of 1973, and the Bangladesh famine of 1974, food output actually was on the increase in these countries.[53] State policies that led to these famines were either incompetent in preventing the agricultural supplies from reaching the starving population, inefficient in transporting the right amount of food, or simply corrupted by allowing rich farmers to hoard supplies to raise the price of food. This is why the nature of state institutions plays a crucial role in Sen's calculus of preventing food crises. He believes that only a democratic government accountable for respecting human rights can become a meaningful prerequisite for a stable country where all people have adequate access to food.

Sen observes that people suffer from hunger when they cannot secure their entitlement to food. He breaks down the concept of entitlement into three distinct determinations. First, it is the issue of endowment, which is defined as the ownership over productive resources like labor and land. Second, production possibilities and their use become an important consideration, especially with respect to technology, including the new generation of genetically modified seeds. Third, exchange conditions also play a fundamental role in ensuring entitlement. The ability to freely sell and buy agricultural goods at a fair price despite the changing external conditions is something that must be protected by rule of law. Taken together, these three sets of influences provide a more complete scenario surrounding the issue of food security, because "understanding the causation of hunger and starvation calls for an analysis of the entire economic mechanism, not just an accounting of food output and supply."[54]

Consistent with his work on freeing human potential by creating an enabling sociopolitical environment, Sen argues that the priority should be placed on developing the entitlement approach to famine analysis, which "concentrates on the ability of people to command food thorough the legal means available in the society."[55] The entitlement approach "has the effect of emphasizing legal rights" and other relevant factors, including ownership rights, contractual obligations, and legal exchanges operating in a transparent way according to the rule of law.[56] In recognizing the limitations of the entitlement approach, he lists the following factors: looting and illegal distribution of food, diseases and epidemics, and inter-family distribution systems. There is always a possibility of unforeseen circumstances impacting the assessment. In summary, Sen's research has shown that undernourishment, starvation, and famine are often consequences of inept economic and political systems where the poor have no rights and their capabilities are compromised. As a result, famines do not occur in democracies, because democratic governments are accountable to the people and have to produce results in the time of crisis. Democracies are also characterized by free press, eager to scrutinize policy-making processes.

However, while Sen's argument holds strong when examining the incidence of famine, Sen also recognizes that democracies are not immune to chronic hunger, and for this reason he supports food imports for the purpose of stockpiling and reducing the costs for the vulnerable poor. For example, his native India has not experienced famine since becoming a democracy, but a large portion of its population continues to suffer from chronic hunger. During the 1980s, approximately 300 million poor people lacked food security in India.[57] According to the most recent FAO Food Insecurity Report, India has the second highest estimated number of malnourished people in the world: 194.6 million, or 15.2 percent of its population, as measured during the 2014–2016 period. On the positive side, these numbers are decreasing, since during the 2005–2007 period there were 233.8 million undernourished people in India, which represented 20.5 percent of the whole population.[58] To address the problem, India, and also China and Pakistan, are known for high rates of government procurement for the purpose of stockpiling food. In fact, stockpiling accounts for more than 20 percent of the total production of wheat in each country. India also procured 30 percent of all rice production in the country.[59] Arguing that this kind of stockpiling was undertaken to ensure food security for its vulnerable population, India fought to extend the existing exemption from the WTO rules for developing countries to maintain such programs, as previously discussed.

India has made remarkable progress over the last two decades in its fight against poverty, but the uneven development across the country remains problematic. There are 28 states in the federation, each differing in size, resources, economic performance, and capacity. Success stories, such as Kerala, which has the lowest rate of malnutrition among children, and other southern states such as Andhra Pradesh, and Karnataka, which have reduced the poverty level, and stories of failure, such as Bihar, Orissa, Uttar Pradesh, and West Bengal, which have seen increases in the number of the poor during the same period, mark the drastic difference between regions within India.[60]

The main response to the issue of food insecurity by the Indian government since independence has been the public distribution system (PDS), which is a large-scale rationing program designed to provide the poor with sustained access to basic food. The program was first launched in 1947 and underwent a number of reforms over the years to address complaints about its uneven and sometimes ineffective distribution. For example, between the years 1986 to 1996, Kerala, one of the most successful Indian states, received 63 kg of grain per person per year from the PDS, while in contrast Bahir, one of the most disadvantaged and poorest regions of the country, received only 6 kg per person per year.[61] This kind of discrepancy has led to the transformation of the PDS system in 1997 into a targeted program, focused on those who need assistance the most. The new targeted distribution system obliged every state in India to come up with transparent arrangements to identify the poor. There was also a requirement for greater accountability when it comes to the distribution of food grains. Further reforms in 2000 by the central government, which still provides 90 percent funding for the system, enhanced the program with a new undertaking called "Antyodaya Anna Yojana" (AAY). This initiative promised to eradicate chronic hunger in India by expanding the number of eligible households. The reforms also increased the amount of rice and wheat given to the poor per month. Since 2002, the AAY program was further expanded three times. In May 2017, the Ministry of Consumer Affairs of India reported that the AAY covered approximately 25 million households considered to be desperately poor. In 2013, India enacted the National Food Security Act, meant to formalize the government's commitment to provide food security for all its citizens.[62] India's PDS has been operating for 70 years, with millions of people receiving food grains to survive, although the program continues to be criticized for its inefficiencies. According to the 2015 FAO report, India still has the second-highest estimated number of undernourished people in the world.[63]

A unique mixture of policy measures has long characterized China's approach to agriculture. Over the last two decades, China has made outstanding progress in reducing food insecurity. In the early 1990s, China had 289 million hungry people (or 23.9 percent of its population), during the 2005–2007 period this number was reduced to 207.3 million. China managed to reduce this further to 133.8 million undernourished people in 2014–2016, or 9.3 percent of its population.[64] In the past, however, many mistakes were made. Archival historical research suggests that at least 45 million people died prematurely as the country experienced the Great Chinese Famine of 1958–1962. This man-made catastrophe resulted from the disastrous collectivization of farms, economic mismanagement, and incompetent regulations pushed for by the Mao communist regime.[65]

Lessons from that dreadful period influenced the subsequent food policy in China. From the first wave of pro-liberalization reforms, the Chinese government prioritized rural development, beginning with the 1978 program, which in effect leased the state-owned land to famers on a contractual basis for 15 years. The reforms were successful, the agricultural productivity significantly increased, and the land contracts were extended by an additional 30 years in the late 1990s. As the reforms progressed, the government eliminated most of its direct intervention programs in parallel with reducing average import tariff for all agricultural products from 42.2 percent in 1992 to 21 percent in 2001 and 17 percent in 2004. Gradually Chinese farmers have gained from increased allocative efficiency based on market prices, and by the mid-2000s most agricultural commodity prices in China were reflecting the international prices. Another long-running policy initiative focused on supporting research on genetically modified seeds and other bio technological inventions for the benefits of domestic farmers. Overall, there was consistent investment in water control (irrigation), land improvement, agricultural technology, and transportation infrastructure.[66] China's experience supports the argument about the importance of agricultural policies to enhancing development and economic growth in low income developing countries.

However, in the recent years the Chinese government has been facing growing challenges when it comes to ensuring food security and achieving sustainable agricultural development. For one, the urban–rural income gap rose sharply between 1997 and 2003. This trend continued, although at a slower pace, given the spectacular growth of manufacturing in the urban areas. The income gap reached about ¥20,000[67] (or US$3,200) in 2015, with nearly 60 million people living in poverty in the rural areas. In 2004, China shifted from being a net food exporter to a net foot importer. As a result, the government launched the largest agricultural subsidy program in the world. The program eliminated taxes in the agricultural sector and introduced direct grain, seed, and technology subsidies, as well as access to credit, insurance, and land conservation services. China also introduced price policy support, which since 2012 led to the growing discrepancy in prices between the domestic and international markets. To prevent market distortions and falling prices, the government started to purchase and store massive amounts of agricultural products. Such policy interventions have further undermined the competitiveness of Chinese farmers, whose income stagnated. In addition, concerns over the sustainability of resources are also growing as the environmental degradation has intensified.[68]

Agricultural policies are increasingly viewed in the broader context of socio-economic development. Most recently, combined programs have been introduced among some of the poorest rural areas of the world. These programs merge agricultural interventions (such as subsidized credit, investment grants, access to market, natural resource management, distribution of seeds, subsidized machinery, technology training) and social protection (cash transfers, access to health care, public works) into one program. Their key objective is to graduate poor rural families from extreme poverty and help them toward becoming self-sustaining households, often by participating in microfinance programs. The first combined programs were launched in Bangladesh, Pakistan, Honduras, Ghana, Ethiopia, Peru, Lesotho, and India. The results varied, but all types of programs have demonstrated positive impacts on the following aspects of development: investment in productive assets, savings and access to formal credit, more stable, permanent and profitable sources of income, self-employment (particularly for women), and/or more profitable and decent employment, food security, rising income and consumption level, poverty reduction.[69] In Lesotho, a pilot initiative called the Linking Food Security to Social Protection Program aimed at improving the food security of poor households by providing vegetable seeds and comprehensive training on gardening and proper nutrition. The targeted households were those who were also eligible for social cash transfers under the Child Grants Program. The results again were largely positive in terms of increased production of vegetables and enhancing food security. The study also showed, however, the limited supply of labor steered households towards reliance on child labor, which could be a potentially unintended consequence of such programs.[70]

In the attempt to deal with growing concerns over the sustainability of agricultural production in China, the government has introduced maize market reform in 2016. It combines market principles in determining maize prices with separate income support provided to farmers. The idea that farmers need social protection in parallel with governmental policies that see investment in new technologies and sustainable development remains an important principle behind ensuring food security in many parts of the developing world.[71]

The 2016 Human Development Report calls for effective actions on the issue of rural poverty as it registers certain systemic patterns of deprivation in the rural communities of the world. Although there is variation across regions, people in rural areas tend to be more multi dimensionally poor than people in urban areas (29 percent versus 11 percent). For example, "nearly half of people in rural areas worldwide lack access to improved sanitation facilities, compared with a sixth of people in urban areas. And twice as many rural children as urban children are out of school."[72]

Notes

1 Jenny Clover (2003) "Food Security in Sub-Saharan Africa," *African Security Review*, Vol. 12, No. 1, p. 5.
2 Fabio Veras Soares, Marco Knowles, Silvio Daidone, Nyasha Tirivayi (2017) *Combined Effects and Synergies between Agricultural and Social Protection Interventions: What is the Evidence So Far?*, Rome, Italy: FAO.
3 FAO (2015) *The State of Food Insecurity in the World 2015*, Rome, Italy: FAO.
4 FAO (2005) "Voluntary Guidelines – to Support the Progressive Realization of the Right to Adequate Food in the Context of National Food Security" Adopted by the 127th Session of the FAO Council, November 2004, Rome.
5 Ibid.
6 Fabio Veras Soares, Marco Knowles, Silvio Daidone, Nyasha Tirivayi (2017) *Combined Effects and Synergies between Agricultural and Social Protection Interventions*, op. cit.
7 FAO (2015) *The State of Food Insecurity in the World 2015*, op. cit., p. 1.
8 For more information see FAO website, online, available at: www.fao.org/.
9 FAO (2004) *The State of Food Insecurity in the World 2004*, Rome, Italy: FAO, p. 8.
10 Per Pinstrup-Andersen (2003) "Food Security in Developing Countries: Why Government Action is Needed," *UN Chronicle*, Vol. 40, No. 3, September/November, p. 65.
11 FAO (2015) *The State of Food Insecurity in the World 2015*, op. cit., pp. 9–10.
12 Ibid.
13 Lisa C. Smith, Amani E. El Obeid, and Helen H. Jensen (2000) "The Geography and Causes of Food Insecurity in Developing Countries" *Agricultural Economics*, Vol. 22.
14 Jenny Clover (2003) "Food Security in Sub-Saharan Africa," op. cit., p. 9.
15 Howard Handelman (2000) *The Challenge of Third World Development* (2nd edition), Upper Saddle River, NJ: Prentice Hall, p. 105.
16 Ibid., p. 106.
17 Jane Kennan, Sheila Page (2011) *CAP Reform and Development: Introduction, Reform Options and Suggestions for Further Research*, London, UK: ODI (Overseas Development Institute), May 14.
18 Economist Intelligence Union (2017) *A Future for the EU's Common Agricultural Policy*, February.
19 UNDP (2003) *Human Development Report 2003 – Millennium Development Goals: A Compact among Nations to End Human Poverty*, New York NY: UNDP, pp. 154–155.

20 Anna Lanoszka (2009) *The World Trade Organization,* Boulder, CO: Lynne Rienner Publishers, pp. 81–95.

21 WTO (2001) *Doha Ministerial Declaration* (WT/MIN(01)DEC/1), November 20.

22 Ruosi Zhang (2004) "Food Security: Food Trade Regime and Food Aid Regime" *Journal of International Economic Law,* Vol. 7, No. 3, p. 569.

23 Ibid., p. 566.

24 Oxfam (2009) *Broken Promises – What Happened to "Development" in the WTO's Doha Round?* Oxfam Briefing Paper 131, July 16.

25 WTO (2005) *Hong Kong Ministerial Declaration* (WT/MIN(05)DEC), December 22.

26 International Center for Trade and Sustainable Development (ICTSD) (2005) *Bridges* (WTO/MC6), December 19.

27 WTO (2013) *Bali Ministerial Declaration* (WT/MIN(13)/DEC), December 11.

28 ICTSD (2016) *Public Stockholding for Food Security Purposes: Options for a Permanent Solution,* Geneva, November 21.

29 Theresa Phillips (2008) "Genetically Modified Organisms (GMOs): Transgenic Crops and Recombinant DNA Technology," *Nature Education,* Vol. 1, No. 1, p. 213. More details and the full text of the Convention on Biological Diversity are online, available at: www.cbd.int/convention/.

30 Anne Ingeborg Myhr (2007) "The Precautionary Principle in GMO's Regulations," in T. Traavik and L.C. Lim (eds.), *Biosafety First,* Trondheim, Norway: Tapir Academic Publishers.

31 Nicholas A. Ashford (2007) "The Legacy of Precautionary Principle in US Law: The Rise of Cost–Benefit Analysis and Risk Assessment as Undermining Factors in Health, Safety, and Environmental Protection," in Nicolas de Sadeleer (ed.), *Implementing the Precautionary Principle: Approaches from the Nordic Countries, the EU and the USA,* London, UK: Earthscan, pp. 352–378.

32 Article 1 Para 3. of TRIPS lists the following: the Paris Convention (patents), the Berne Convention (copyrights), the Rome Convention (broadcasting), and the Washington Treaty (integrated circuits).

33 Keith E. Mascus (2000) *Intellectual Property Rights in the Global Economy,* Washington, DC: Institute for International Economics.

34 WTO (2001) *Declaration on the TRIPS Agreement and Public Health* (WT/MIN/[01]/DEC/W/2), November 14.

35 Roman Grynberg and Roy Mickey Joy (2000) "The Accession of Vanuatu to the WTO – Lessons for the Multilateral Trading System," *Journal of World Trade,* Vol. 34, No. 6, pp. 159–173.

36 Convention on Biological Diversity (2017). See the text of the Convention on Biological Diversity, online, available at: www.cbd.int/convention/.

37 FAO (1989) *Report of the Conference of FAO – Twenty-Fifth Session* (C/1989/REP), Rome, November, 11–29.

38 Stephen B. Brush (2005) *Farmers' Rights and Protection of Traditional Agricultural Knowledge,* CAPRi Working Paper 36, Washington, DC: International Food Policy Research Institute, January, p. 26.

39 As of March 2017.

40 Traditional Knowledge Digital Library (2017). Online, available at: www.tkdl.res.in.

41 Ibid.

42 Samir K. Brahmachari (2011) *Traditional Knowledge Digital Library (TKDL) – An Effective and Novel Tool for Protection of India's Traditional Knowledge Against Bio-Piracy,* Document (SS-132/SF-132/12.08.2011), New Delhi, India: Department of Scientific and Industrial Research.

43 Sam Savage (2008) "Enola Bean Patent Claim Rejected by US Patent Office," *redOrbit,* May 1.

44 Ibid.

45 WTO (2001) *Declaration on the TRIPS Agreement and Public Health* (WT/MIN(01)/DEC/2), November 20.

46 A geographical indication is a sign used on products that have a specific geographical origin and possess qualities or a reputation due to that origin. See WIPO website, online, available at: www.wipo.int/geo_indications/en/.

47 WTO (2001) *Doha Ministerial Declaration* (WT/MIN(01)DEC/1), November 20, paragraphs:17–19.

48 WTO (2005) *Hong Kong Ministerial Declaration* (WT/MIN(05)DEC), December 22, paragraph: 29.

49 WTO (2013) *Bali Ministerial Declaration – Decision on TRIPS Non-violation and Situation Complaints* (WT/MIN(13)/31_WT/L/906) December 11.

50 WTO (2015) *Nairobi Ministerial Declaration – Decision on TRIPS Non-violation and Situation Complaints* (WT/MIN(15)/41_WT/L/976) December 21.

51 Amartya Sen (1999) *Development as Freedom*, New York, NY: Random House p. 161.

52 Ibid., pp. 164–170.

53 Amartya Sen (1998) "Ingredients of Famine Analysis: Availability and Entitlements," in *Resources, Values and Development*, Cambridge, MA: Harvard University Press, pp. 452–484.

54 Amartya Sen (1999) *Development as Freedom*, op. cit., p. 164.

55 Amartya Sen (1998) "Ingredients of Famine Analysis: Availability and Entitlements," op. cit., p. 452.

56 Ibid., p. 480.

57 Lisa C. Smith, Amani E. El Obeid, and Helen H. Jensen (2000) "The Geography and Causes of Food Insecurity in Developing Countries," op., cit., p. 204.

58 FAO (2015) *The State of Food Insecurity in the World 2015*, op. cit., Annex, table A1, p. 46.

59 ICTSD (2016) *Public Stockholding for Food Security Purposes: Options for a Permanent Solution*, op. cit., p. 8.

60 UN (2009) *Rethinking Poverty – Report on the World Social Situation 2010* (ST/ESA/324), p. 41.

61 Jos Mooij (1999) "Food Policy in India: The Importance of Electoral Politics in Policy Implementation," *Journal of International Development*, Vol. 11, No. 4, pp. 626–628.

62 Government of India (2017) Ministry of Consumer Affairs, Food, and Public Distribution Department of Food and Public Distribution Website. Online, available at: http://dfpd.nic.in/public-distribution.htmm accessed May 10, 2017.

63 FAO (2015) *The State of Food Insecurity in the World 2015*, op. cit., p. 15.

64 FAO (2015) *The State of Food Insecurity in the World 2015*, op. cit., Annex, table A1, p. 46.

65 Frank Dikötter (2011) *Mao's Great Famine: The History of China's Most Devastating Catastrophe, 1958–1962*, London, UK: Bloomsbury.

66 Jikun Huang and Guolei Yang (2017) "Understanding Recent Challenges and New Food Policy in China," *Global Food Security*, Vol. 12, p. 122.

67 ¥ – Chinese Yuan Renminbi. US$1 = ¥6.9 as of May 2017.

68 Jikun Huang and Guolei Yang (2017) "Understanding Recent Challenges and New Food Policy in China." op. cit., p. 125.

69 Fabio Veras Soares *et al.* (2017) *Combined Effects and Synergies between Agricultural and Social Protection Interventions*, op. cit.

70 Silvia Daidone *et al.* (2017) "Linking Agriculture and Social Protection for Food Security: The Case of Lesotho," *Global Food Security*, Vol. 12, pp. 146–154.

71 Jikun Huang and Guolei Yang (2017) "Understanding Recent Challenges and New Food Policy in China," op. cit., p. 125.

72 UNDP (2016) *Human Development Report 2016 – Human Development for Everyone*, New York NY: UNDP, pp. 54–55.

Suggested further reading

R. Ford Denison (2012) *Darwinian Agriculture: How Understanding Evolution Can Improve Agriculture*. Princeton, NJ: Princeton University Press.

Thomas DuBois and Huaiyin Li (eds.) *Agricultural Reform and Rural Transformation in China Since 1949*, Leiden, The Netherlands: Brill.

Tukumbi Lumumba-Kasongo (ed.) (2017) *Land Reforms and Natural Resource Conflicts in Africa: New Development Paradigms in the Era of Global Liberalization*, New York, NY and London, UK: Routledge.

Njoki Nathani-Wane (2014) *Indigenous African Knowledge Production: Food-Processing Practices among Kenyan Rural Women*, Toronto, Canada: University of Toronto Press.

Prabhu Pingali and Gershon Feder (eds.) (2017) *Agriculture and Rural Development in a Globalizing World: Challenges and Opportunities*, London, UK: Routledge.

Matin Qaim (2016) *Genetically Modified Crops and Agricultural Development*, New York, NY: Palgrave Macmillan.

Katar Singh and Anil Shishodia (2016) *Rural Development: Principles, Policies, and Management* (4th edition), New Delhi, India: Sage Publications.

Urban development and challenges of migration

9

I Conceptualizing urbanization

Historically, urbanization has been closely linked with industrialization. It is generally agreed that as a country advances and diversifies its economy, there is also a coinciding growth of urban centers. Cities are the centers for economic activity and the source of stable employment. With the changing global economy, more and more rural of the population has moved to urban areas. In fact, the phenomenon of urbanization is perhaps the most notable characteristic of twentieth-century economic development. In 1800, only 3 percent of the world's population lived in urban areas. By 1900, this figure grew close to 14 percent, and 12 cities already had more than one million inhabitants. In 1950, the number of cities with million-plus people increased to 83, and 30 percent of the world's population lived in urban centers. Humanity reached a peculiar threshold in 2007, when for the first time in history the world's urban population exceeded the global rural population. In 2014, about 54 percent of the global population was urban and more than 400 cities had million-plus populations. The same year also saw 28 megacities with populations that exceeded ten million. Researchers from the Economic and Social Affairs Department of the United Nations predict that by 2050 nearly 66 percent of the global population will be living in cities, and most of the urban growth will occur in the developing world.[1]

Already today, most large cities are located in the Global South. China alone has six megacities, and ten cities with more than five million dwellers. Five of India's large cities are projected to become megacities in the coming years. However, the fastest growing urban agglomerations are medium-sized cities (those between 500,000 and one million people) in Africa and Asia. Since different definitions are used when measuring the size of urban populations, it is helpful to rely on one approach. We define urban agglomeration based on two criteria recently used in the OECD report assessing urbanization dynamics in West Africa. These two criteria relate to the land use and the quantity of the population: 1. An agglomeration is a continuously built-up and developed area, with less than 200m between two buildings; 2. An agglomeration is considered urban if it has a minimum of 10,000 agglomerated inhabitants.[2] The data in Table 9.1 is taken from a UN

Table 9.1 Top five largest urban agglomerations in 2014, including comparative numbers
for 1990 and projections for 2030 (in millions)

1990		2014		2030	
1. Tokyo, Japan	32.5	1. Tokyo, Japan	37.8	1. Tokyo, Japan	37.2
2. Delhi, India	9.7	2. Delhi, India	24.9	2. Delhi, India	36.1
3. Shanghai, China	7.8	3. Shanghai, China	22.9	3. Shanghai, China	30.7
4. Mexico City	15.6	4. Mexico City	20.8	4. Mexico City	23.8
5. São Paulo, Brazil	14.7	5. São Paulo, Brazil	20.8	5. São Paulo, Brazil	23.4

report, and shows the five largest urban agglomerations in 2014, including comparative numbers for 1990 and projections for 2030 (in millions):[3]

The UN report ranks Mumbai (Bombay) India sixth, with 20.7 million people, followed by Osaka, Japan with 20.1 million, then Beijing, China with 19.5 million, next is the New York-Newark urban area in the United States with 18.5 million people, and Cairo, Egypt is tenth with 18.4 million inhabitants.[4] These urban centers, as enormous as they are, continue to expand.

In recent years, massive migration of people from depressed rural areas to cities has accelerated in the developing world. As cities grow, the costs of good quality urban living are rising, while social needs are expanding. Even in the developed world, many cities are struggling to provide sufficient services since local governments are habitually constrained by limited resources. Most western cities are no longer absorbing people from the countryside, but they are still facing a different kind of migration; they are trying to accommodate international economic migrants and political refugees. In general, cities everywhere are confronting escalating demands regarding access to health care, education, food, and clean water, while struggling to ensure an adequate level of public safety, housing, transportation, and energy security for their growing populations. The problems are compounded by air pollution, fresh water shortages, crime, illicit activities, and decaying infrastructure. Some cities are expanding so fast that urban development is consuming what previously was good farmland, adding to the long-term problems of sustainability.

No city in the world is immune to the problems associated with an unstoppable global force of urbanization. Business and community leaders ubiquitously work with various levels of governments on improving the quality of life in the city. The move towards balanced urban development has been steadily gaining momentum.[5] The concept is difficult to define since different cities face different priorities and dilemmas. However, the need for addressing multiple challenges of urbanization without compromising the well-being of future urban dwellers is widely shared by cities around the world.

Both in the developed and developing countries, cities have become a primary destination for poverty-stricken people looking for better lives. Whether at issue is rural–urban or international migration, most people who leave behind the land of their birth move to new urban centers. Just as the adaptive capacity of cities is tested, given the transformative force of migration, the newcomers encounter multiple challenges upon arrival, including the psychological costs of adaptation. Notwithstanding the problems, urbaniza-tion is considered essential for opening opportunities that allow a great number of people escape from poverty.[6]

II Unstoppable cities as a force and prize of progress

Upward urban concentration within developing countries brings forth questions around the presumed link between development and urbanization and the sustainability of this trend. The growth of urban centers is widely considered as an inevitable step on the ladder of development. Jeffrey D. Sachs illustrates the main thrust of this argument by showing that at the beginning of the new millennium poor countries had a large portion of their populations living in the rural areas. This was in sharp contrast to the developed world:

> in Malawi, 84 percent of the population lives in rural areas; in Bangladesh, 76 percent; in India, 72 percent; and in China, 61 percent. In the United States, at the other end of the development spectrum, it is just 20 percent.[7]

According to Sachs, modern economic growth started with the Industrial Revolution in Britain. It has been accompanied by five major transformations, with urbanization leading the way, followed by social mobility, changing gender roles, changing family structure, and increased division of labor.[8] There is a natural progression on the path of economic development that moves from subsistence agriculture toward light manufacturing and urbanization, and ending up with high-tech services economy. This is how Sachs sees the growing city in one of the poorest countries in the world:

> I was up at dawn one morning in Dhaka, Bangladesh, to see a remarkable sight: thousands of people walking to work in long lines stretching from the outskirts of Dhaka and from some of its poorest neighborhoods. Looking more closely, I noticed that these workers were almost all young women, perhaps between the ages of 18 and 25. These are the workers of a burgeoning garment industry in Dhaka who cut, stitch, and package millions of pieces of apparel each month for shipment to the United States and Europe.... These young women already have a foothold in the modern economy that is a critical, measurable step up ... from the villages of Bangladesh where most of them were born.[9]

Consistent with this view, the governments in emerging economies view the economic activity in urban centers as necessary for the country, and hence they even provide initi-atives for people to move to cities: "Policymakers tend to focus on the greater output per worker in larger cities and ignore the higher opportunity costs of labor and land, creating a policy bias toward locating heavy, often polluting industries in larger cities."[10] China is an interesting example of this kind of urban centered bias. The Chinese liberalizing eco-nomic reforms, which started in the late 1970s, produced spectacular economic growth over the last three decades, but they also exacerbated regional inequalities and created huge and vastly polluted urban areas.

The rural–urban divide in China occurs between coastal and inland areas. The geographical advantages of coastal areas are palpable from a transportation point of view. As part of the initial reform process, the government designed a development strategy and implemented a coastal-based policy that led to establishing special economic zones; the first of which was established in Shenzhen in 1980. To further boost industrialization and foreign investment, the Chinese state also created economic and technology development zones, and free trade areas in and around coastal cities, and provided favorable tax breaks to coastal provinces.[11] These open zones constituted what became known as a "golden coastline" in the East of the country.

With respect to China's specific development strategy with regard to the coastal areas, six policies were highlighted: coastal regions were receiving more subsidies from the state; they were enjoying higher foreign exchange retention rates; these regions had greater fiscal autonomy; workers were receiving higher wages; these regions were also getting lower prices of primary products; and were receiving greater freedom in currency circulation. One of the most controversial of these six policies concerns lowering the prices of primary goods. This state initiated preferential price policy further assisted the development of coastal regions at the expense of rural areas, since inland provinces were dependent on selling primary products but have to sell them at low preferential prices. Yet, when it came to purchasing finished goods from coastal provinces, the prices were set too high and the inland provinces were not receiving any subsidies to meet these prices.[12] These types of policies have achieved phenomenal growth of trade and investment in the coastal areas and have generated prosperity in these regions. However, these policies have also caused inequality to emerge between the coastal and inland areas of China.

Primarily because of the economic success and opportunities that exist in the coastal areas, there has been a large migration of people towards the eastern coastline. For example, in Shenzhen, the total population increased from 0.875 million in 1990 to almost 11 million people in 2014.[13] In the mid-1990s alone around 150 million people had migrated to coastal urban areas and this movement of has continued until present. The actual numbers became difficult to estimate since they were mostly undocumented migrant workers from rural areas in search of work in large cities. It is important to know that China has maintained a system of residency registration known as *hukou*, by which government permission is needed to formally change one's place of residency. A similar system existed in South Korea but was abolished in 2008. While internal migration within China had largely been suppressed by the central government, in recent years it has become quietly accepted, and unregistered migration has become more accessible.

There are millions of such undocumented people living in Chinese cities, although they have long been denied access to education, state-run health care, and other social services. Often termed the *"floating population,"* these illegal migrants have been subject to discrimination, considered a drain on urban resources, and treated as lower-class citizens by permanent urban residents.[14] Further, there is also a peculiar *brain drain* taking place simultaneously, as educated and younger individuals move to more prosperous regions, contributing to the concentration of human capital there. Thus, not only have China's coastal cities benefited by receiving government support, but they also receive the most able individuals from the inland regions. The contrasts between the coastal and the rural regions of China became staggering by the end of 1990s. As of 2003, there were an estimated 170 million rural residents living in poverty, with some scholars warning that rising inequality could counter the positive effects of China's overall economic success.[15] In addition, the deepening divide between prosperous and poor regions and between permanent and illegal city dwellers would create resentment and lead to violent hostilities.

On the positive side, massive movements of internal migrants to cities became a transformative force of the Chinese economy. They contributed to the rise of increasingly diversified manufacturing operations in China. Most economists agree that one of the primary reasons behind income inequality in the developing world is a stunted manufacturing sector.[16] In contrast, China's urban bias had a hugely positive impact on growth in manufacturing and on rising wages. These city-made wages were shared with the families left behind. It is estimated that a substantial inflow of remittances sent from family members working in the cities to the rural areas worked as a substitute for farm incomes. In fact, among households with lower marginal labor productivity in agriculture, migration to the city resulted in an increase in rural incomes. Internal migration can then work to reduce inequality. Several conditions, however, have to be met: people who migrate cannot work or be profitable as agricultural workers, and most importantly, once they move to the city and find a job they consistently send remittances back to the family left in the countryside.[17]

To facilitate the positive effects of urban to city migration, in December 2015, the Chinese government announced a major reform. The *hukou* system was relaxed by allowing migrant workers to apply for a permit to live in the city, with access to public services and education for children, subject to a valid employment contract. This move was consistent with the new urbanization policy unveiled in March 2014 that plans to formally move 250 million additional people from the country's poor farming regions into cities by 2026. This ambitious project involves the mass uprooting of rural residents, supported by destruction and reconstruction, and building new homes, office facilities, and infrastructure.[18] Another far-reaching developmental project designed by the Chinese government was inaugurated in May 2017. The "One Belt, One Road Project" that will link Asia, Africa, and Europe, focuses on building a massive global infrastructure network, promoting connectivity and trade among diverse nations.[19]

China's economic reforms focusing on urbanization have produced an enormous wealth and developmental progress for the country, elevating China to the position of a global power. The economic progress, however, has resulted in growing socio-economic inequality among Chinese citizens and came with a prize. Presently China is facing a magnitude of severe environmental problems, which include air pollution in the cities on a level that is hazardous to human health. In fact, 20 of the world's 30 most polluted cities are in China, and pollution is reducing life expectancy there. In October 2013, the ten million-plus city of Harbin was shut down for days because air pollution particulates were 40 times the level considered safe by the WHO. In addition, water pollution is said to affect more than half of China's rivers, lakes, and urban ground water. According to a report by the State Oceanic Administration in 2014, some 38 out of 71 rivers that entered the ocean had water quality levels assessed at the lowest possible category, which is considered unsuitable even for human touch. There also other entrenched problems, such as deforestation, erosion, desertification, soil contamination, salinization, loss of arable land, acid rain, biodiversity loss, and clean water shortage. Notwithstanding the progress made by the country since the late 1970s, given the severity of China's environmental problems, the key question is whether this economic development is sustainable.[20]

In contrast to an overall positive view of cities, this is how Robert D. Kaplan describes his impression of urbanization:

> And the cities keep growing. I got a general sense of the future while driving from the airport to downtown Conakry, the capital of Guinea. The 45-minute journey in heavy traffic was through one never-ending shantytown: a nightmarish Dickensian spectacle to which Dickens himself would never have given credence. The corrugated metal shacks and scabrous walls were coated with a black slime. Stores were built out of rusted shipping containers, junked cars, and jumbles of wire mesh. The streets were one long puddle of floating garbage. Mosquitoes and flies were everywhere. Children, many of whom had protruding bellies, seemed as numerous as ants. When the tide went out, dead rats and the skeletons of cars were exposed on the mucky beach. In 28 years Guinea's population will double if growth goes on at current rates. Hardwood logging continues at a madcap speed, and people flee the Guinean countryside for Conakry.[21]

Dire circumstances, such as those described by Robert D. Kaplan, which he experienced through his travels in Guinea, continue to be found in many parts of the Global South. Guinea is located in West Africa, a region that includes 17 countries.[22] In 2010, 133 million people in West Africa lived in 1,947 urban agglomerations with at least 10,000 inhabitants, which was 530 more agglomerations than in 2001. In 1970 there were only 493 such agglomerations in West Africa, Presently, there are 22 cities with million-plus populations as compared to 10 in 2000. It is estimated that these 22 cites together have 54.4 million inhabitants.[23] The total population of West Africa went from 62 million in 1950 to 329 million in 2010. Over the period 2000–2010, the region experienced one of the highest average annual population growths in the world of 3.2 percent. The impact of demographic pressures on urbanization in West Africa is important for two reasons: 1. the growth of large cities in West Africa is mainly driven by their natural population increase; 2. the strongest urban growth occurs in the most densely populated areas.[24]

Thus, even if we assume that urbanization is a necessary step on the path of development, once a city begins to grow the processes of urbanization accompanied by a strong population growth and/or migration from the rural villages can have previously unforeseen consequences. Apart from the environmental problems, regionalism sometimes referred to as dualism can also be problematic. The concept describes an uneven level of economic development within the same geographical area. It is a serious problem as many less developed countries are unable to reconcile internal divisions or regions operating as dual economies. Dualism consists of a revenue generating urban economy, on the one hand, and a largely subsistence-based rural economy on the other. Larger economies such as China have resources and actively work on reducing the inequality between different regions of the country. In the case of smaller poor countries, the problems persist and such internal divisions can be exacerbated by the migration of desperate people escaping rural poverty.

When in the past, former agricultural laborers moved into cities and found employment in the manufacturing or services sector they became high-productivity workers who contributed to countries' higher economic growth rates. Recently, however, a large number of displaced people worldwide find it difficult to secure decent jobs anywhere. The report of the International Labor Organization points out the growing challenges to employ the available capital and workers.[25] The uneven economic recovery after the 2008 financial crisis saw more people being unemployed worldwide. In 2013 there were 202 million unemployed, with the biggest increases registered in the East and South Asia followed by sub-Saharan Africa. If current trends continue, it is estimated that by 2018 there will be more than 215 million job seekers globally.[26] The ILO also estimates that, given the tight job market globally, a total of 23 million people became discouraged and dropped out of the labor force altogether in 2013. The report projects that global job prospects will change very little in the near future, with the number of permanently discouraged workers rising to 30 million by 2018.[27]

The large numbers of displaced and unemployed people worldwide and the accelerated pace of urbanization have had profound implications on the way of living in contemporary urban centers. The 2016 Report on World Cities notes that "cities are currently operating in economic, social, and cultural ecologies that are radically different from the outmoded urban model of the twentieth century."[28] The report identifies the most important urgent issues that have emerged during the last 20 years and categorizes them as either persistent issues or emerging issues. Persistent issues include the unstoppable growth of urban areas, problems of governance and funding – which is linked to inadequate social and public services –, and a growing number of people living in slums and informal settlements. Emerging urban issues include environmental problems, exclusion and rising inequality, rising insecurity, and an upsurge in international migration.[29]

Population growth coupled with large migrations to the urban centers is occurring through much of the developing world, leading to the nightmarish conditions with which the state has been unable to cope. Since city dwellers cannot grow their own food to sustain themselves, the economic and political migrants that move to cities around the world often end up homeless or on social assistance, suffering social exclusion and extreme poverty. Among all the people who come to urban areas looking for economic opportunities, only some are able to better themselves by finding permanent jobs, while the rest must resort to informal employment to survive. Social benefits are rare in the developing world; however, those who are employed in the formal economy do have certain rights, such as access to health and welfare programs, or a guaranteed minimum wage. For those surviving in the informal economy, such rights are non-existent. As parts of the cities turn into slums, basic human needs are not met. Another important consequence of urban slums has to do with health and the prevention of diseases. Poor sanitary facilities, desperate living conditions, pollution, and other environmental factors account for a large number of diseases in developing countries. These include infectious diseases (such as measles), parasitic diseases (such as malaria), and respiratory diseases (such as tuberculosis), and perhaps most importantly, the spread of AIDS. The evidence shows that migrants are disproportionally represented among the urban poor living in slums.[30]

The incredible growth of cities has not been met by corresponding institutional developments that allow local governments to secure sufficient levels of revenue. In many countries, despite the decentralization, cities still really on the state for essential funding. Consistent budgetary problems experienced by governments often lead to politicizing the funding formula. Cities that are strategically located on the electoral map of the country tend to receive a disproportionate share of the money. With respect to people living in slums, or informal settlements, there is some positive news. Recent estimates show that the proportion of the urban population living in informal settlements in the developing world decreased from 39.4 percent in 2000 to 29.7 percent in 2014. However, in terms of absolute numbers, there was actually an increase. In 2000, approximately 791 million people lived in slums, compared with 881 million in 2014. Many countries continue to experience large gaps between slum dwellers and the rest of the urban population. Regional discrepancies continue to characterize the delivery of basic services such as water and sanitation in the developing world. The situation is particularly dire in sub-Saharan Africa and Southern Asia. The problems resulting from lack of space and fresh water in the cities are further aggravated by air pollution as cities are known for a high level of energy consumption and greenhouse gas emissions.[31]

The accelerated reduction of green species is one of the main reasons behind the deterioration of the ecological conditions of many cities. Consider Dhaka, the capital of Bangladesh, a megacity previously citied as an example of economic progress. Due to such factors as lack of developmental policy, absence of urban planning, prevalence of politically motivated construction contracts, there has been an almost complete destruction of green spaces in Dhaka. In the 1960s, almost 80 percent of the land in the Greater Dhaka region was non-urban and consisted of vegetation, open spaces, wetlands, and cultivated lands. By 2005, this figure had reduced to 40 percent.[32] Most recent estimates further bring this number down to less than 10 percent. The fast-shrinking green space has had many consequences, such as air pollution, increase in infectious diseases, deterioration of water systems, and rising energy costs of cooling buildings, just to name a few.

In addition, some estimate that almost half of Dhaka's current 15 million inhabitants live in the informal shantytowns that can be found in any open area of the city. An urban geographer who was born in Dhaka calls it "the megacity of the poor" and claims that the majority of its dwellers make less than US$200 a month.[33] The city is notorious for a lack of clean water, poor infrastructure, substandard sanitation, and inferior transportation services. On April 24, 2013, the Rana Plaza building in Savar outside Dhaka collapsed, killing 1,138 textile workers, who worked in horrible conditions. This was a factory used by one of the local subcontractors in the global textile supply chain. Despite promises made by the government to take action on factory safety and labor standards, the progress has been slow. Because the accident highlighted the horrible situation of workers in the garments industry, which in the first quarter of 2016 accounted for 76 percent of Bangladesh's US$9.7 billion exports, the improvements were mainly concentrated there. In September 2016, another industrial accident at a packaging factory outside Dhaka resulted in scores of fatalities.[34] Urbanization as a sign of economic progress in Bangladesh has come at a heavy price, begging the question of how sustainable it is.

III Displacement in the age of unprecedented migration

The 2016 Report on World Cities considers expansion in international migration as an emerging urban issue.[35] Poverty and lack of economic opportunities in many parts of the developing world, together with armed conflicts and political prosecution, have resulted in increased cross-border movements of people seeking places to live better lives. While economic factors seem to most directly affect international migration, violence and authoritarianism have been a significant determinant of population displacements. The statistics show that in 2015 close to one in five migrants ended up living in one of the 20 biggest cities in the world. The same year, international migrants constituted over a third of the total population of London (UK), Singapore, Auckland (New Zealand), and Sydney (Australia), and more than half of the population of Brussels (Belgium), and Dubai (United Arab Emirates). Of Canada's 6.8 million foreign born population, 46 percent live in the Greater Toronto Area. In 2015, at least one in four residents of Frankfurt, Paris, and Amsterdam were foreign born. However, approximately 70 percent of migrants, from 45 of 63 low and middle-income countries, who are escaping conflicts, disasters, and other shocks, go to the cities of neighboring countries.[36]

There are many issues that elude easy definitions when it comes to the international movements of people. One contested concept relates to forced migration. Poverty, environmental degradation, and armed conflicts often exacerbate each other – thus ensuring the conflict continuance. As a researcher of migration, Richard Black states that these linkages confuse the very sources of forced displacement, and thereby pose greater challenges in understanding the situation in terms of development. Interestingly enough, development itself can be at the root of forced displacement. Probing the connection between refugees and development, Black further writes that:

> A first clue is provided by the one form of forced migration that does not fit neatly into the "downward spiral" hypothesis of poverty, violence and displacement – namely so-called "development-induced displacement." Those who have been displaced by dams and other major public works bear many of the hallmarks of the refugee – they are forced to leave their homes and land, often at short notice and with little or no compensation. Yet by definition their flight is caused not by a lack of development, but by "development" itself, albeit a particular vision of what development should be.[37]

The interesting point raised by Black begs the question whether those forced to leave their homes "at short notice" and compensated only a little, if at all, are qualified to be called refugees. It is possible that some of the people who are asked to reallocate because of the construction of new government projects, may be placed in a situation of utter economic deprivation and personal insecurity because they are being forcefully removed from their communities. They cannot stay as they could lose their lives. Consequently, one can argue that such people do meet the criteria of being refugees.

To be sure, the 1951 Geneva Convention Relating to the Status of Refugees defines "refugee" as someone who,

> owing to well-founded fear of being persecuted for reasons of race, religion, nationality, membership to a particular social group, or political opinion, is outside the country of his nationality and is unable or, owing to such fear, is unwilling to avail himself of the protection of that country; or who, not having a nationality and being outside the country of his former habitual residence, ... is unable or, owing to such fear, unwilling to return to it.[38]

This legalistic definition may first appear unambiguous. However, two concepts, persecution and lack of protection, are subject to interpretation among politicians and legal experts. In a case where someone is fleeing a country due to legal prosecution, it is sometimes difficult to establish whether or not the reasons for which are accepted in the definition. Another concept inherent in the term "refugee" is the lack of protection afforded to the individual in his home country. What is meant by national protection here, however, demands clarification. A popular definition of this concept is that it means protection with which the state shields the person from the feared persecution. Any absence of such protection is due to either an inability or unwillingness of the state. In fact, the countries of origin may purposely give up their responsibility for the citizens considered to be hostile to the government in power. On the other hand, states often do not only provide security and protection to its individual members within their territories. They seek to provide diplomatic protection by ensuring their citizens' rights when they are abroad through international law and bilateral treaties. As a result, they may be a disproportionate protection given to the individuals connected with the political elites, while protection is refused to those who are hostile towards the political elites. Refugees then become in a way legal orphans at the mercy of international law.[39]

Refugees constituted a significant portion of displaced people in the past century and continue to do so. However, economic causes are often difficult to separate from the context of social crisis and political violence. According to the UNHCR 2015 Global Trends Report, we are currently witnessing the highest levels of human displacement on record. An unprecedented 65.3 million people have been forcefully displaced, and among them are 21.3 million refugees. Some 53 percent of all refugees come only from three countries: Somalia (1.1 million), Afghanistan (2.7 million), and Syria (4.9 million).[40] Many of the studies concerning the impact on the psyche of the refugee by war, migration, and even acculturation, have shown that refugees tend to demonstrate a prevalence of high rates of post-traumatic stress disorder and affective disorders. Trauma and depression can occur as a result of experiencing war and brutality firsthand and the struggle to adjust to a new life outside one's country of origin. Conditions can vary between refugee camps, as some are adequately supplied with food and basic drugs while other camps, unfortunately, are execrable in comparison.

Life in refugee camps was particularly challenging for Afghan women. A study into the condition of Afghan refugee woman described the situation as thus:

> Refugee women experienced war trauma, sexual violence, deepening poverty, and increased household responsibilities as a result of the conflict. They also witnessed the erosion of their legal and social status in their homeland, with direct consequences for their status in camp communities. Camp organization ... was deeply politicized and ultimately detrimental to women's interests.[41]

Pakistan was the initial target country for many Afghan refugees, and increasingly was unable to cope with the situation. Eventually its borders with Afghanistan were officially sealed off. However, a country does not have to share borders with Afghanistan to experience the consequences of a refugee problem in Afghanistan. In August 2001, a Norwegian freighter named the MV Tampa rescued over 400 people from the Indian Ocean who had attempted to reach Australia from Indonesia to seek asylum. Following a diplomatic dispute over the status of the asylum seekers, many of which were Afghanis, an arrangement was made to transport the refugees to either New Zealand or Nauru. In response to this event, the Australian policy called the Pacific Solution was developed, whereby refugees would be held in offshore detention centers found in nearby countries in the Pacific while their applications for refugee status were being processed in Australia.[42]

From the start, controversy has surrounded the Pacific Solution. The costs for running the detention centers were enormous, while conditions in the detention camps had also been found by independent observers to be substandard, and even appalling. In 2008, the newly elected government dismantled the policy in response to a mounting criticism. However, in the years that followed, the world has seen a growing number of displaced people seeking refuge and economic opportunities in the developed world. The Australian government was unable to address the growing problems in accommodating a large number of asylum seekers who attempted to reach the shores of Australia. In August 2012, the government announced a similar policy to the one known as the Pacific Solution by reopening two detention centers on the Pacific island-states of Nauru and Manus for the offshore processing of asylum seekers. It was argued that offshore processing was necessary to act as a deterrent after thousands of people drowned at sea before the policy was reintroduced. In the spring of 2017, the new US Trump Administration struck a deal with Australia concerning its refugee policy negotiated under President Obama. Under this modified deal, US Homeland Security officials have started "extreme vetting" interviews at Australia's detention centers to honor a promise to offer refuge to up to 1,250 asylum seeks. In exchange, Australia agreed to take refugees from an American detention center in Costa Rica.[43] These convoluted measures put in place by the United States and Australia testify to the inability of western governments to successfully deal with growing flows of migrants.

However, the most difficult refugee crisis began to unfold in Europe following the Syrian war. It is estimated that 340,000 migrants tried to get to Europe in the first seven months of 2015, a threefold increase over 2014. In addition to a large number of Syrians, there were Afghans trying to flee the ongoing civil war, Roma faced with discrimination in Kosovo, as well as asylum seekers from Northern Africa escaping violence and terrible conditions in countries such as Eritrea, Somalia, Libya, just to name a few. Despite the existing institutions, the EU members failed to agree on a common asylum system. Greece for example, placed asylum seekers on several of its islands, while encouraging them to go to Macedonia and creating a refugee crisis there in the process. France and the Netherlands decided to toughen its refugee laws, while Germany attempted to become more open, Hungary tended to put most refuges into downgrading detention centers or send them away while its laws classified most of them as "economic migrants."[44] The United Kingdom became engulfed in fears over the uncontrolled flows of migrations from other European countries. This strengthened the argument for withdrawing from the EU altogether, which become the reality after the June 2016 referendum. The same year, general elections in Lithuania and Poland gave power to distinctively anti-emigration parties.

As politicians struggled to deal with the crisis, most discussions in Europe centered on how to fairly distribute hundreds of thousands of new migrants throughout the EU. Ironically, most of the Syrian refugees ended up in Jordan, Lebanon, and Turkey. Only around 4 percent of displaced Syrians attempted to reach Europe, while around 60 percent of them (more than six million people) remained in Syria, since Jordan and Lebanon, unable to cope with the influx of refugees, effectively closed their borders in 2014.

For much of the twentieth century, migration from rural areas to the new urban centers fueled economic growth, as did the massive movements of poor immigrants escaping poverty from the old to the new world. For example, between 1871 and 1915, 36 million people left Europe. The inflows of immigrants brought substantial economic growth to the quickly developing host countries, especially the United States. The countries left behind also experienced productivity gains since surplus labor was eliminated.[45] However, this approach no longer seems to be working. Presently, the historically unprecedented number of migrants testify about the persistent global inequality between those who have access to economic opportunities to enjoy a peaceful and prosperous life and those who do not.

> International migration is a powerful symbol of global inequality, whether in terms of wages, labor market opportunities, or lifestyles. Millions of workers and their families move each year across borders and across continents, seeking to reduce what they see as the gap between their own position and that of people in other, wealthier, places.
>
> In turn, there is a growing consensus in the development field that migration represents an important livelihood diversification strategy for many in the world's poorest nations. This includes not only international migration, but also permanent, temporary and seasonal migrations within poorer countries, a phenomenon of considerable importance across much of Africa, Asia and Latin America.

Yet it is also clear that migration – and perhaps especially international migration – is an activity that carries significant risks and costs. As such, although migration is certainly rooted, at least in part, in income and wealth inequalities between sending and receiving areas, it does not necessarily reduce inequality in the way intended by many migrants.[46]

Today, most of the migrating people who are escaping violence and poverty face very bleak future scenarios. If they survive the trip by sea in overcrowded unlawful boats, they will remain stuck for years in poorly run refugee camps and degrading detention centers. Other desperate migrants seek informal, often high-risk, employment in seamy parts of big cities. The lack of legal protection for economic migrants sometimes results in activities such as human trafficking. Human trafficking, an act forbidden by international conventions, is a reality for thousands of people each year. As the United Nations Protocol to Prevent, Suppress and Punish Trafficking in Persons defines the term, trafficking in persons means:

> the recruitment, transportation, transfer, harbouring or receipt of persons, by means of the threat or use of force or other forms of coercion, of abduction, of fraud, of deception, of the abuse of power or of a position of vulnerability or of the giving or receiving of payments or benefits to achieve the consent of a person having control over another person, for the purpose of exploitation. Exploitation shall include, at a minimum, the exploitation of the prostitution of others or other forms of sexual exploitation, forced labour or services, slavery or practices similar to slavery, servitude or the removal of organs.[47]

Trafficking has many negative consequences on society. Socially, the trafficking of human persons devastates social structures such as families and the immediate community itself. Furthermore, even when victims of trafficking and exploitation return to their communities, they are often avoided and may be stigmatized. In the period between 2010 and 2012 some 53 percent of the detected victims were trafficked for sexual exploitation and 40 percent were trafficked for forced labor. In fact, the most recent UN report on the issue indicates over the last decade a steady rise in trafficking for forced labor including sectors such as manufacturing, construction, cleaning, restaurants, catering, domestic work, and textile production.[48]

Forced labor is a problematic area of economic activity connected with the issue of international migration, and increasingly with the problem of human trafficking. It relates to the inhuman conditions in which some of the newly arrived legal aliens as well as illegally smuggled migrants work to sustain themselves. The so-called *sweatshops* exist both in the developed and the developing world. The US Department of Labor defines *sweatshop* as a factory that violates two or more labor laws. In such factories, working conditions are poor and the practices toward laborers are exploitative, such that there are long hours and there are no living wages. Labor standards and human rights are nonexistent as workers are often subject to abuse and thus censor themselves from protest in fear. Moreover, sweatshops are known to employ children. Faced with the grim reality of work in hazardous conditions, some illegal migrants are lured into illicit activities organized by armed gangs. As a result, the policies that strive to make growing cities sustainable should include an urban development model for the displaced that centers on labor market initiatives.

IV The search for a balanced and inclusive urban development model

The consistent rise in global flows of city-bound migrants inspired the name of the most recent report by the International Organization for Migration (IOM): Migrants and Cities: New Partnerships to Manage Mobility. The report examines migration at the city level and probes the connection between migration and urban development. Specifically, the report looks at diversity in urban settings in view of recent migration patterns and examines the specific types and circumstances of vulnerability experienced by urban migrants. In seeing migrants as agents of international development, the study also proposes some strategies for urban partnerships to manage mobility for the mutual benefit of the city and its diverse populations.[49] We are also reminded to think about the three groups of people: the migrants themselves, the people left behind in the country of origin, and the indigenous population of the host country."[50] These groups constitute three uniquely interrelated perspectives in need of examination before effective policy concerning urban development can be developed.

A considerable amount of work was done in the past by the IOM in terms of policy recommendation. Over a decade ago, a series of meetings were held under the IOM framework with the purpose of enhancing cooperation between countries on the subject and to prevent illegal migration and the exploitation of migrants. Several key themes were identified: 1. the necessity of a comprehensive approach to migration, which takes into consideration its dimensions, historical ties, root causes, and its social, economic, and cultural consequences in countries of origin and destination; 2. the necessity to guarantee full respect for the human rights of all migrants without regard to immigration status, including the rights of migrant workers and their families; 3. the necessity to continue to combat irregular migration and migrant smuggling and human trafficking; 4. the necessity to analyze the forms and means of managing migratory flows, including irregular migration; 5. the importance of remittances as a significant source of income for many countries, emphasizing that all efforts should be made to facilitate transfers; 6. recognition of the contribution of migrants to the economic development and social and cultural life of countries of origin and of destination. In the past, there was a tendency to examine the factors that trigger international migration in the countries of origin and to overlook the effects of the policies in countries where the migrants end up settling. The list above can provide some guidance as to how matters of international migration may be addressed in a more balanced way.

In view of recent upsurge in international migration, a new approach to refugee resettlement has been proposed.[51] It is called the zonal development model and begins with an understanding that all migrants, including political refugees who are escaping armed conflicts and political oppression at home, not only seek safety but also hope for better economic prospects. These are often skilled, able, and young individuals. The UN Refugee Agency (United Nations High Commissioner for Refugees or UNHCR) reported in June 2016, that over half of an estimated 21.3 million refugees were under the age of 18. These displaced people represent a potential labor force that is largely forgone when the next step on their life journey is a detention refugee camp or a life spent as illegal aliens hardly surviving in a foreign city.

Consider the recent case of Syrian refugees. A large majority of them did not go to refugee camps or Europe but ended up in Amman, Beirut, and other Middle Eastern cities working illegally. Approximately 83 percent of all refugees in Jordan live in cities, with about 170,000 refugees in the capital Amman.[52] Calls for integrating these refugees into the societies of the host countries are met with resistance. Many governments consider refugees to be a threat to domestic employment and a burden on already overstretched budgets. An alternative approach proposes the creation of special economic zones where refugees could be offered jobs and autonomy.

Effective economic zones in Jordan could develop Syrian businesses and people's skills in preparation for the end of the civil war, while invigorating Jordan's plans for industrial development. Presently, Jordan's economy has a weak manufacturing sector as it lacks the geographic concentration of firms to reap the benefits of economies of scale: "to industrialize, then, Jordan needs a small number of major businesses and a large number of skilled workers to relocate to manufacturing clusters."[53] The refugee crisis offers an opportunity in this context. Refugee camps together with certain urban areas can be reorganized as industrial economic zones where the displaced can be offered legitimate jobs and training. The developmental agencies should be encouraged to provide financial and other incentives for different firms to establish their operation within such economic zones. Although Jordan has already established several special economic zones, some of them are underutilized because businesses lack proper incentives to move there.

One potential problem relating to this kind of zonal and export-oriented model of development for the displaced is the need for land to accommodate operations and workers. It is known that special economic zones on the outskirts of major developing cities, resulted in negative peri-urbanization or the tendency to encroach on rural land as part of the expansion of economic activity connected with the city. Attracted to the economic opportunities in the special economic zones outside the cities, such as in Shenzhen (China), Jakarta (Indonesia), and Bangkok (Thailand), migrants tend settle around them by building low-income informal settlements.[54] Apart from problems related to the delivery of basic services to the peri-urban settlements, the loss of peri-urban agricultural land can also reduce the capacity of cities to have a reliable source of food supply.[55]

The idea of creating special economic zones for the displaced also invites criticism from human rights activists, who worry about the possibility of low labor standards leading to the outright exploitation of refugees. Yet nothing stops the host country from regulating such zones to ensure the ethical and lawful treatment of people working there. Because of the Syrian refugee crisis, Jordan was uniquely positioned to launch the zonal model of development. However, this model could be operationalized in other countries. Given the growing problem of illicit and forced labor in the cities that are known to have a large population of illegal migrants, governments together with the development community can work on adapting the model to provide legitimate opportunities to the displaced. For example, the diaspora communities can help facilitate the establishment and operation of special economic zones for refugees and migrants from their home countries.

One of the consequences of mass displacement and international migration is the creation of *diasporas*. Diaspora, meaning the "sowing of seeds" in Ancient Greek, was favorably looked upon as it meant to the Greeks the expansion of their territories, and in effect colonization. In modern usage, the word has come to mean the dispersal, either induced or forced, of a people who share a common historical memory and feel solidarity to another land wherein they continue to develop and practice their culture.[56] In the context of development, it is argued that socio-economic progress may be aided by diaspora because the community may provide intellectual, economic, or political support to relatives and friends left behind in the homeland. The diaspora community often becomes a source of external income by way of remittances (formal and informal) received by the families in the native countries.

Diaspora might also form institutions, formal or otherwise, and non-governmental organizations (NGOs), whose activities include the promotion of their homeland and its development. The African Foundation for Development (AFFORD) is a good example of how an NGO can serve as a channel through which a diaspora community seeks to assist in the development of their countries of origin. AFFORD was formed in 1994 by a group of Africans living in the United Kingdom to elevate Africa's standing among the developmental initiatives in order to encourage wealth and job creation in Africa. Its financial arm brings together a number of programs, grants, and services aimed at harnessing relevant investment opportunities.[57]

Above all, however, diaspora communities are known for creating networks among business firms as well as social and cultural organizations. Diasporas contain many support groups that have historically helped arriving migrants and refugees to settle in their new cities. The existing diaspora networks can become knowledgeable and culturally sensitive partners in promoting development in economic zones for the displaced. Diaspora and migrant communities collaborate closely with city officials and business and civil society leaders in several Canadian and American cities on urban integration issues. The recent initiatives try to foster participatory economic growth in those cities experiencing a large influx of new migrants.

To prevent the unmanageable growth of socio-economically unequal and environmentally unsustainable cities, the 2016 World Cities Report calls for all governments to give prominence to a new urban agenda in national policies. This new agenda suggests creating conditions that enable a paradigm shift towards a new model of urbanization based on the five principles that have emerged from an examination of ongoing trends and a growing awareness of new challenges. These five principles call for a shift in urban policy thinking with the aim of: 1. ensuring that the new urbanization model contains mechanisms and procedures that protect and promote human rights and the rule of law; 2. ensuring equitable urban development and inclusive growth; 3. empowering civil society, expanding democratic participation, and reinforcing collaboration; 4. promoting environmental sustainability; 5. promoting innovations that facilitate learning and the sharing of knowledge.[58]

The report further articulates three fundamental components of the new universal urban agenda:[59]

1   Working on rules and regulations aimed at strengthening urban legislation and the system of governance.
2   Reinvigorating city planning and urban design in view of the future expansion.
3   Adapting financial and economic mechanisms to changing needs of the urban population.

These three frameworks for actions reflect the interrelated challenges stemming from continuing urbanization. The focus is on building a more effective legal framework supported by accountable governability processes and guarded by inclusive institutions. The strategy for tomorrow is to envision more inclusive, socially integrated, and geographically connected cities. This is only possible when adequate laws and transparent regulations ensure good planning practices. The work aims at a positive view of a city as an organically developing and well-managed system that expands its economic potential in a sustainable way that improves the lives of all its citizens.

It will be up to individual governments to implement these frameworks for actions. Rapidly expanding cities in the developing world would find it particularly challenging to adopt the new urban agenda, given their focus on providing job opportunities for their growing populations.

The types of employment available in cities of the developing countries are often grueling jobs, which are in stark contrast to the safe and rewarding employment people hope for. Many transient workers live in harsh conditions and work long hours. Such workers have little, if any, social protection and the safety standards are routinely compromised. The contra-argument is that such jobs are better than the alternative – an extreme poverty. While such arguments are usually based on economic considerations, a broader, more inclusive conception of development greatly questions this scenario.

However, it is very difficult to reach an international agreement ensuring high labor standards for all workers. In addition, most of the incentives regarding universal labor rights come from the developed countries, and because of this, such incentives are normally met with great skepticism on the part of the developing world. Part of the problem rests on the perceptions (justified or not) within developing countries about the true motives behind the initiatives. Many poor countries see the developed states' concern about high labor standards and prohibition of child labor to be an attempt of protecting their own industry at the expense of developing countries. They worry that the developed countries would come up with an agreement on labor standards only so that they would be able to stop trading with those poor countries found in violation of such agreement. Consequently, the developing countries greatly resent these initiatives and view them as hypocrisy aimed at protectionism.

While the developed world wishes to extend labor rights internationally, the developing world is slow in advancing such standards, as the case of Bangladesh has demonstrated. While much of the current labor issues fall under the watch of the ILO, which was created in 1919 with the objective of improving labor conditions worldwide, the ILO lacks the necessary enforcement mechanism to provide solutions. Most importantly, as some scholars have pointed out, not only are there disagreements regarding whether international labor standards should be universally enforced, there is also a general absence of agreement on which labor practices should even be included in such an agreement.

> We have a near-universal consensus only in favor of prohibiting forced labor. On other issues, like the appropriate rules to regulate collective bargaining or child labor or discrimination, we have a mixture of good intentions, some blood-curdling stories about undoubted abuses in extreme cases, and great uncertainty over what the appropriate labor standards should be.[60]

A good example of this kind of disagreement revolves around the issue of the terrible working conditions in the global textile industry. While generally accepted that working conditions in such jobs are appalling, scholars and practitioners are divided on whether these jobs are good or bad for a country's development. Some, like Sachs, defend any employment opportunities because of their arguably important role in economic development. Opponents quickly attack such arguments because they believe that by justifying abysmal working conditions, we allow them to persist. Further, those critics often invoke the rights aspect, arguing that all employees deserve to be treated fairly and humanely. The collapse of the factory in 2013 in Dhaka reignited this debate. The position of the labor activists is that "the nations of the world ought to be able to agree on some set of universally accepted human rights regarding working conditions that would apply to all nations."[61] This argument rests on the belief that there are certain aspects of labor standards that should be considered natural rights and part of humanitarian principles.

A view of many governments in developing countries is that the initiatives behind a global agreement on high labor standards is merely another form of protectionism designed by the rich countries to limit exports from the poor countries. As these countries are trying to slowly bring themselves out of poverty by relying upon the comparative advantage of cheap labor, the argument goes that the developed world should not stop them by demanding the introduction of higher standards, labor, environmental, etc. It is important to recognize that enforcing such international rules, even if they came into effect, would be very difficult. The ILO does not have any effective enforcement mechanisms, and hence it would not be able to enforce such standards. The WTO is reluctant to deal with labor issues and in fact it is not institutionally prepared to handle them. Even those who acknowledge that global labor standards are a good idea, state that the WTO is the wrong place to institutionalize them.[62]

Perhaps due to low political motivation, multilateral efforts to create an enforceable agreement on labor standards have quieted down. Paradoxically, the continued weaknesses in the global economy and widening income gap highlight the importance of good quality jobs. Over 46 percent of total employment today (impacting 1.5 billion people) is considered vulnerable, as characterized by a high level of precariousness.[63] Poor job quality means that many workers have only limited access to social protection programs, the wages remain low, and there is no job security. Cities in developing and developed countries have also experienced a spurt in informal employment. According to the 2016 World Cities Report, undocumented workers in Los Angeles constituted 65 percent of the city's labor force, while in the United Arab Emirates 95 percent of workers in the private sector are migrants.[64] Cities and migrants can mutually benefit each other, however, effecting urban governance that includes local decentralized partnerships must be developed in conjunction with a multilayered urbanization policy that takes into account the principles of sustainability.

The 2015 World Migration Report summarizes some of the main issues concerning urban migration and economic development. Its main findings alert us to the fact that migrants who make the decision to move to foreign cities in search of better lives are particularly vulnerable and in danger of exclusion. The host communities are routinely unable to cope with a large number of newcomers, especially when the economic situation in the city is highly competitive, causing tensions among the existing population. Even if the cities are economically vibrant, migrants routinely: "face specific barriers and obstacles that result in specific patterns of marginalization. As a consequence, they often end up over-represented among the weakest, most vulnerable social groups within urban communities."[65]

In order to prevent marginalization of migrants, the cities have to work with different levels of governments and with the leaders of the host communities on designing inclusive policies aimed at training and educating the newcomers. Such policies have to be supported with an adequate level of funding. Harnessing the potential and the skills of migrants can have hugely beneficial outcomes. Historically, migration has contributed to enriching the cosmopolitan makeup of emerging cities. Migrants have brought relentless energy, cultural diversity, and economic industrious drive. However, as the report points out, in order to ensure the vitality of new settlements and positive integration of newcomers, the dynamics of growing cities must be part of a multilevel national strategy building. The productive strategy takes into account the regional impact of expanding cities and anticipates the issues arising from increased labor mobility. In this context, an accountable and inclusive process of urban planning conducted in the spirit of sustainability is critical. Such positive local urban strategy and "effective national mobility management policies are therefore essential not only to prevent the potential vulnerabilities linked with movement into cities, but also to leverage the potential of building the resilience and increasing the well-being of migrants."[66]

The report is also valuable in pointing out the transformative global patterns of labor and social mobility resulting in ongoing "urban transition." The expanding cities are part of a larger international dynamic accelerated by migration. This is why the successful integration of migrants into legal labor markets is critical. After all, the "urban transition" includes:

> the concurrence of urban and rural development as they are closely linked through increasing mobility. Therefore, at a policy level, internal and international migration and urbanization trends need to be factored into labor market strategies as they affect both rural and urban development.[67]

Notes

1 UN (2014) *World Urbanization Prospects – The 2014 Revision*, New York, NY: UN DESA Population Division.

2 François Moriconi-Ebrard, Dominique Harre, and Philipp Heinrigs (2016) *Urbanization Dynamics in West Africa 1950–2010 – Africapolis I, 2015 Update*, Paris, Franc: OECD, p. 20.

3 UN (2014) *World Urbanization Prospects –The 2014 Revision*, op. cit., Annex, table II, p. 26.

4 Ibid.

5 Basant Maheshwari, Vijay P. Singh, Bhadranie Thoradeniya (eds.) (2016) *Balanced Urban Development: Options and Strategies for Livable Cities*, Basel, Switzerland: Springer International Publishing.

6 Paul Collier (2015) *Exodus: How Migration is Changing Our World*, Oxford, UK: Oxford University Press, p. 175.

7 Jeffrey D. Sachs (2005) *The End of Poverty: Economic Possibilities for Our Time*, op. cit., p. 18.

8 Ibid., pp. 35–37.

9 Ibid., p. 11.

10 Vernon Henderson (2002) "Urbanization in Developing Countries," *World Bank Research Observer*, Vol. 17, No. 1, p. 96.

11 Xiaobo Kanbur and Kevin H. Zhang (2003) "How Does Globalization Affect Regional Inequality Within a Developing Country? Evidence from China," *Journal of Development Studies*, Vol. 39, No. 4, p. 96.

12 Cindy C. Fan (1997) "Uneven Development and Beyond: Regional Development Theory in Post-Mao China," *International Journal of Urban and Regional Research*, Vol. 24, No. 1, pp. 624–627.

13 UN (2014) *World Urbanization Prospects – The 2014 Revision*, op. cit., Annex, table II, p. 26.

14 Nancy E. Riley (2004) "China's Population: New Trends and Challenges," *Population Bulletin*, Vol. 59, No. 2, p. 28.

15 Ibid.

16 Alice H. Amsden (2007) *Escape from Empire – The Developing World's Journey through Heaven and Hell*, Cambridge, MA: The MIT Press, p. 140.

17 Nong Zhu and Xubei Luo (2008) *The Impact of Remittances on Rural Poverty and Inequality in China*, Policy Research Working Paper 4637, Washington, DC: The World Bank.

18 This policy initiative is called the "National New-type Urbanization Plan (2014–2020)." See Chris Weller (2015) "Here's China's Genius Plan to Move 250 Million People from Farms to Cities," *Business Insider*, August 5.

19 Zhao Bingxing (2017) *China's "One Belt, One Road" Initiative: A New Silk Road Linking Asia, Africa, and Europe*, Quebec City, Canada: Centre for Research on Globalization, May 10.

20 Judith Shapiro (2016) *China's Environmental Challenges* (2nd edition), Cambridge, UK: Polity Press, chapter 1.

21 Robert D. Kaplan (2000) *The Coming Anarchy: Shattering the Dreams of the Post-Cold War*, New York, NY: Vintage Books, pp. 17–18.

22 Guinea-Bissau, Ghana, Chad, Nigeria, Gambia, Burkina Faso, Liberia, Mali, Benin, Cape Verde, Togo, Senegal, Côte d'Ivoire, Mauretania, Niger, Sierra Leone, and Guinea.

23 François Moriconi-Ebrard, Dominique Harre, and Philipp Heinrigs (2016) *Urbanization Dynamics in West Africa 1950–2010*, op. cit., p. 37.

24 Ibid., p. 52.

25 ILO (2014) *Global Employment Trends 2014: Risk of a Jobless Recovery?* Geneva, Switzerland: International Labor Organization (ILO).

26 Ibid., p. 11.

27 Ibid., p. 17.

28 UN-Habitat (2016) *World Cities Report 2016 – Urbanization and Development: Emerging Futures*, Nairobi, Kenya.

29 Ibid.

30 IOM (2015) *World Migration Report 2015*, Geneva, Switzerland: International Organization for Migration (IOM), pp. 44–46.

31 UN-Habitat (2016) *World Cities Report 2016 – Urbanization and Development: Emerging Futures*, op. cit., pp. 1–6.

32 Talukder Byomkesh, Nobukazu Nakagoshi, and Ashraf Dewan (2012) "Urbanization and Green Space Dynamics in Greater Dhaka, Bangladesh," *Landscape and Ecological Engineering*, Vol. 8, pp. 45–58, February.

33 Erik German and Solana Pyne (2010) "Dhaka: Fastest Growing Megacity in the World," *Global Post*, September 8.

34 Simon Mundy (2016) "Bangladesh Factory Fire Kills 25 and Injures Dozens," *Financial Times*, September 11.

35 UN-Habitat (2016) *World Cities Report 2016 – Urbanization and Development*, op. cit.

36 IOM (2015) *World Migration Report 2015*, op. cit.

37 Richard Black (2002) "Refugees," in Vandana Desai and Robert B. Potter (eds.), *The Companion to Development Studies*, New York, NY: Arnold and Oxford University Press, p. 438.

38 UNHCR (1951). See the text of the convention at the UN Refugee Agency (UNHCR) website, online, available at: www.unhcr.org.

39 Antonio Fortin (2001) "The Meaning of Protection in the Refugee Definition," *International Journal of Refugee Law*, Vol. 12, No. 4, pp. 548–558.

40 UNHCR (2017). See the UNHCR Statistical Database, online, available at: http://popstats.unhcr. org/en/overview.

41 Ayesha Khan (2002) "Afghan Refugee Women's Experience of Conflict and Disintegration," *Meridians*, Vol. 3, No. 1, p. 96.

42 Penelope Mathew (2002) "Australian Refugee Protection in the Wake of the Tampa," *American Journal of International Law*, Vol. 96, No. 3, pp. 661–663.

43 Colin Packham (2017) "Exclusive: US Starts 'Extreme Vetting' at Australia's Offshore Detention Centers" *Reuters*, May 23.

44 Lukas Kaelin (2015) "Europe's Broken Borders: How to Manage the Refugee Crisis," *Snapshot Foreign Affairs*, September 2.

45 James Harold (2002) *The End of Globalization*, Cambridge, MA: Harvard University Press, Electronic Edition, Location 166 of 3454.

46 Richard Black, Claudia Natali, and Jessica Skinner (2005) *Migration and Inequality*, World Bank Background Paper, Washington, DC: The World Bank.

47 UNODC (2000). See the Protocol on the United Nations Office on Drugs and Crime website, online, available at: www.unodc.org/.

48 UNODC (2014) *The Global Report on Trafficking in Persons*, New York, NY: United Nations Office on Drugs and Crime (UNODC), p. 9.

49 IOM (2015) *World Migration Report 2015*, op. cit.

50 Paul Collier (2015) *Exodus: How Migration is Changing Our World*, op. cit., p. 22.

51 Alexander Betts and Paul Collier (2017) "Help Refugees Help Themselves – Let Displaced Syrians Join the Labor Market," *Foreign Affairs*, Vol. 94, No. 6, pp. 84–92.

52 Ibid., p. 85.

53 Ibid., p. 86.

54 IOM (2015) *World Migration Report 2015*, op. cit., pp. 42–43.

55 Basant Maheshwari, Vijay P. Singh, Bhadranie Thoradeniya (eds.) (2016) *Balanced Urban Development: Options and Strategies for Livable Cities*, op. cit., pp. 591–593.

56 Giles Mohan (2002) "Diaspora and Development," in Robinson, Jenny (ed.), *Development and Displacement*, New York, NY: Oxford University Press, pp. 80–82.

57 AFFORD (2017). The African Foundation for Development (AFFORD) website, online, available at: www.afford-uk.org/.

58 UN-Habitat (2016) *World Cities Report 2016 – Urbanization and Development: Emerging Futures*, op. cit., p. 36.

59 UN-Habitat (2016) *World Cities Report 2016 – Urbanization and Development: Emerging Futures*, op. cit., p. 38.

60 Drusilla K. Brown (2003) "Labor Standards: Where Do They Belong on the International Trade Agenda?," *Journal of Economic Perspectives*, Vol. 15, No. 3, p. 96.

61 Ibid., p. 89.

62 Jagdish Bhagwati (2000) *The Wind of the Hundred Days – How Washington Mismanaged Globalization*, Cambridge, MA: The MIT Press, pp. 273–283.

63 UN-Habitat (2016) *World Cities Report 2016 – Urbanization and Development: Emerging Futures*, op. cit., p. 16.

64 Ibid., pp. 15–16.

65 IOM (2015) *World Migration Report 2015*, op. cit., p. 104.

66 Ibid.

67 Ibid., p. 150.

Suggested further reading

Tom Angotti (ed.) (2017) *Urban Latin America: Inequalities and Neoliberal Reforms*, Lanham, MD: Rowman & Littlefield.

Michael A. Burayidi (ed.) (2015) *Cities and the Politics of Difference: Multiculturalism and Diversity in Urban Planning*, Toronto, Canada: University of Toronto Press.

Karen Chapple (2015) *Planning Sustainable Cities and Regions: Towards More Equitable Development*, London, UK: Routledge.

Niko Frantzeskaki, Vanesa Castán Broto, Lars Coenen, and Derk Loorbach (eds.) (2017) *Urban Sustainability Transitions*, London, UK: Routledge.

Partha S. Ghosh (2016) *Migrants, Refugees, and the Stateless in South Asia*, New Delhi, India: Sage Publications.

Steve Kayizzi-Mugerwa, Abebe Shimeles, and Nadège Désirée (eds.) (2016) *Urbanization and Socio-Economic Development in Africa: Challenges and Opportunities*, New York, NY and London, UK: Routledge.

Florence Padovani (ed.) (2016) *Development-Induced Displacement in India and China: A Comparative Look at the Burdens of Growth* (foreword essay by Michael M. Cernea), Lanham, MD: Lexington Books.

In conclusion   **10**

This book is dedicated to all my students, with the hope of introducing you to the intricate array of subjects concerning the socio-economic development of societies. The experience of spending my childhood under a dysfunctional and oppressive regime helped me realize what it means to be denied any hope of future prospects. Only later did I discover that my situation paled in comparison with the acute poverty and inequality that persists in the world. When writing the book, I constantly worried whether I would be able to address the complexity of the field called *development*. I know that given the constraints of space and time many themes are missing and many issues are not dealt with in a proper depth. It will be up to you to explore them further.

Some of you will be working for a major international development agency, others will become advisers to governments and political leaders. When dealing with problems of poverty and deprivation, your first goal is to see the situation from the point of view of the individuals you are trying to help. We can only make progress with eliminating poverty if we work to ensure that every desolate person on the streets of a small village or a big city is empowered by having rights and access to knowledge. Such an empowered individual gains incentives to contribute to the collective well-being of community. Having fundamental rights and education are necessary prerequisites to living a fulfilling and dignified life.

This kind of progress starts with a call for transparency and accountability in implementing foreign aid programs. The move towards democracy is always welcomed. Nobody wants to live in an autocratic country where the arbitrary decision making is conducted by a self-interested clique. The push for institutional reforms can be encouraged with proper incentives, because even authoritarian regimes worry about their long-term prospects. Having said that, the processes of democratization have to unfold organically. Democratic institutions cannot be designed in a reductionist style during the meetings of the donor nations.

It is appropriate, I believe, to finish this work with a few pointed observations by scholars and activists who have spent their lives helping the poor escape poverty while teaching us to better understand the multiple aspects of development.

Wangari Maathai: Too often, the term "democracy" has simply become a bromide offered during voting, rather than a means of enhancing the capacities of governmental and non-governmental institutions, providing basic services to the people, and empowering them to be active partners in development.

All political systems, institutions of the state, and cultural values (as well as pathways toward, and indicators of, economic growth) are justifiable only insofar as they encourage basic freedoms, including human rights, and individual and collective well-being. In that respect, democracy doesn't solely mean "one person, one vote." It also means, among other things, the protection of minority rights; an effective and truly representative parliament; an independent judiciary; an informed and engaged citizenry; an independent fourth estate; the rights to assemble; practice one's religion freely; and advocate for one's own view peacefully without fear of reprisal or arbitrary arrest; and an empowered civil society that can operate without intimidation. By this definition, many African countries – and, indeed, many societies in both the developing and developed worlds – fall short of genuine democracy....

The responsible and accountable management of resources, as well as the sharing of them, equitably, can be accomplished only if there is democratic space, where rights are respected. In a dictatorial or one-party system, resources cannot be shared equitably and sustainably, because the political leaders tend to apportion them among themselves, their cronies, and their supporters. Since only the elite has access to the wealth of the country, the vast majority of the population is excluded and dissenting voices have little power to bring about change. Where democratic space has been created, however, cultures of peace are more likely to be built and to flourish; when such space is constrained or nonexistent, peace will likewise be elusive and conflict more likely.[1]

Muhammad Yunus: I always feel that eliminating poverty from the world is a matter of will, rather than of finding ways and means. Even today we don't pay serious attention to the issue of poverty, because we ourselves are not personally involved in it. We are not poor. We distance ourselves from the issue by saying that if the poor worked harder they wouldn't be poor. When we want to help the poor, we usually offer them charity. Most often we use charity to avoid recognizing the problem and a finding a solution for it. Charity becomes a way to shrug off our responsibility. Charity is no solution to poverty. Charity only perpetuates poverty by taking the initiative away from the poor. Charity allows us to go ahead with our own lives without worrying about those of other people.

But the real issue is creating a level playing-field for everybody, giving every human being a fair and equal chance. Human society has tried in many ways to ensure equal opportunities, but the poverty issue remains unsolved. The poor are left to the state to take care of. The state has created massive bureaucracies with their rules and procedures to look after the poor....

Changes are the product of intense effort. The intensity of effort depends on the felt need for change, and the resources that are mobilized to bring about the changes desired. In a greed-based economy, obviously, changes will be greed-driven. These changes may not always be socially desirable. Socially desired changes may not be attractive from the greed perspective.

That is where social-consciousness-driven organizations are needed. The state and civil society must provide the support of financial and other resources behind such socially-conscious organizations.[2]

Ha-Joon Chang: Many of the formal rules restricting equality of opportunity have been abolished in the last few generations. In this struggle against inequality of opportunity, the market has been a great help.... Given all this, it is tempting to argue that, once you ensure equality of opportunity, free from any formal discrimination other than according to merit, the market will eliminate any residual prejudices through competitive mechanism. However, this is only the start. A lot more has to be done to build a genuinely fair society.... Unless we create an environment where everyone is guaranteed some minimum capabilities through some guarantee of minimum income, education and healthcare, we cannot say that we have fair competition.[3]

William Easterly: However benevolent an autocrat may appear for the moment, unrestrained power will always turn out to be the enemy of development. It is time at last for the debate that never happened to happen. It is time at last for the silence on unequal rights for rich and poor to end. It is time at last for all men and women to be equally free.[4]

Amartya Sen: The case for taking a broad and many-sided approach to development has become clearer in recent years, partly as a result of the difficulties faced as well as successes achieved by different countries over the recent decades. These issues relate closely to the need of balancing the role of the government – and of other political and social institutions – with the functioning of markets.

They also suggest the relevance of a "comprehensive development framework." [It] involves rejecting a compartmentalized view of the process of development (for example, going just for "liberalization" or some other single, overarching process). The search for a single all-purpose remedy (such as "open the markets" or "get the prices right") has had much hold on professional thinking in the past....

Individuals live and operate in a world of institutions. Our opportunities and prospects depend crucially on what institutions exist and how they function.... Even though different commentators have chosen to focus on particular institutions (such as the market, or the democratic system, or the media, or the public distribution system), we have to view them together, to be able to see what they can do or cannot do in combination with other institutions. It is in this integrated perspective that the different institutions can be reasonably assessed and examined....

In the context of developing countries in general, the need for public policy initiatives in creating social opportunities is crucially important. [In] the past of the rich countries of today we can see quite remarkable history of public action, dealing respectively with education, health care, land reforms and so on. The wide sharing of these social opportunities made it possible for the bulk of the people to participate directly in the process of economic expansion.[5]

Notes

1 Wangari Maathai (2009) *The Challenge for Africa*, New York, NY: Pantheon Books, pp. 55–57.
2 Muhammad Yunus with Alan Jolis (2001) *Banker to the Poor – The Autobiography*, Dhaka, Bangladesh: The University Press Limited, pp. 283–284.
3 Ha-Joon Chang (2012) *23 Things They Don't Tell You About Capitalism*, New York, NY: Bloomsbury Press, pp. 213, 220.
4 William Easterly (2013) *The Tyranny of Experts: Economists, Dictators, and the Forgotten Rights of the Poor*, New York, NY: Basic Books, p. 351.
5 Amartya Sen (1999) *Development as Freedom*, New York, NY: Random House, pp. 126–127, 142, 143.

Index